Construction Contracts

Also available from Taylor & Francis

Understanding JCT Standard Building Contracts 8th edition

D. Chappell

Pb: ISBN 978–0–415–41385–5

Construction Contracts Questions and Answers

D. Chappell

Pb: ISBN 978–0–415–37597–9

Dictionary of Property and Construction Law

R. Pickering-Hardy *et al.*

Hb: ISBN 978–0–419–26100–1
Pb: ISBN 978–0–419–26110–0

Development and the Law: a guide for construction and property professionals

G. Bruce-Radcliffe

Hb: ISBN 978–0–415–29021–0

Information and ordering details

For price availability and ordering visit our website
www.tandfbuiltenvironment.com
Alternatively our books are available from all good bookshops.

Construction Contracts
Law and management

Fourth Edition

John Murdoch and Will Hughes

Taylor & Francis
Taylor & Francis Group

LONDON AND NEW YORK

First edition published 1992 by E & F N Spon
Second edition published 1996
Third edition published 2000 by Spon Press
Fourth edition published 2008 by Taylor & Francis
2 Park Square, Milton Park, Abingdon, Oxon OX14 4RN

Simultaneously published in the USA and Canada
by Taylor & Francis
270 Madison Ave, New York, NY 10016, USA

*Taylor & Francis is an imprint of the Taylor & Francis Group,
an informa business*

Publisher's Note
This book has been prepared from camera-ready copy provided
by Will Hughes

British Library Cataloguing in Publication Data
A catalogue record for this book is available from the British Library

Library of Congress Cataloging in Publication Data
Murdoch, J. R.
Construction contracts : law and management / John Murdoch and
Will Hughes. -- 4th ed.
p. cm.
Simultaneously published in the USA and Canada.
Includes bibliographical references and index.
ISBN 978-0-415-39368-3 (hardback : alk. paper) --
ISBN 978-0-415-39369-0 (pbk. : alk. paper)
1. Construction contracts--Great Britain. I. Hughes, Will, Ph. D. II. Title.
KD1641.M87 2007
343.41′078624--dc22
2007017713

ISBN10: 0–415–39368–X Hbk
ISBN10: 0–415–39369–8 Pbk
ISBN10: 0–203–96574–4 ebk

ISBN13: 978–0–415–39368–3 Hbk
ISBN13: 978–0–415–39369–0 Pbk
ISBN13: 978–0–203–96574–0 ebk

Contents

Introduction

This book is aimed primarily at students for whom the study of building or civil engineering contracts forms part of a construction-based course. We have had in mind the syllabus requirements for first degrees in Building, Civil Engineering, Architecture, Quantity Surveying and Building Surveying, as well as those of postgraduate courses in Construction Management and Project Management. We have also assumed that such students will already have been introduced to the general principles of English law, especially those relating to contract and tort. As a result, while aspects of those subjects that are of particular relevance to construction are dealt with here, the reader must look elsewhere for the general legal background.

In producing this fourth edition, we have again been greatly assisted by the many helpful comments made by reviewers and users of its predecessor. We are particularly indebted to Jan-Bertram Hillig, who has patiently and diligently guided us through recent changes in standard-form contracts and, in so doing, has also made many other helpful suggestions. Nonetheless, our basic aim is identical to that which underpinned the first edition: to provide an explanation of the fundamental principles of construction contract law, rather than a clause-by-clause analysis of any particular standard-form contract. As a result, while we draw most frequently upon JCT SBC 05 for our illustrations of particular points, this merely reflects the pre-eminent position occupied by that particular form of contract in the UK construction industry.

We conclude by repeating our previous warning as to the dangers inherent in a little learning. Neither this book, nor the courses for which it is intended, seek to produce construction lawyers. The objective is rather to enable those who are not lawyers to resolve simple construction disputes before they become litigious, and to recognize when matters require professional legal advice. It should be the aim of every construction student to understand the legal framework sufficiently that they can instruct and brief specialist lawyers and this book is designed to help them towards that understanding.

Table of cases

Statutes and statutory instruments

Glossary

ACA	Association of Consulting Architects
ACA/2	ACA Standard Form of Building Contract, 2nd edition
ACE	Association of Consulting Engineers
ADR	Alternative Dispute Resolution
BEC	Building Employers' Confederation (now CC)
CCSJC	Conditions of Contract Standing Joint Committee
CD 98	JCT Standard form of Building Contract with Contractor's Design, 1998
CDM	Construction (Design and Management) Regulations 2007
CDPS	Contractor's Design Portion Supplement for use with JCT 98
CECA	Civil Engineering Contractors Association (CECA)
CEDR	Centre for Effective Dispute Resolution
CESMM3	Civil Engineering Standard Method of Measurement (3rd edition)
CIB	Construction Industry Board
CIC	Construction Industry Council
CIOB	Chartered Institute of Building
CM 02	JCT Construction Management documentation, 2002
CWa/F	JCT Collateral Warranty for Funding Institutions
CWa/P&T	JCT Collateral Warranty for Purchasers and Tenants
DAB	Dispute Adjudication Board
DB 05	JCT Design and Build Contract, 2005
DOM/2	BEC Domestic Sub-contract for use with JCT CD 81
DRB	Dispute Resolution Board
FA 05	JCT Framework Agreement, 2005
FASS	Federation of Associations of Specialists and Sub-contractors
FBSC	Federation of Building Specialist Contractors
FCEC	Federation of Civil Engineering Contractors
FEIC	European International Federation of Construction
FIDIC	Fédération Internationale des Ingénieurs-Conseils (International Federation of Consulting Engineers)
GC/Works/1	General Conditions of Government Contract for Building and Civil Engineering Works (now PSA/1)
GC/Works/5	Standard-form contract in two parts: General Conditions for the Appointment of Consultants and supporting Model Forms, both 1998
IC 05	JCT Intermediate Building Contract, 2005
ICE	Institution of Civil Engineers
ICE 7	ICE Conditions of Contract, 7th ed., measurement version
ICSub/A 05	JCT Intermediate Sub-Contract Agreement, 2005

ICSub/C 05	JCT Intermediate Sub-Contract Conditions, 2005
ICSub/NAM/A 05	JCT Intermediate Named Sub-Contract Agreement, 2005
ICSub/NAM/C 05	JCT Intermediate Named Sub-Contractor Conditions, 2005
ICSub/NAM/E 05	JCT Intermediate Named Sub-Contractor/Employer Agreement, 2005
IFC 98	JCT Intermediate Form of Building Contract, 1998
JCT	Joint Contracts Tribunal
JCT 63	JCT Standard form of Building Contract, 1963
JCT 80	JCT Standard form of Building Contract, 1980
JCT 98	JCT Standard form of Building Contract, 1998
LADs	Liquidated and Ascertained Damages
MC 98	JCT Standard form of Management Contract, 1998
MP 05	JCT Major Project Construction Contract, 2005
MTC 06	JCT Measured Term Contract, 2006
MW 05	JCT Minor Works Building Contract, 2005
NEC 3	Engineering and Construction Contract, 3rd edition, 2005 (originally entitled New Engineering Contract)
NFBTE	National Federation of Building Trades Employers
NHBC	National House Building Council
NJCC	National Joint Consultative Committee for Building
NSC/A	JCT Nominated Sub-contractor agreement
NSC/C	JCT Nominated Sub-contractor conditions of contract
NSC/T	JCT Nominated Sub-contractor form of tender
NSC/W	JCT Warranty between Nominated Sub-contractor and Employer
OR	Official Referee
PACE	Property Advisors to the Civil Estate
PCC 06	JCT Prime Cost Building Contract, 2006
RIBA	Royal Institute of British Architects
RICS	Royal Institution of Chartered Surveyors
RM 06	JCT Repair and Maintenance Contract (Commercial), 2006
SBC 05	JCT Standard Building Contract, 2005
SCWa/E	JCT Sub-Contractor Collateral Warranty for Employer
SECG	Specialist Engineering Contractors' Group (formerly CASEC)
SFA/99	RIBA Standard Form of Agreement for the Appointment of an Architect, 1999
SMM7	RICS Standard Method of Measurement (7th edition)
TCC	Technology and Construction Court
TeCSA	Technology and Construction Solicitors' Association
TNS/1	JCT Form of Tender for a Nominated Supplier
TNS/2	JCT Form of Warranty for a Nominated Supplier

1   UK construction industry context

This book is about construction contracts. The purpose of this introductory Chapter is to place such contracts in their proper context by describing the shape of the UK construction industry in terms of the general groupings of those who take part in the process. Although most readers will already have a comprehensive knowledge of the industry, our aim in this Chapter is to step back from the detail and to develop an overview which is less dependent on the interests of specific professional groups than is usually found in books about the construction sector.

1.1   THE NATURE OF THE INDUSTRY

Construction projects can best be understood in the context of the whole industry. Technological complexity ranges from the familiar, well-known materials and trades through to highly complex facilities involving multiple interacting sub-systems. Regardless of its technological complexity, any reasonably sized project involves a high level of organizational complexity. This arises because there are many specialized skills and professions with a useful contribution to the process. Most who study the industry do so from the point of view of the profession to which they aspire. Because of this, there are many different descriptions of the construction sector, drawn from different specialist disciplines. This produces a certain amount of confusion, which is compounded by the fact that construction involves such a wide range of activity that the industry's external boundaries are also unclear. The term 'construction' can include the erection, repair and demolition of things as diverse as houses, offices, shops, dams, bridges, motorways, home extensions, chimneys, factories and airports. Many different firms carry out specialist work relating to particular technologies, but few firms are confined to only one building type or one technology. Thus, the industry (and issues that affect construction projects) are difficult to comprehend fully because:

- The relationships between the parts are not always clear.
- The boundary of the industry is unclear.

The fragmentation of construction into a large number of diverse skills is an inevitable consequence of the economic, technological and sociological environment: there is an extraordinary diversity of professions, specialists and suppliers. It is important to approach construction contract problems in an organized, rational way. Although each professional discipline likes to focus upon its own contribution and the way that it relates to other project team members, a deeper understanding can be gained by considering how the industry provides a service to clients and to society at large. There are many texts that already focus on the needs of one professional group or another. In order to provide a more meaningful context, therefore, we begin by separating people into five groups: builders, designers, regulators, purchasers and users of buildings. Each of these groups is increasingly sub-divided into specialist interests such that any building project will bring together a large number of different specialists. The way they combine is specific to each project. This uniqueness arises from the individual demands of the project coupled with the continuing evolution of specific roles. This is why it is not sufficient simply to know the contents of standard contracts. The specific details of each project and the continuing evolution of changing roles demand that students of construction can understand the importance of contract structure and the options open to those who choose project strategies.

The construction industry, and the contracts that are used, only make sense in the context of changing circumstances and in the wider context of how the industry provides a service to its clients and users. As a first step, each of the five groups is introduced below, with a brief account of their context and their relationships to construction projects and to each other to put the current patterns into a wider perspective.

1.1.1   Builders

Although construction is not a new activity, the most significant developments have taken place since industrialization (Hughes and Hillebrandt 2003). Before the Industrial Revolution construction involved only a handful of technologies – such as bricklaying, carpentry, thatching and stonemasonry. Some projects were sufficiently important to justify the appointment of an architect but few projects employed other than craft skills.

In the absence of a designer, buildings simply evolved, involving slight modifications as each new project applied the lessons from experience. Pre-industrial projects were totally organized by a master mason or an architect. The interactions between the few trades were predictable. Each craftsman had a detailed knowledge of a particular technology and knew what to expect of the other trades. Thus, organization and management were simpler than they are today.

The Industrial Revolution led to the emergence of new materials and ways of working. These led to more adventurous and innovative buildings. For example, the use of steel beams enabled larger spans to be achieved. In parallel with the developments to the technology of materials, the transportation network became more sophisticated enabling the rapid spread of new technologies. Thus, sites became more complex, involving increasing numbers of specialist trades.

As the technological complexity of any process grows, so the demands for integration and co-ordination increase (Lawrence and Lorsch 1967). In the case of the construction industry, this demand led to the emergence of the general contractor (see Chapter 3), a role first undertaken by Thomas Cubitts of London in the early 19th century (Spiers 1983). Before this, clients would have entered into a series of separate trade contracts with the people who were doing the work. The general contractor fulfilled a need by employing and providing all the necessary skills, providing all of the materials, plant and equipment and undertaking to build what the client had had designed. Thus, in a general contract, the basic premise is that the client takes the responsibility for design and the contractor takes the responsibility for fabrication. Although this process is often referred to as traditional general contracting, it is a tradition that only goes back to the 19th century. One important, but confusing piece of terminology is the use of the word 'employer' to indicate the client for the purpose of many standard-form building contracts. Employer denotes the organization or person who pays the building contractor.

The task of builders is to fabricate the products of the industry. Recent developments have caused some building firms to move away from a focus of fabrication and more towards a focus of management and co-ordination of others (trade contractors). Some builders respond to the market by specializing in narrower fields (whether technical specialization or management): others respond by offering wider, integrated packages (such as design and build). Perhaps the only constant is that the builder must ultimately ensure that work takes place on a site.

1.1.2 Designers

The advancing technological complexity of the industry also led designers to embrace new techniques. There grew a demand for specialist designers who understood the new technologies. It is useful to think of design as a once whole discipline that has been successively eroded by more specific disciplines. As explained above, the need for co-ordination of building work led to the emergence of general contractors. Further, the need for measurement and valuation of work in progress and for cost planning led to the emergence of quantity surveying; the need for a specialized understanding of new technologies led to the emergence of structural engineering and services engineering; the need for overall control of the process led to the emergence of project management. In simple projects, some of these disciplines have little involvement, but their roles can be very significant in the case of complex buildings.

Like builders, designers face an increasingly complex management problem. The co-ordination of information from tens or sometimes hundreds of specialists is a very real problem (Gray and Hughes 2000). Good design requires a clear policy for the project that provides a basis for all design decisions. Although architectural training may cover management issues, the skills of good leadership will not necessarily be found in all architects (Hawk 1996). Clients who feel that architectural leadership may not be forthcoming will look to alternatives, such as the appointment of a project manager or the use of a procurement system that plays down the role of the architect.

The progressive erosion of the architect's role leads to the question of whether an architect should lead a project or should be just one of the consultants managed by a project leader. The view taken on this question depends upon what architecture is believed to be. The debate can be resolved down to two alternatives: it is either art or science. An art involves the artist exercising subjective and personal choice with little need to rationalize or explain the output. This contrasts sharply with the view of architecture as a science, which involves rational choices based upon objective techniques that can be explained and justified. Architecture as art cannot be subjected to external management: indeed it can only occur if the architect is in complete control of the process. Architecture as science can be subjected to external controls because output can be measured against some predetermined objective set by the architect. Reality is rarely so simple: real projects involve a complex and difficult tension between these views and such a debate is rarely exercised at the outset of a project, when it is most needed.

1.1.3   Regulators

Buildings and structures affect everyone who comes into contact with them and very few people believe that the freedom to erect structures should be unfettered. There are many instances where a structure can threaten the freedom, privacy or rights of an individual. Thus, legislation of many types has evolved to regulate the activities of those who wish to build. Planning legislation controls the appearance of buildings; building control legislation controls safety of finished buildings; health and safety legislation controls safety of the process of building; and so on.

Planning control in the UK originally arose from the Town and Country Planning Act 1947 which basically set up a process of locally based plans that describe the views of the local planning authority on how the area will develop. Today, the applicable planning control acts are the Town and Country Planning Act 1990 and the Planning and Compulsory Purchase Act 2004. Additionally, all building work requires permission before it can go ahead. In this way, proposals for building and alteration work can be gauged against the local development plan. An applicant who fails to get permission has a right of appeal.

Building control is intended to identify certain minimum standards defined nationally but enforced locally. The Building Act 1984 refers to Approved Documents, which contain advice on how to satisfy the functional requirements of the regulations. This advice is not compulsory, but if it is not followed, then it must be proved to the satisfaction of the Building Inspectorate that the building satisfies the functional requirements of the regulations.

The Health and Safety at Work Act 1974 (under which the Construction Design and Management Regulations 2007 are issued) and the Occupiers' Liability Acts 1957 and 1984 also have an impact on the organization and management of construction.

1.1.4   Purchasers

All construction work is ultimately undertaken for the benefit of a client. But even within one project, not everyone works for the same paymaster. Clearly, someone is paying for the work and such a person is best thought of as the purchaser. The concept of client is wider than this as it includes end users, workforce, etc. On some projects, users and purchasers are different and become involved in the project in different ways. On other projects one organization, or even one person undertakes the two roles. Everyone involved, whether designer or builder, is involved through a contract. This will connect a purchaser with a supplier or with a provider of services. The document may be a building contract, a sub-contract or a professional's appointment document. Although many of the participants will be professionals, the basic commercial nature of the process cannot be denied. Contracts are records of business transactions and the courts will approach them with the same rules that apply to all commercial contracts.

The importance of purchasers cannot be over-emphasized. Construction is about providing a service. This provision is complex because of the nature of the product, the duration of the project and the involvement of so many different people. But this should not distract us from the idea that those who pay have expectations. If their expectations are not met, then dissatisfaction is bound to follow.

Purchasers do not fall into a discrete category. The word covers everyone who buys construction work, from a householder buying a garage or a multi-national corporation buying a factory complex, to a national government buying a nuclear power facility. In other words, it is wrong to categorize purchasers into one group. There are very few generalizations that apply across the board and it is important always to be clear about the type of purchaser for a project.

1.1.5 Users

Finally we turn to the users of buildings. Although there is an overlap with the preceding category, users of buildings are a much wider group. All of us pass by buildings, enter buildings and live in buildings. The products of the construction industry affect everyone; therefore a decision is needed in each project about the extent to which people outside the immediate project team should be involved. As mentioned earlier, one of the purposes of legislation is to oblige construction project teams to take account of the impact of a project upon society. However, any firm involved in the property development process risks alienating potential clients or potential public support. Because so many people are affected in so many different ways, it is important to develop approaches that take account of some of these effects. For example, surveys among workers and customers might help to reveal possible problems; feedback from earlier projects may shed light on potential problem areas; public consultation processes may avoid protests and boycotts of controversial developments. There are all sorts of ways in which the users of buildings might be involved at an early stage of a project but the benefits of such early involvement are far reaching.

1.2   THE NATURE OF PROFESSIONALISM IN CONSTRUCTION

The professional institutions in construction differ from those in other industries primarily by their sheer number. The various skills embodied by the institutions grow increasingly specialized and institutions proliferate. Before considering the impact of so many professions, it is worthwhile considering the concept of professionalism.

1.2.1 Professionalism defined

The literature on professionalism (for example Elliott 1972) identifies four basic defining characteristics in the way that the term is used: a distinct body of knowledge, barriers to entry, serving the public and mutual recognition.

Distinct body of knowledge

First, each of the professions has its own distinctive competence, which is embodied in an identifiable corpus of knowledge. Typically, in construction, this will be represented by a professional institution's library, by its active participation in an area of research and by a close involvement with academic courses leading to qualifications. All of the professional institutions in construction make demonstrable commitments in these areas.

Barriers to entry

Professions seek to regulate who can enter. One of their roles is to obstruct those who are not properly qualified to practice. They do this through entrance examinations and other qualifying mechanisms which seek to ensure the relevant level of skill and conduct in those seeking to join. Typically, professional institutions require academic qualifications first, and then a period of approved training before being admitted to full membership. The academic qualification will usually derive from university education to degree level on a course of study recognized and approved by the institution. The period of professional training typically involves working for two years under the guidance and supervision of a qualified professional who takes responsibility for ensuring that the novice is exposed to a wide range of professional practice.

Service to the public

A further distinction drawn by professionals is that they seek to serve the public first and foremost. This concept underlies many of the aims and objectives of modern professional institutions. It means that the true professional places the public good before mere financial reward. This phenomenon is usually apparent in a code of conduct for members. Any member who breaches the code will normally be asked to resign from the profession. A professional who places public service above profit will act in the wider interest, not just the interest of the immediate client. It is doubtful whether this would work in practice unless a professional has a certain financial independence from any particular client.

Mutual recognition

A profession that is not recognized by other professions has no real status. Mutual recognition is an essential part of defining a professional institution. Without this, an institution is merely a collection of like-minded people who do not fit into the established network of professions. Clearly, a professional institution needs to belong to an infrastructure of institutions; otherwise there is no reason for it to exist. In the UK, the granting of Royal Charters through the Privy Council involves a process of consultation with other chartered groups, to ensure that any new holder of a charter is both meaningful and unique.

1.2.2   The problem of institutionalism

There are disadvantages in professionalization. The institutions that result are inherently conservative because of the burgeoning full-time staff and the inevitable bureaucracy that surrounds every decision. Membership does not guarantee excellence, as there will always be good, mediocre and bad people in any group.

An institutionalized framework of roles can cause particular problems for construction projects. Roles may become defined in terms of a practitioner's relationship with the institution so participants come to construction projects with preconceptions about their roles and about the roles of others in the team. This fuels further demand for integrating the various professional contributions. This is because such roles are defined for the purposes of the institution whereas each project has unique objectives and involves unique combinations of skills. The inherent inflexibility of institutionalism has been criticized in the past for preventing appropriate control procedures from being developed and appropriate tasks from being identified. These problems were alluded to by Latham (1994: 37) in his call for a *wholly integrated package of documents, which clearly define the roles and duties of all involved*. When flexibility is needed in the organization of construction projects, uncertainty, and hence insecurity, drive participants towards their familiar terms of engagement and fee scales. Thus, just when the need for adaptability is greatest, the likelihood of its emerging is least (Bresnen 1990).

1.3   THE NATURE OF PROJECTS

Projects involve commercial risks and they involve people. These two aspects are the most significant defining characteristics of projects and project strategies.

1.3.1   The nature of risk in construction

Risk is defined as *(noun) hazard, danger, chance of loss or injury; the degree of probability of loss...; (verb transitive) to expose to risk, endanger; to incur the chance of (an unfortunate consequence) by some action* (Schwarz 1993). The definition shows that there is more here than mere chance. Taking a risk involves a hazard combined with volition or will. In relation to construction projects, one usually takes risks by choosing from among a range of risks, deciding which are acceptable and laying off those that are not. This is very different to the concept of uncertainty, which some commentators conflate with risk.

The precise nature of the types of risk in construction projects is considered more fully in Chapter 7. A few basic points will illustrate the principles. Risk management is concerned with identifying the salient risks, assessing their likelihood and deciding how best to manage the project efficiently in the light of this information. In entering into a contract, parties face a choice about how to deal with the risks inherent in the venture. The emphasis should be on the process of identifying the nature of the particular risks for a construction project and deciding where these risks should lie within the project team. Different types of building contract will allocate risks in different quarters. There are several mechanisms for achieving different risk distributions, chief among which are the methods available for calculating and making payments.

In allocating a risk, we are concerned with the eventual payment and responsibility for the cost of the event, should it eventuate. The main point about contractual risks is that the contract apportions these between the parties, whether expressly or otherwise. Even if the contract is silent on a particular risk, that risk will still lie with one party or the other. The contract may seek to transfer a risk by making one party financially liable should the eventuality take place. In this way, risks are translated into financial equivalents so that they may be transferred or otherwise dealt with. Clearly contractual risk is to do with what happens when some mischance occurs. Before the mischance occurs someone may have predicted it in some way. Predictions will estimate the likely magnitude of cost and the statistical probability of it happening. The point about contractual risk is that, if one is repeatedly involved in construction, anything that *can* go wrong eventually *will*.

The frequency with which someone engages with a risk is a crucially important aspect in deciding how to respond to it. A client who only builds once may be fortunate enough to avoid some or even all of the risks involved. A contractor, developer, architect or surveyor, on the other hand, is statistically bound sooner or later to meet some of these disasters. To illustrate this point, consider the insurance of a home against fire. Suppose there is a 1 in 10,000 risk of a house burning down. The cost of replacing it is too much to be borne by householders, who already spend most of their working lives paying for the house in the first place. Therefore, the risk of fire is insured by paying an insurer 1/10,000th of the cost of re-building, plus a premium for overheads and profit. If the insurer holds 10,000 such policies, then the householders' statistical uncertainty is converted into the insurer's certainty. The probability of fire occurring in one of the houses is 100%. Provided that the original estimate of 1 in 10,000 is correct, the insurer will survive. Now, by the same calculation, if someone owns 10,000 buildings, it becomes a statistical certainty that one of them will burn, although which one is still unpredictable. There is no statistical risk to insure against because the probability of fire occurring *somewhere* is a statistical certainty. Insurance policies carry no financial benefit in this case: the owner is well advised to become a self-insurer. By the same process, clients who repeatedly build may wish to retain responsibility for certain risks and keep the financial benefit. It is this potential statistical inevitability that makes the consideration of risk so important on a project-by-project basis.

In dealing with risks, then, we are dealing not with mere uncertainty, but with the uncertainty related to future events and financial liability for them. This liability may be dealt with rationally by considering for each identified risk the three elements of magnitude, probability and frequency. Understanding the magnitude and probability will indicate the scale of the contingency to be considered and understanding the frequency will indicate who ought to be assuming responsibility for the risk.

1.3.2   Risk and price

Before turning to consider procurement in more detail, a final introductory point is worth making. When a project is first under consideration, decisions will be needed about where financial liability for a whole range of risks is to lie. Should they lie with the contractor, the designers, other consultants or the owner?   Mis-management of construction risks happens when such factors are not considered by employers (clients) and tendering contractors, for the following reason.

If the cost of a possible eventuality is to be borne by the contractor, then the price submitted to do the work should include an element for this contingency. This ought to be spread across a series of contracts, because the item at risk will not occur on them all. (If it will occur on all of them, then it is not a risk but a certainty.) Alternatively, it may be decided that the cost of the eventuality should be borne by the client. In that case, the offered price should be correspondingly lower but the final price will be increased if the eventuality comes to pass. Any extra risk to be carried by the contractor should therefore be reflected in the price charged for the work.

This discussion leads to an inevitable conclusion, a basic principle to be observed in the letting of all construction contracts: wherever risk is transferred from the contractor to the owner, this should be reflected in the price for the work, to balance the risk assumed by the owner (Wallace 1986). This basic principle remains unappreciated by many who work in the industry. Its observance would undoubtedly remove a lot of the ambiguity that surrounds construction contracts and would probably therefore avoid many of the disputes that occur. Indeed, Uff (2003) has suggested that lawyers have an ethical duty to avoid creating unbalanced risks for any party, and that they should 'avoid the creation of a risk which cannot be practically and financially borne'.

Recognition of the essential relationship between risk and price has another important effect. It renders unnecessary any discussion of such emotive (but misleading) issues as whether the passing of the risk to one party or the other contractor is 'unfair' or 'immoral'. As Wallace (1986) points out, *any discussion about whether or not a particular risk should be so included in the price is a discussion of policy, and not of 'fairness', 'morality' or 'justice'.*

1.3.3   The involvement of participants

An important aspect of the way that people belong to groups is the partial involvement of participants. Any member of an organization is typically a member of many organizations simultaneously. People have interests outside their work: they may be members of professional institutions; their project membership may arise as a consequence of their belonging to a firm; and so on. In other words, it would be wholly wrong to expect exclusive and total devotion to a project from those who take part. It is inevitable that participation in any organization, project or firm, is partial (Scott 1981).

The piecing together of each of the specialist skills that are needed produces an organizational structure and a pattern of relationships that are temporary. This feature is another of the distinguishing features of construction project management and is sometimes referred to as the creation of a temporary multi-organization (Cherns and Bryant 1984). This phrase indicates not simply the transience of a project but also the fact that people become involved by virtue of their membership of another organization: whether professional institute, firm, partnership or other group. In order to understand the way that contractual relationships are regulated in construction projects, it is important to note that everyone who becomes involved with a project presumably intends to detach from it at some point. For this reason, organizational structures in construction projects are constantly changing as different people come and go from the collective effort.

1.4   PROCUREMENT METHODS

This introductory Chapter has described in very general terms the types of participant in the construction process. It has highlighted some of the reasons that people become involved and outlined some of the formative influences of the major interest groups. The fact that each of the participants has his or her own reasons for becoming involved is important for those who seek to control contractual relationships and avoid disputes. The characteristic patterns of participants' involvement, and the disposition of risk among them, constitute the procurement method, or procurement system for a project. A range of procurement methods is described in subsequent Chapters, after a discussion of the major roles typically involved.

2 Roles and relationships

The roles and responsibilities of each of the members of the project team should be considered carefully, if we are to avoid some of the problems associated with assembling temporary teams of professionals. There are dangers in over-simplifying the problems associated with the need to enter into complex and difficult relationships. Research into the terminology of construction project roles reveals a complex array of disciplines and sub-disciplines, with many different terms for similar roles. Professional terms of appointment use these concepts in tying a project team together, and it is important to ensure that the consultants appointed to a project team are appointed on a basis that matches with the main building contract and its sub-contracts.

2.1 COMMON PROBLEMS

The process of building procurement involves a series of different specialists in contributing to the work at different times. These people have widely differing skills; they often work for different organizations, in different geographic locations and at different times. The level of understanding between them is often less than would be desirable. There are several perennial problems that can stand in the way of effective team building in construction projects.

2.1.1 Professional pride

There is a wide variety of professional consultants in the construction industry. Undergraduates may be studying on vocational courses, expecting to become professionals. It is very easy to fall into the trap of believing that any profession other than one's own is somehow inferior. The attitude that sometimes seems to prevail in the industry is that members of 'other' professions are greedy, self-righteous, dim or prima donnas. One must constantly remember that they too have been educated and trained to at least the same level. The intelligence or skill of one's colleagues in construction should *never* be underestimated. Pride has its place, but when it becomes conceit it can be very destructive. An associated problem is the way that increasing specialization of roles can lead to certain areas of responsibility falling between the clearly defined roles of the professions, unless attention is paid to the requirements of a project, rather than the needs of the professions.

2.1.2   Overlaps between stages

There are many stylized representations of the process of building procurement. The simplest of these involves a sequence of three discrete stages of briefing, designing and then constructing. This view is widespread, but projects rarely happen like this in practice. It may be convenient to teach design as a process separated from construction, but the world is not that simple. It is unwise to develop a brief and then freeze it before design starts, although many people would advise this. A system cannot function properly without feedback and the construction process is a system. Thus it is important that the brief is developed and refined in parallel with the design process. In fact, as the design and construction processes evolve, a sophisticated design brief may not be seen as comprehensive until the project is completed.

Similarly, as the fabrication activities take place on site, information is continuously passing back and forth between designers and fabricators. This consists of clarifications, revisions, shop drawings and 'as-built' drawings. It is often essential for designers to leave some details until fabrication is under way. Furthermore, the client's organization is subject to change as time passes – and construction projects can occupy significant passages of time. Indeed, this is one characteristic of construction contracts that distinguishes them from other contracts. During this time new processes, equipment or materials may emerge, changing the client's attitude to what was originally in the brief. Alternatively, the economic situation may change or even wipe out the financial viability of a project. Therefore, it is very important to ensure that there is a method for accommodating changes to what was originally specified.

The iteration between the various processes and stages in construction is sometimes acknowledged in the literature on construction management, but its importance can be overlooked by specialists whose own objectives may not be quite in tune with those of the client organization.

2.1.3   Extended project participation

Another common, but erroneous assumption is that the project team consists of half a dozen main consultants who come together at the start of the project and work as a team, interacting with each other to decide everything necessary to put a building together. This is a simplified picture that takes little account of the large number of other participants.

Even on a small project there may have to be a large number of consultations, as well as approvals, by building inspectors, planning officers, environmental health officers, specialist sub-contractors, suppliers and so on. Project meetings may involve a wide range of specialized advisors and various people carrying out some kind of statutory role. An attendance list from a routine design team meeting for the project may include any of the following: architects, building control officer, electricity board representatives, electrical services engineers, environmental health officer, fire officers, gas board representatives, health officer, heating and ventilation engineers, planning officer, shopfitters, structural engineers, telecommunication engineers, traffic engineers, water board

Table 1: Structure of responsibilities in construction projects

Client/ employer	Representative	Client project manager, client's representative, employer's representative/project manager.
	Advisor	Advisory group, feasibility consultant.
	Stakeholder	End-user, general public, tenant, workforce.
	Supplier	Client's direct contractor, Preferred supplier
Advisor/ consultant	Design leadership	Architect (management function), design leader, lead consultant, lead designer.
	Management	Consultant team manager, design manager.
	Design	Architect (design function), architectural designer. Designer, specialist advisor, engineer, consultant (etc).
	Administration	Architect, contract administrator, supervising officer. Planning supervisor, Project administration.
	Site inspector	Clerk of works, Resident engineer.
	Financial	Cost advisor, cost consultant, quantity surveyor.
Constructor	Overall responsibility	Builder, contractor, lead contractor, general contractor, main contractor, principal contractor, design-build contractor, design contractor, management contractor.
	Constructor's staff	Construction manager, construction planner, contract manager, person-in-charge, site agent.
	Partial responsibility	Engineering contractor, package contractor, specialist, specialist contractor, specialist sub-contractor, specialist supplier, specialist trade contractor. Domestic sub-contractor, labour-only sub-contractor, named sub-contractor, nominated sub-contractor, nominated supplier, specialist sub-contractor, supply-only sub-contractor, sub-contractor, trade contractor, works contractor.

representatives. Even if not present at meetings, their expertise and participation in decision-making will be required at some point in the process. The communication and co-ordination problems among such a diverse group is exacerbated by the fact that key people within such a group may leave and be replaced by others, either because they move to another job, or because as the project advances, responsibility for contributing to it tends to move from senior to junior people within each organization. Chapman (1998) has shown that this is a very real risk in construction projects and one that is rarely planned for in advance.

Research has shown that even on relatively small projects, as many as two hundred people can be involved in the decisions on construction projects before they reach the site (Hughes 1989). The involvement of such large numbers of participants is not always obvious. Moreover, once the wide range of participants is exposed, the relationships between them are far from clear. Table 1 shows how some structure may be imposed on a wide variety of participants, showing how they may be grouped. This shows an enormous range of terms that describe various project participants, some of which are roughly synonymous with each other, and all of which are in current use. It also shows how the traditional role of the architect has become fragmented, with different specialists taking on some aspects of the work. This Table is extracted from a major research project on roles in construction project teams carried out for the UK's Joint Contracts Tribunal (Hughes and Murdoch 2001). The point of this Table is that there are many diverse terms for roles that are essentially very similar. But each of the terms for these roles has a specific context, and to transfer a term into an unusual context will often result in confusion.

2.1.4   Temporary multi-organizations

One result of the specialization and professionalization of the many tasks in the construction process is fragmentation. The reasons for this fragmentation are associated with the fact that a variety of different people come together on a project, temporarily, each with his or her own set of objectives and expectations, each from a separate firm. The reason for the wide variety in the skills needed on a construction project originates from the technical complexity of the problems associated with procuring buildings, creating a demand for detailed specialist knowledge. The need of people to feel as though they belong to a group encourages these technical specialists to group together into professional organizations such as institutions, strengthening the differences between them. These specialists tend to be employed by specialist firms, further reinforcing their differences. This phenomenon is not confined to the construction sector and has long been acknowledged in the organizational literature (as far back as Thompson 1967, for example). Because of this, the divisions brought about by technical complexity are exacerbated by organizational complexity. Finally, because each of the contributing organizations is involved in the project through a contract, the differences in their orientation and allegiances are brought into sharp relief. The fact that these firms come together solely for the purposes of completing a particular project results in a temporary multi-organization (see Section 1.3.3). Understanding the cumulative effects of technical complexity leading to the proliferation of professional institutions, professionalization leading to organizational complexity, and organizational complexity leading to legal complexity, is crucial in developing an appreciation of the problems deeply embedded in construction industry practice.

2.1.5   Conflict in project teams

The characteristic grouping of project teams very often results in conflict at a variety of levels. Each project participant has particular aims and objectives and, in the past, it has been rare to find contract structures that encourage harmony among these aims. Until recently, project participants have *expected* to enter into confrontations with each other and with the client. The purpose of contracts and appointment documents is to regulate these confrontations and to provide a basis whereby one party can enforce the promise of another. It is not their purpose simply to avoid or prevent tension and conflict. However, recent developments have helped to modify expectations. All of the major contract-drafting bodies have moved to a much greater emphasis on integrated packages of documents. While it is not yet clear whether the cause is a new approach to contract drafting, the state of the economy or the introduction of statutory adjudication schemes, the fact is that the industry is a lot less litigious than it was a few years ago, and there is a growing awareness of the need for team-working and the development of long-term business relationships between organizations. Partnering, framework agreements, joint ventures and consortia are growing in their use, and many people feel that this is a great improvement. There are some signs that the increasing use of these approaches is merely a market response to certain high-profile clients who only contract with those who will work along these lines (Gruneberg and Hughes 2005). Moreover, there is always going to be a tension in public sector projects between the need to develop long-term relationships and the expectations of transparency and accountability.

One of the purposes of assembling a team of people from different professions is to harness a variety of views. It should be expected that each person brings his or her own criteria for decision-making; they all have their own agendas and this is as it should be. The tensions thus generated should precipitate debate and dialogue so that clear choices can be made. In other words, controlling conflict is not the same as eliminating it. To quote Tjosvold (1992):

> *The idea that conflict is destructive and causes misery is so self-evident that it is seldom debated. Employees fight about many issues, but the wisdom of avoiding conflict is too often not one of them. However, it is the failure to use conflict that causes the distress and low productivity associated with escalating conflict. Conflict avoidance and the failure to develop an organization equipped to manage it, not conflict itself, disrupt. Open, skilful discussion is needed to turn differences into synergistic gains rather than squabbling losses.*

Interestingly, the problem of handling conflict was also addressed decades ago by the organizational researchers, Lawrence and Lorsch (1967). They identified the task of an *integrator* and stated that such a person, in order to succeed, must be able to deal with conflict. They identified three methods for dealing with conflict: confrontation (choosing, after discussion, a solution from those put forward for consideration.), smoothing (avoiding conflict) and forcing (the naked use of power). They found that the most effective integration was achieved in organizations that used confrontation, supplemented as necessary by forcing behaviour to ensure that issues were properly confronted. Smoothing was the least effective method.

Clearly, the problem is not merely a question of avoiding conflict and eliminating disputes, but a more definite problem of how to take advantage of the potential benefits of conflict without removing necessary sanctions.

2.2   CONSULTANT ROLES

Professional institutions continue to proliferate, and some have not been in existence long enough to have established any traditions. Nonetheless, there is often an emphasis on 'traditional' by those who wish to give the impression that professionals occupy a clear and unassailable position. While this may be a laudable aim, it is an unfortunate trait of our professions that their high ideals of protecting the interests of clients and the public sometimes take second place to the survival of the professional institution. The position of a professional institution, therefore, should always be open to question. The preferred role of a professional may not always match up with what is required for any given project. Indeed, professions often seek to develop a professional identity that is uniform and consistent. The difficulty with this is that the pursuit of a professional identity can get in the way of providing a proper service to a client. Andrews (1983) observed that *as specialized institutions proliferate, institutional survival matters more than the appropriateness to changing circumstances of fiercely protected roles*, which is as true now as it was in the 1980s. The professions tend to establish a baseline of services to be offered by their members and (historically) recommended fee scales often accompanied this. In this way, the professions manage to limit the type of work undertaken by their members. These days, a more important limiting factor is the professional indemnity insurance policy which underwrites the activities of a professional. The insurers will only underwrite specific risks and this generally means that professionals who step outside the boundaries of their institution's typical work will be uninsured. Generally, clients would be unsettled about employing professional advisors who had no indemnity insurance cover. The expected roles, then, can be discovered by looking at the relevant institutions' published 'professional services agreements' or 'terms of appointment', phrases used to describe the contract between consultant and client.

2.3   PROFESSIONAL SERVICES AGREEMENTS

The current trend in documentation for appointing consultants to project teams was propelled by the momentum that began to gather pace with the publication of the Latham Report (1994). There are a few basic principles that seem to underpin the modern consultants' contract. Most have dropped the title 'conditions of engagement' and are now called variously professional services agreement, professional services contract or simply agreement. There have been considerable efforts on the part of those who draft standard-form contracts to create the kind of integration between the various participants' contracts called for by Latham. This has resulted in a common pattern of defining the services separately from the contractual conditions that govern the provision of those services. It must always be remembered that the conditions governing the relationship are separate and distinct from the specific duties that a consultant will undertake. Thus, the same set of conditions, at least in theory, could cover a variety of consultants' appointments, with each having a different set of duties and responsibilities within the project.

Professional services agreements serve various purposes. First, they contain the detailed conditions of a contract between two parties (but these conditions do not in themselves form a legal contract; this is generally contained in a memorandum of agreement which incorporates the conditions). In this process, standard conditions are convenient to use because appointments can be made with them simply by referring to them. There is no need to draft a specific appointment agreement. Further, so long as there is scope for amending them, they can also provide a useful starting point for negotiations about duties for a particular project.

From the point of view of the professional institutions, standard conditions provide a framework for identifying the duties of the professional. By unifying appointments, similar services are offered by all members of the profession. In the sense of providing a set of rules for conduct, these documents fulfil the most laudable aims of a profession. They discourage unscrupulous practitioners from offering second-rate services. Such unification enables institutions to negotiate professional indemnity insurance policies on behalf of all those who employ the standard conditions.

Another function fulfilled by a formal agreement for services is that of attributing fees to various aspects of the work. By listing the activities to be done sequentially, those activities can be grouped together so that parts of the fee can be paid at each of the major steps in the consultants' involvement.

A potential problem arises when different professions each have their own professional services agreements, and this may lead to confusion. It can be unproductive and cause wasted work or delay because the purpose behind writing the standard terms has been to standardize the relationship between the client and the consultant. This misses the significance of teamwork in construction. It is essential that the team acts as a whole, and the fragmentation created by differing terms of appointment can compromise this (Latham 1994).

As stated above, these standard professional services agreements are contracts and thus subject to the rules of contract law. The terms within them relate to the obligations of the parties, and detail the services to be carried out and the payments to be made for them. These terms need to be incorporated into the contract between the parties, for they are not a contract in themselves. The general situation is that they will not be automatically incorporated (i.e. implied) unless the client is familiar with the usual relationship between client and architect and has engaged similar professionals before on similar terms.[1] Otherwise, the terms must be expressly included by reference to them from the document that forms the contract. This can be done by using a memorandum of agreement, signed by the parties, which refers to and expressly incorporates the conditions of engagement or professional services agreement. Indeed, most standard agreements currently in circulation include articles of agreement, recitals and an *attestation* or a *sample form of agreement*.

It is becoming common practice for professional institutions to publish their documents as forms, containing the memorandum of agreement (or articles of agreement) as well as the conditions. These forms are completed and signed by the parties, incorporating the conditions into the agreement.

At one time, professional institutions specified mandatory conditions of engagement for their members, along with minimum scales of charges. But the Monopolies and Mergers Commission declared that such minima were not in the public interest and should be abolished. In consequence, mandatory conditions of engagement were abolished in 1982.

---

[1] *Sidney Kaye, Eric Firmin & Partners v Bronesky* (1973) 4 BLR 1

2.4   ARCHITECT

The accepted role of the architect has long been to design the building, advise on the selection and appointment of other consultants, manage the design, select and appoint the contractor and/or sub-contractors, and generally represent the client's interests as far as possible. Coupled with these responsibilities is the duty to act as an independent certifier under certain standard forms of contract, where the architect has to judge certain issues impartially.

There are some apparent and inherent conflicts in this combination of responsibilities, which are examined in more detail in Chapter 18. It has been suggested that the skills necessary to be a good designer are not the same as the skills necessary to be a good manager (Honeyman 1991). This, it should be said, flatly contradicts the idea that the architect can only design effectively when he or she has the highest authority in the project (Gutman 1988). Unfortunately, many people have come to interpret the former notion as meaning that the skills necessary for designing are *incompatible* with management skills and that one person cannot possibly do both. This is clearly a very different view from the latter. The two roles are different, but not necessarily incompatible. However, not all people can develop both sets of skills with equal facility, and it has become clear that there are some architects who are better at design than management, others who are better at management, and still others (the most valuable type) who are equally competent at both. Similar arguments apply to the role of engineers and their responsibilities in civil engineering projects.

An example of the conflicts within the roles of the architect is the task of managing the design process. There is a view that *design* is the total process of integrating all of the constraints on a planned building project in such a way as to optimize as many of them as possible. This explains why design and management should be integrated in one person. There is another view, equally compelling, which sees management as the function that integrates design information from all of the constraints and other designers. This view suggests that the interaction between groups of professionals, successfully managed by one who is not 'designing' will result in the optimization of as many constraints as possible. This is not the place to indicate the validity of one view or the other. The aim here is simply to raise these issues so that the reader may recognize the arguments.

The standard architect's professional services agreement is produced by the RIBA; here we refer to the standard form of 1999 (SFA/99) in its updated version of 2004 (Royal Institute of British Architects 2004). The use of written terms of appointment for architectural services is strongly advocated for all members of the RIBA in the body's code of professional conduct (Royal Institute of British Architects 2005a). The actual conditions to be used are not specified. However, the use of the RIBA standard documentation is recommended in the annex to guidance note 4 of the code of professional conduct. In line with most packages of standard contract documentation, the standard form is available in several different versions, all based upon the core conditions contained in SFA/99 (all updated in 2004; a new suite of RIBA forms of appointment is scheduled for 2007):

- SFA/99: Standard form of agreement for the appointment of an architect.
- CE/99: Conditions of engagement for the appointment of an architect.
- SW/99: Small works.

- SC/99: Form of appointment for a sub-consultant.
- DB1/99: Employer's requirements.
- DB2/99: Contractor's proposals.
- PS/99: Form of appointment as a planning supervisor.
- PM/99: Form of appointment as a project manager.

As mentioned above, it is the memorandum of agreement that forms the contract, the conditions being secondary terms to add detailed terms to the agreement. The memorandum can be executed as a deed or as a simple contract. RIBA members are asked to use SFA/99 as a *standard*, rather than as a *model*. The distinction is that a standard form should rarely, if ever, be amended whereas a model form is intended to form only the starting point for detailed negotiations about terms and conditions.

2.5 TYPICAL TERMS IN PROFESSIONAL SERVICES AGREEMENTS

While there are many standard agreements for the appointment of consultants, the aim here is merely to summarize the common threads in them, using SFA/99 as the basis and highlighting the major differences exhibited in other documents. Other agreements mentioned are the NEC Professional Services Contract 3rd edition (Institution of Civil Engineers 2005c) and GC/Works/5, which is published in two documents, the General Conditions for the Appointment of Consultants and the supporting Model Forms (Property Advisors to the Civil Estate 1998a and 1998b).

2.5.1 Defining the appointment

Standard terms of appointment consist of different documents. These appointment documents are nothing more or less than a contract.

SFA/99 (updated 2004) contains a memorandum of agreement, four schedules and the conditions of engagement. The memorandum leaves space to add the names of the parties and it contains the articles in which the parties shall record the project-specific data, such as the law that applies to this contract, time limits, limit of liability, the amount of professional indemnity insurance, and the names of the adjudicator and arbitrator. Most articles include default provisions. At the end of the memorandum the parties sign the contract, either as a simple contract or as a deed. The schedules then contain the detail of the services to be performed by the architect. Schedule 1 describes the client's requirements regarding the project (see Section 2.5.2 below), schedule 2 deals with the kinds of services that the architect will perform (see Section 2.5.2 below), schedule 3 identifies the fees and expenses and schedule 4 (very usefully) identifies the other appointments made by the client for the purposes of the project. This last schedule leaves spaces to insert the names and addresses of the project manager, planning supervisor, quantity surveyor, structural engineer, building services engineer, site inspector and others. It also identifies which elements of the building are to be designed by participants other than the architect.

A different approach is taken in GC/Works/5. Its first part contains the general conditions for the appointment of consultants which are not supposed to be amended by the parties. In contrast, the supporting model forms are designed to be completed with the project-specific data. This second part of GC/Works/5 begins with a single section called appointment particulars, which seems to be more detailed than its SFA/99 counterpart and which include the description of services and the particulars of other consultants appointed by the client. This documentation also includes considerable detail on the tendering procedure to be adopted for the appointment to the project, and the relationship between price and quality for the purposes of each project. Fees and expenses are identified as well as the documents and other information that comprise the tendering documents. Interestingly, a significant amount of detail is called for in the model form for the consultant's tender (Model Form 2), which includes the formal *offer* as item 27, incorporating by reference GC/Works/5 General conditions for the appointment of consultants. The memorandum of agreement (Model Form 10) is then a rather small document simply recording that the parties have agreed the particulars and incorporating by reference GC/Works/5 General conditions for the appointment of consultants.

The NEC 3 Professional Services Contract has yet another different structure. According to the guidance notes (Institution of Civil Engineers 2005e: 1) it is designed as a 'shell contract', which requires important information to be provided separately. It does not contain any form of agreement because it is based on the assumption that the creation of the contract can be fulfilled in different ways, for example by the client's acceptance of a tender or by the consultant's acceptance of a counter-offer prepared by the client (Institution of Civil Engineers 2005e: 25). Whatever the form of the agreement is, there must be a reference to the conditions of NEC. For example, such a reference is contained in the 'sample form of agreement' which is published as the first appendix to the guidance notes (Institution of Civil Engineers 2005e: 144). The second item outside the NEC documentation is the description of the services. In NEC's clauses this description is referred to as the 'scope'. The NEC documentation itself contains five components. First, there are the general clauses (called *core clauses*) whose content is described in Section 2.5.4. Second, the *main options* deal with different types of payment mechanism. Third are the *dispute resolution options*. These are followed by the *secondary options*, i.e. a range of options such as price adjustment for inflation, delay damages or limitations of liability. Finally, the *contract data* contain the project-specific data. There are several pages of *data provided by the employer* identifying the client's choice regarding the aforementioned options (main options, dispute resolution options, secondary options), the employer, the adjudicator, the law applicable to the contract, and the language of the contract (these latter items making clear that the agreement is not solely for use in the UK). The people who have access to the construction site are outlined as well as contract specific data such as the timing of certain events and intervals. There are also two pages of *data provided by the consultant* identifying who it is, including named individuals who will perform the work.

2.5.2   Definition of services

As stated earlier, the conditions governing the relationship between consultant and client are generally separate from the detailed description of the precise services to be undertaken under the agreement. SFA/99 typifies the current approach to this, by the use of *schedules* in the agreement. For the definition of the services to be performed by the architect, the first two of these schedules are of importance. The first describes the client's requirements in order to form the basis of the architect's services and fees. The second schedule provides for two options: the parties can either specify the details of the services directly on the schedule (which is simply a blank page to fill in) or specify the details in the supporting supplement. In this supplement, the parties can determine the roles that the architect will perform by choosing from designer, design leader, lead consultant, and contract administrator. The scope of the design and management services associated with each role is identified in some detail by means of default provisions. For the sake of clarity the supplement also lists a range of services that are not normally undertaken by the architect, unless specifically identified. The guide (Phillips 2004: 65) advises that the numerous default settings of the supplement can be tailored, deleted or replaced depending upon the circumstances of the project.

As mentioned above, GC/Works/5 utilizes a detailed schedule of *appointments particulars* (Model Form 1), which includes reference to the services to be provided, which will be listed in an annex and incorporated by reference. As in SFA/99, there is also reference to the services that will *not* be provided.

The NEC Professional services contract includes in the *data provided by the employer* two entries of relevance here. One entry states *the services are...* and another states *the scope is in...* As regards the first entry, the guidance notes (Institution of Civil Engineers 2005c: 147) suggest that the services should be described briefly, apparently more for the purposes of identification, rather than obligation. The scope is to be identified in a separate document, merely incorporated here by reference. This reference is specified by the client in the second entry (*the scope is in ...*), usually by citing the title or number of this document (for example 'document XY 007'). The guidance notes provide a useful table collating all the references from the contract to the scope of the consultants' work (Institution of Civil Engineers 2005c: 26).

2.5.3 Typical conditions

The conditions governing these agreements are laid out in very different ways, even though they broadly cover the same topics. By way of example, the main headings of SFA/99 (updated 2004) are used in the following brief descriptions of the key points covered by such agreements.

- *General*: Under this initial heading, basic issues are determined such as the insignificance of the headings for the interpretation of the clauses, the choice of law, and the parties' duty to execute any communication among them in writing. One clause demands both parties to act in a spirit of mutual trust and co-operation, adding particular obligations such as the architect's duty to advise the client about the need for appointments of others. The parties are also *contractually* bound to meet their respective statutory obligations arising from the Construction (Design and Management) Regulations 2007.

- *Obligations and authority of the architect*: The standard of liability is identified as reasonable skill and care and the extent of authority is identified. An important point is that the architect acts on behalf of the client and he or she has to report to the client on the progress in the performance of the services. As to the limits of the architect's obligations, one clause provides that the architect does not warrant that the services will be completed in accordance with the timetable or the budget cost for construction work, that statutory approvals will be granted, that others will perform adequately or that others will remain solvent. The architect is called upon to co-operate with other appointed consultants, provide them information where needed, comment where appropriate on their work and integrate the various contributions into his or her own work.

- *Obligations and authority of the client*: A range of clauses impose obligations on the client that enable the project to progress, obliging the client to supply information, make decisions, give approvals, instruct the architect, instruct applications for planning legislation, building control and so on. In terms of work to be performed by others, the client will not hold the architect responsible for their management and operational methods. Apart from this, the client is at liberty to hold the architect responsible for not detecting the others' failures to perform (given that the architect should have reasonably detected such failures). To this extent, the clause is in line with the *Investors in Industry* case[2] in which the Court of Appeal set out the architect's overall duty to check for compliance (see Section 13.3.1). The client will also require other participants of the project to co-operate with the architect and to provide the architect with information. An interesting clause obliges the client to seek legal advice when needed in connection with any dispute between the client and other parties involved in the project.

- *Assignment and sub-contracting*: Assignment of the whole or part of this agreement is not permitted by either party without the express consent of the other. The architect is prevented from sub-contracting any part of the services without the consent of the client, which shall not be unreasonably withheld.

- *Payment*: Detailed clauses govern the calculation and timing of payments, including, among other things, part-payment for services not completed, refund of expenses and disbursements, payment of monthly accounts, the accrual of interest on late payments, and payment in the event of suspension or termination.

- *Copyright and use of information*: The architect owns the copyright on material that he or she has produced and grants a licence to the client to use this for the sole purposes of the project. These purposes are delineated and certain provisions regarding the use are added. The architect may publish photographs of the project subject to the consent of the client which shall not be unreasonably withheld. For both parties, the disclosure of confidential information is exceptionally allowed in specific circumstances, for example if necessary for the performance of the service.

---

[2] *Investors in industry Ltd v South Bedfordshire DC* [1986] 1 All ER 787.

- *Liabilities and insurance*: A time limit is placed on the period within which proceedings under this contract may be commenced. This period is entered into the memorandum of agreement, with a default of six years (or twelve if executed as a deed). This period begins to run from the date of the last service performed under this contract, or practical completion of the building contract, whichever is the earlier. Also, the architect's overall liability may be limited by inserting a sum in the memorandum of agreement and the architect's liability should be just and equitable in relation to the responsibilities and liabilities of others involved in the project. The architect is obliged to maintain professional indemnity insurance cover.
- *Suspension and termination*: Both parties may suspend the performance of any of the services, but the architect may do so only as a response to non-payment by the client or if the client fails to comply with the Construction (Design and Management) Regulations 2007. Either party may determine, by stating the grounds and which of their obligations are affected. Reasonable notice is required, except in the case of insolvency of either party, or the death or incapacity of the architect, when immediate notice is sufficient. Such termination is not intended to prejudice any other rights and remedies of either party.
- *Dispute resolution*: In the event of disputes between the parties, negotiation and mediation are encouraged. In particular, the parties may attempt to settle any dispute by use of RIBA's mediation service. Either party may make reference to adjudication at any point (which is a statutory right arising from the Housing Grants, Construction and Regeneration Act 1996), using the Model Adjudication Procedures of the Construction Industry Council. There is also an arbitration agreement.

2.5.4 Other standard agreements

GC/Works/5 covers much the same material as SFA/99, although with a very different structure, under the following headings:

- *Definitions, documents and representatives*: These provisions are in line with similar conditions in SFA/99.
- *General obligations*: Broadly in line with the general provisions of SFA/99, with the addition of clauses covering access to Crown establishments, discrimination, changes in the consultant's business and commercial interests that may conflict with the consultant's services.
- *Security*: Under this heading are extensive clauses that reflect the origin of this contract in government work, dealing with the Official Secrets Act 1989 and a general confidentiality clause. There are also clauses dealing with corruption, providing for the contract to be determined by the client in the event that there is reasonable cause to do so.
- *Controls and programming*: Restricting the consultant's ability to commit the client to exceed any limits on expenditure and restricting the consultant's power to issue variations without the prior written consent of the client. There are requirements for the consultant to provide a

consultant's programme and to obtain the client's authority before proceeding between work stages. An optional requirement obliges the consultant to comply with the codes of procedure known as Co-ordinated Project Information (see Section 10.1.2).

- *Notices, instructions, additional services and payments*: Covering the requirements for the issue of notices, for the consultant to comply with all instructions issued by the client, including changes to the services. The consultant will submit monthly accounts broadly consistent with the provisions in SFA/99. There are provisions for the consultant to be reimbursed in the event that a contractor is not appointed for the project.
- *Particular powers and remedies*: Covering suspension, termination, negotiated settlements, adjudication, arbitration and choice of law.
- *Assignment, sole consultants, sub-consultants, other consultants and parent company guarantee*: The client seems freer to assign than the consultant, there is provision for the consultancy to act as the sole source for all consultancy work in the project, and a statement that any sub-consultants must be prepared to enter into a warranty directly with the client. The consultant's obligation to co-operate with other consultants is similar in effect to that in SFA/99.

The NEC Professional Services Contract is structured according to the headings of the Engineering and Construction Contract, and does not seem to contain any clauses that impose obligations and rights that are very different from those already discussed. The headings are:

- *General*: Including an obligation to act in a spirit of mutual trust and co-operation.
- *Parties' main responsibilities*: Similar obligations to SFA/99 for both parties.
- *Time*: Including more detail than other forms as to the content and status of the consultants' programme, as well as authority for the client to decide the date of completion and certify it within a week of it happening.
- *Quality*: Including an obligation on the consultant to operate a *quality management system* and to correct defects identified by the client.
- *Payment*: Broadly in line with SFA/99.
- *Compensation events*: The provisions under this clause bear more resemblance to provisions in building contracts than those usually found in professional services agreements, and thus are more detailed than similar provisions in SFA/99 or GC/Works/5.
- *Rights to material*: This seems to attribute rights and obligations similar to those in SFA/99 that grant a licence to the client to use the consultant's material and impose confidentiality restrictions on both parties.
- *Indemnity, insurance and liability*: The consultants' liability seems less restricted than that in SFA/99, otherwise, these provision are broadly similar.
- *Termination*: The termination provisions are broadly in line with SFA/99.

2.6 INTEGRATED DOCUMENTATION

The biggest problem currently facing professional institutions in preparing professional services agreements is the tension between fulfilling demands to develop co-ordinated documentation for the whole team on the one hand, and protecting the interests of their members on the other. While major developer clients often provide their own documents for consultants to use, practice in the industry is diverse. The *New Engineering Contract* package of documentation (in which the building contract is called the *Engineering and Construction Contract*) already provides a fully integrated set of documentation, this being one of the reasons that the Latham Report (1994) recommended the universal adoption of these forms for the industry. The conditions of engagement are referred to as the *NEC 3 Professional Services Contract* (Institution of Civil Engineers 2005d). Another integrated package of contract documents, called GC/Works, has been developed for government clients, with a view to their use for private sector clients (Property Advisors to the Civil Estate 1998b). GC/Works/5 is a two-part Consultant's Agreement intended for use where the main building contract is GC/Works/1.

3 General contracting

General contracting involves the separation of construction from design. A main contractor is employed to build what the designers have specified. Since this form of procurement was developed it has become very common and is often referred to as traditional. Although the popularity of general contracting seems to be waning, it still accounts for most of the construction work in the UK. However, some of the basic defining characteristics of general contracting no longer hold true.

3.1 BACKGROUND

In the UK, General contracting has been around since the early 19th Century, when Cubitts in London first began to offer the services of a general contractor (Spiers 1983). Before that time building work tended to be procured either as a series of direct contracts between client and trade contractors (the essence of what is nowadays called construction management, see Chapter 6), or as a lump-sum design and build package (see Chapter 4).

General contracting was a response to the increasing sophistication of construction technology during the Industrial Revolution. As techniques and materials proliferated, co-ordination problems on building sites became more complex. At the same time, the crafts and trades associated with construction were becoming more formalized. A series of issues combined to make the idea of general contracting a viable proposition; continuity of employment for operatives, economies of scale in the use of plant, the development of the transport infrastructure, and so on. The desire of architects to focus more on design and client-related issues and less on the day-to-day business of construction fuelled demand for a contractor who would shoulder all of the responsibility for building the project.

Another development at that time helped to shape what became a typical pattern of general contracting. The need to measure work in progress for the purposes of paying contractors and the need to prepare documentation for tendering purposes was increasing. The technicians who dealt with these aspects were becoming increasingly dominant in the process of construction. As a result the profession of quantity surveying was born. These people specialized in costs, prices and financial control. The contractor was obliged to build what was documented by the architect, but the process of competitive tendering meant that bills had to be prepared in such a way that contractors could be chosen based upon the lowest bid. In order for this process to be effective, and to ensure that all of the contractors were tendering on the same basis, bills of quantities came to be standardized. Over the years common practice became codified and recorded into

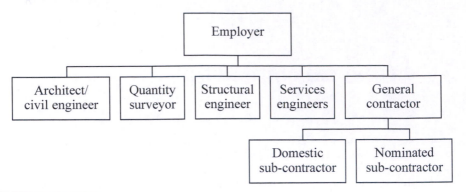

Figure 1: Contractual relationships in general contracting

what is now known as the standard method of measurement, for example SMM7 for building works (Royal Institution of Chartered Surveyors 1998) and CESMM3 for civil engineering works (Institution of Civil Engineers 1991).

The continuing development of science and technology has led to the emergence of many new specialist skills. It is a basic tenet of organizational theory that increasing complexity in the environment of an organization should be matched by an increasingly complex organizational structure (Lawrence and Lorsch 1967). To put it simply, if the technology is complex, more specialists are needed to cope with it. Each of the specialists is likely to have its own training and jargon. Thus, a consequence of the increased use of specialists is a corresponding increase in the demand for co-ordination and integration.

When building projects were totally organized by one architect, or by a master mason in charge of a craft based product, there were only a handful of technologies (Bowley 1966). These interacted in predictable ways. Not only did each craftsman have an intimate knowledge of a particular trade, but also each knew precisely what to expect of the other trades. As a result, integration and co-ordination could take place by means of operatives making mutual adjustments in their own work. In the modern technological building process, very few people can grasp all of the different technologies. The craftsmen cannot be left to sort out the interfaces between work packages; these become the responsibility of the designer. The result is an increased demand for management.

The basic defining characteristic of general contracting is that the contractor agrees to produce what has been specified in the documents. Designers, on behalf of the employer, produce the documents and builders produce the building. In theory, the contractor should be invited to price a complete set of documents that describe the proposed building fully. Such documentation demands that the architect (or lead designer) co-ordinates design advice from a wide variety of specialists. The result is that the contractor has no responsibility for design. The contractor's offer of price is based on the bill of quantities, a document that itemizes and quantifies, as far as possible, every aspect of the work. The bill forms not only the pricing document but also, because of its comprehensiveness, an important mechanism for controlling costs as the project progresses. Therefore, it has a central role in the process and the quantity surveyor, as the author of the bill, is an important contributor. General contracting, then, revolves around the relationships between employer, architects, quantity surveyors and builders. The contractual relationships are summarized in Figure 1.

Structural and services engineers provide specialized design advice which is co-ordinated by the architect. In civil engineering projects, the lead designer will generally be a civil engineer and will co-ordinate the design advice from other specialist engineers. These specialist designers will often take on supervisory duties as well as design. The purpose would be to visit the site and inspect the work to ensure that the work is produced in accordance with the design. There will be problems when work does not concur with the design. Such problems can only be resolved by examining the means by which the contractor was instructed what to build. The importance of the documents used for this purpose is self-evident. It is for this reason that the standard-form contracts tend to oblige the builder to produce what is in the documents.

One of the most important documents in general contracts is the bill of quantities. It cannot be produced if the design is incomplete. It is usually impracticable to prepare a complete design because of the time needed and because sometimes choices need to be left as late as possible in the process. If design is only partially completed then general contracting begins to break down because it is based on the assumption that the contractor prices, and builds, what has been documented. When the documentation is incomplete, there is a high demand for communications and information. These demands render the process very difficult to manage. Similarly, when complex technologies are required, the need for specialist sub-contractors can place too many demands on the co-ordination and information systems. Clearly, there are problems if general contracting is used where it is inappropriate. Therefore, guidelines are needed as to where it is best suited.

3.2   USE OF GENERAL CONTRACTING

Since general contracting is widely used, there are many standard forms that have evolved. In civil engineering, for example, the current standard form is called ICE 7: Institution of Civil Engineers Conditions of Contract 7th Edition 1999 (see Section 8.4.1). There are three widely used forms in the building industry; JCT SBC 05: Joint Contracts Tribunal Standard Building Contract Edition 2005, JCT IC 05: Joint Contracts Tribunal Intermediate Building Contract Edition 2005 and JCT MW 05: Joint Contracts Tribunal Minor Works Building Contract Edition 2005 (a more comprehensive list is given in Chapter 8). The details of each contract differ markedly, but many of the general principles are transferable.

The circumstances under which a traditional general contract should be used are not always clear. This is because general contracting has evolved over many years. The role of the JCT was to keep the standard form up to date with modern practice, removing any difficulties that may arise in connection with its use (Joint Contracts Tribunal 1990). In the context of the history of the development of the JCT, this provides a strong indication of the assumption that there should be one standard form of contract for building work, an idea dealt with in Section 8.1.

Because so little guidance exists about where such contracts ought to be used, it is necessary to examine the circumstances where they are used in practice. This reveals that general contracting is often used as a 'default method'. In other words, it may be used merely because the consultants have failed to consider the issue of appropriateness. The general contracting approach seems to be adopted where no one plans the means of procurement. There have been attempts to describe those circumstances where general contracting is likely to be successful. Indeed, JCT have issued a Practice Note which is intended to help in the selection of the most appropriate form of JCT contract (Joint Contracts Tribunal 2001) but, naturally enough, this fails to take into account the 'competition' from non-JCT contracts. The following list summarizes typical circumstances in which general contracts are used:

1. The employer has caused the design to be prepared and for the purposes of the building contract takes responsibility for it.
2. The employer's designer is sufficiently experienced to co-ordinate and lead the design team and to manage the interface between design and production.
3. The design is substantially complete when the contractor is selected.
4. An independent quantity surveyor will be used to plan and control the financial aspects of the project.
5. The contractor is selected on the basis of the contractor's estimate and carries the risk that the estimate may be wrong.
6. The employer reserves the right to select sub-contractors for certain parts of the work.[1]
7. 'Prime cost sums',[2] including employer-selected sub-contracts, do not form the major proportion of the contract sum.
8. The employer's agents feel that it is important to use an acceptable negotiated form to ensure a fair and familiar distribution of risk.
9. The employer makes no explicit choice and the advisors do not raise the issue.

This list has been developed by considering the types of project where general contracting has been used successfully. It is not proposed as a typical way of arriving at the conclusion to use general contracting. But many of these criteria are observed on most general contracts. If a general contract is let under different circumstances, perhaps a different procurement method should have been used. Each of the circumstances is discussed in more detail below.

3.2.1   Employer commissions design

In general contracting, the designers act on behalf of the employer in converting the employer's requirements first into a brief and subsequently into a workable

---

[1] In ICE 7, employer selected sub-contractors are called *nominated* sub-contractors. In JCT IC 05 they are called *named* sub-contractors (see Chapter 19). There are no provisions for employer-selected sub-contractors in either JCT SBC 05 or the Engineering and Construction Contract.
[2] A 'Prime Cost Sum' is a part of the work that is not priced by the tendering contractor. The tender documents simply tell the contractor how much money to include for such an item so that overheads and profit can be added. When the work is executed the contractor is paid whatever the item ultimately costs (see Section 9.1.2).

design. The processes of briefing and design are complex, difficult and often poorly understood (Gray and Hughes 2000). By developing a brief and a design philosophy for a project, an architect acts not simply as an agent of the client but also as a representative of the architectural profession exercising judgements and guarding the interests of the public (see Section 1.8). In the light of this, the idea that the employer instructs the architect and the architect converts these instructions into a design is a gross over-simplification. The preparation of a brief involves an intensive and comprehensive dialogue between designer and client, in the context of the project's environment. Therefore, the involvement of the designer in the subsequent construction process is inevitable because of the impossibility of converting the overall requirements into a simple set of instructions. If the employer's requirements *can* be given in a simple way, then the designer may be redundant and general contracting is probably unsuitable.

Although civil engineering projects are not generally subjected to aesthetic arguments, the relationship between a civil engineering project and its environment places civil engineers in a similar position. The involvement of the lead designer in the production process is necessary to ensure that the contractor's work is in accordance with the specifications, to revise and refine the design where necessary and to respond to the contractor's requirements for guidance and information.

3.2.2 Experience of the lead designer

The lead designer has much flexibility in choosing the design team. Typically, consultants with significant input will be in direct contract with the employer but often the lead designer will sub-contract parts of the design to specialists. The wide range of design inputs demands good co-ordination and integration skills, as well as good communication skills. In addition, the lead designer must develop a clear understanding of the employer's requirements and, in the case of commercial projects, a clear understanding of the way that the project fits into the employer's business strategy. Such a complex role demands a wide range of skills. Although many architects and engineers have proved themselves worthy of this task, the employer must be clear about the kind of person that is needed and must use careful selection procedures. Clients should be wary of design practices that use the senior partners to impress clients and earn commissions, but more junior staff to actually undertake the work.

3.2.3 Complete design

It is customary to choose a contractor by opting for the one with the lowest price. As stated earlier, one of the reasons for a standard method of measurement is to ensure that bills of quantities are prepared in a consistent way. This bill ensures that all bidders calculate their prices on the same basis. However, a *comprehensive* bill (i.e. **fully measured** bill) can only be prepared from a complete design. But there are techniques to deal with incomplete design. For example, quantities need not be precise, in which case a bill of **approximate quantities** can be used. There is a version of JCT SBC 05 available for this. In civil engineering contracts such as

ICE 7 approximate quantities are the norm. Because the final quantities are not known until after the work is executed, this is termed a **remeasurement** contract: the work is specified in approximate measure for the purposes of tendering and after execution is measured again for the purposes of payment. By their nature, civil engineering contracts, unlike building contracts, cannot be specific about everything. For example, ground conditions are not clear until the ground is actually excavated. Finally, there is also a 'without quantities' version of JCT SBC 05 for use where the design is so undeveloped that there are no measured quantities at all. Although this version enables bidding to proceed on incomplete information, it should not be taken on lightly because the without quantities version makes it difficult for a contractor to estimate economy or diseconomy of scale and to calculate an offer accordingly. Hence, the use of alternatives should always be explored.

3.2.4   Involvement of quantity surveyor

The importance of the bills of quantities underlines the need for quantity surveying expertise in general contracting. Quantity surveyors contribute cost planning, cost control, procurement planning and contractor selection expertise. In combination, these enable the choice of contractor to be dictated by lowest price bidding. The training and experience of quantity surveyors enables them to focus on financial aspects above all others. Indeed, clients for whom finance is a priority sometimes appoint a quantity surveyor as a project manager to oversee design and construction. However, it is important to remember that there is more to construction than finance. The project team must be balanced in a way that matches the priorities for a project. The particular tension between design, finance and workmanship, represented by general contracting, is only one of many. A client who wishes to have quantity surveying skills in a central role, balancing these against design and craft criteria, is well placed to take advantage of general contracting. Other balances require different approaches.

3.2.5   Price as the basis of selection

There are many ways of configuring the pricing of work in a general contract. For example, as mentioned in Section 3.2.3, it is typical for ICE 7 projects to involve remeasurement in that the quantities in the bills are intended only as a guide. However, if firm quantities are used (as in JCT SBC 05 **with quantities**), then the work is not remeasured after execution but the contractor is paid on the basis of the quantities in the bill, *not* the quantities actually executed. The only exception to this is work that has been subject to a variation instruction, i.e. the specification has been changed in some way after the contractor starts work (see Section 15.3). JCT IC 05 may be used with bills, but not necessarily. In general, these contracts can be used with any of a wide range of pricing mechanisms. However, one thing remains constant: in all cases, the contractor offers to do the work for a price, not for reimbursement of cost. A contractor who estimates too low is held to the bargain, even if the job runs at a loss. Similarly, for a contractor who estimates too high, the bargain still holds, despite the employer paying over the odds. This principle underlies some important decisions in the courts that demonstrate reluctance to intervene in what may turn out to be a bad bargain (see Section 10.4.2). It applies as much to every rate in a bill as it does to the whole contract sum. This is one reason that the bills play such an important role in the management of construction contracts. An employer who does not wish to have the contract governed in this way should use an alternative procurement method.

3.2.6   Employer-selected sub-contractors

There are several circumstances under which an employer would wish to select a sub-contractor with whom the main contractor must enter into a contract (Hughes, Gray and Murdoch 1997). In some cases, the employer requires control over the selection of certain specialist sub-contractors (see Chapter 20 for a fuller discussion of these points):

- To ensure the chosen sub-contractor has a proven track record for good work.
- To use a proprietary system chosen by the employer.
- To use a sub-contractor with whom the employer has developed a long term business relationship.
- To base the selection of the sub-contractor on a basis other than lowest bid.

In other cases, the employer may make the selection because of the needs of the design team or because of the nature of the construction process:

- Some specialist sub-contract work requires a longer lead time than the project construction programme would allow, therefore, such work must be started before the main contractor has been chosen.
- The design team may wish to ensure the quality of design input from a specialist sub-contractor.
- The designers and quantity surveyors may wish simply to increase their professional influence over the details of the project.

A contractor will be paid whatever nominated sub-contract work costs. Such work falls into the category of prime cost, or cost reimbursement. For this reason, the risks for the contractor are different to those associated with the employment of domestic sub-contractors (see Chapter 19). The latter involve simply an arrangement between the contractor and the sub-contractor and the contractor pays them from within the contract price. Nomination must be used only with extreme care because it fundamentally alters the balance of risk between employer, contractor and sub-contractor. Wallace (1986) goes as far as suggesting that nomination should never be used, referring to it as a conspiracy against employers! It is interesting, in the light of this that, in the JCT SBC 05 contract, all the clauses associated with nomination, developed over many years of evolution of the JCT standard forms, have been dropped.

3.2.7   Proportion of prime cost sums

Nominated sub-contracts are one example of a category of bill items called prime cost sums. Contractors do not price these; they are simply given an indication of the amount so that they can price their overheads and profit and plan their supervision workload. General contracting is based on the assumption that the contractor puts in a price against a complete design. The clauses in general contracts apportion the risks accordingly. If a large proportion of the work is in prime cost sums, the contract will be unsuitable because the contractor's role is reduced to that of a co-ordinator and accountant, rather than a builder, and the clauses do not reflect or give effect to such a role.

3.2.8   Negotiated contract

Most standard forms, particularly the ICE and JCT forms, are negotiated contracts. This means that representatives of the various interest groups have contributed to the drafting process. When this happens, the parties are treated in law in the same way that they would be had they individually negotiated the particular terms of their own bargain. Therefore, in the event of a dispute over interpretation of the conditions of contract, the courts will interpret any ambiguities fairly between the parties.[3] If, on the other hand, a contract has been prepared by only one of the parties, who may use superior bargaining power to get the other party to accept it, then the legal principle of **contra proferentem** might apply (see Section 10.2). This means that ambiguous clauses would be construed against the party who put the contract forward.

3.2.9   In the absence of any other suggestion

If no consideration is given to the method of procurement, it is most likely that traditional general contracting will be used. For many clients of the construction industry, the reflex action when considering the procurement of a building is to appoint an architect. Architects, in turn, will typically expect to design something to be built by a contractor and may also encourage their clients to engage a quantity surveyor. These reflexes inevitably lead to general contracting. Therefore, if a procurement method is not immediately clear, and the project does not clearly demand some specific approach, then perhaps a general contract is probably best. The choice of which form of contract to use depends upon a further set of criteria. Consideration must be given to the following:

- The amount of design to be done before the contractor is selected.
- The level of nomination required.
- The duration of the contract.
- The need for speed.
- The susceptibility of the contractor's costs to market fluctuations.
- The overall size and complexity of the project.

---

[3] *Tersons Ltd v Stevenage Development Corporation* (1963) 5 BLR 54.

- The susceptibility to change of the client's requirements.
- The method by which the contractor should be selected.
- The ability of the client and/or architect to manage and co-ordinate.
- The novelty of the project.
- The skill and experience of the consultants being engaged for the work.

Reference should be made to JCT Practice Note 5 (Series 2) to decide upon which of the JCT main contracts should be used (Joint Contracts Tribunal 2001).

3.3 BASIC CHARACTERISTICS

The range of main contracts available for general contracting has already been commented upon. This section highlights typical features of such contracts. All of these points are considered at length elsewhere in this book; the aim here is to provide an overview.

3.3.1 Design and workmanship

As mentioned above, the key feature is the involvement of the architect in a design and co-ordination role, the quantity surveyor in a cost planning and control role, and the contractor in a production role. This means that the contractor is not, theoretically, liable for design, but only for workmanship. In practice, however, the division between design and workmanship is not always clear-cut. The design must be documented in order for the contractor to be able to tender for the work. However, the documentation will not cover everything. Many detailed aspects are within the skill and knowledge of a competent contractor. Therefore, the exact position of individual nails in a floor, or the location of joints between different pours of concrete and their formation, are matters which are often left to the contractor. There are also many factors which are not documented simply because no one thought of them before the site work started. In these cases, it is essential for the contractor to seek clarification from the architect. A contractor who makes assumptions (even when based upon common sense and experience) will incur design liability for those choices. This leaves the employer exposed because the contractor will probably not carry indemnity insurance for such design decisions.

3.3.2 Contractor's obligations

This leads to the next distinctive feature, which is that the contractor undertakes to do the work described in the documents. It is rare to find a building contract that places upon the contractor the obligation to supply a particular building. This type of obligation (called *entire contract*) would carry with it a fitness for purpose warranty and an obligation to complete the whole of the building in its entirety. Such obligations are covered by design and build contracts (see Chapter 4). But general contracting usually obliges the contractor to carry out the works shown and described in the contract documents. The contracts also state that the quality and the quantity of the work is that which is described in the contract. Bills of quantities, if used, will usually be required by the contract to be prepared according to the relevant standard method of measurement. In building contracts, certain items specified as such in the bills, have to be done to the reasonable satisfaction of the architect, whereas in civil engineering contracts the contractor must ensure that **all work** is in accordance with the contract, to the satisfaction of the engineer. Some contracts go as far as saying that the contractor must comply with any instruction of the contract administrator, on any matter concerning the works whether mentioned in the contract or not! Clearly, there is a wide variety of practices in the way that standard contracts deal with the contractor's obligations. These are dealt with in more detail in Chapter 11.

3.3.3   Nominated sub-contracting

An important feature of general contracting is the nomination of sub-contractors. It is an excellent mechanism for ensuring that the contractor employs sub-contractors of adequate standing. Without nomination, contractors might use high calibre sub-contractors when tendering, in order to keep the bid price high, but on winning the job they may re-negotiate to find the cheapest possible sub-contractor, in order to maximize the return on the contract. Nomination forces the contractor to use a particular sub-contractor. The reasons why nomination was developed have already been discussed. However, the abuses of this mechanism have been frequent (Huxtable 1988). One typical 'abuse' is to use nomination to relieve the architect and/or quantity surveyor of the burden of having to detail the design and specification fully. By nominating a sub-contractor, the design team can get away with minimal documentation. However, this is bad practice, and it will generate problems during the construction stage. If everything is measured and/or specified adequately, then most of the problems can be overcome. Sub-contracting is dealt with in more detail in Chapter 19, and nomination in Chapter 20.

3.3.4   Variations

The power to change the specification, known as a **variation**, is a feature of general contracts (see Chapter 15). This gives the contract administrator the power to change the work required of the contractor. The 'recitals' to the contract are a part of the documentation used to provide background and context for the agreement and would usually provide a brief description of the whole project. Any material alteration to these would go to the root of the contract, and therefore could be challenged by the contractor. This is despite the common practice in standard-form contracts of stating that no variation can vitiate or invalidate a contract. If a change makes fundamental alterations to the contractor's obligations, and it could not have been foreseen at the time the contract was entered into, it is beyond the scope of a variation clause.[4] Such clauses give the contract administrator the

---

[4] *Thorn v London Corporation* (1876) 1 App Cas 120; *Sir Lindsay Parkinson & Co Ltd v Commissioners of Works and Public Buildings* [1949] 2 KB 632, [1950] 1 All ER 208, CA; *Blue Circle Industries plc v Holland Dredging Co (UK) Ltd* (1987) 37 BLR 40, CA; *McAlpine Humberoak Ltd v McDermott International Inc (No 1)* (1992) 58 BLR 1, CA.

power to issue variations on the nature of the work and to the contractor's methods in terms of access to site and so on. There are usually detailed provisions for valuing the financial effect of variations, and these are based, as far as possible, upon the contractor's original price. The principle here is that the contractor should be paid according to what would have been included in the bills, had the contractor known about the varied work at the time of tendering. Only in exceptional circumstances should the basis of the payment to the contractor be total cost reimbursement (day-work). Variations clauses enable the client's design team to refine the design as the contract progresses, but the provisions are often abused by careless clients who see the opportunity to make arbitrary changes to the works as they proceed. This practice leaves the client dangerously exposed to claims from wily contractors who can demonstrate all sorts of consequent effects, which would attract extra payment under the contract.

3.3.5 Payment

Payment provisions are typically based on the assumption that the contractor will be paid in instalments for the work (see Chapter 15). Under most forms these instalments are based on the value of the work executed to date and are regulated by the use of certificates. Interim certificates will state how much money is due the contractor. Other certificates may be issued to record non-completion, completion, making good defects and so on. The contract administrator issues all certificates. A small amount of money on each interim certificate is retained by the employer in a retention fund. This fund provides the client with finance in the event of contractor insolvency during the progress of the work, and encourages the contractor to remedy any defects in the contract after completion.

3.3.6 Completion

Completion is rarely a clear-cut affair. When the work is substantially complete, a certificate is issued, known variously as a certificate of substantial completion, certificate of practical completion or taking-over certificate. At this point a period commences during which time certain defects may be the responsibility of the contractor. This period is variously known as defects liability period, rectification period, defects correction period or maintenance period (this last being a misnomer because there is rarely the intention that the contractor will maintain the facility). This period is usually six or twelve months. During this period, the contractor has the right to remedy any defects that become apparent in the completed building. Without such a right, the employer would be entitled to employ other people, and charge the contractor for the work (see Chapter 14).

3.3.7 Extensions of time and liquidated damages

The contract period may be extended for various specific reasons. Such an extension will relieve the contractor of liability for liquidated damages for that particular period. Liquidated damages are an estimate of the employer's likely losses, which may be incurred by late completion. An amount of money is inserted in the appendix or contract particulars. This amount has to be estimated before the contract is executed. It is not necessary for the employer to provide evidence of such financial loss, because the amount of money levied against the contractor is not related to actual loss; only to the loss that could have been reasonably foreseen at the time of tender. If the amount of money is not a realistic pre-estimate, or if it is punitive, then it is deemed to be a penalty and can have no effect in English law (see Chapter 21).

3.3.8   Fluctuations

Another financial provision is that related to changes in market prices of the contractor's resources. These may be dealt with in a fluctuations clause. On a contract with a short duration, it is unlikely that there would be much alteration in the price of the contractor's supplies. However, in periods of high inflation, or on long contracts, the risk of price rises in such resources is very high. It would be unwise to impose such a risk on the contractor. It is more economical to the employer to absorb these risks, for which purpose the fluctuations clauses are used, effecting changes to the contract sum based upon the market prices for labour, materials and other contractors' costs.

3.3.9   Back-to-back contracts

The publishers of standard forms generally also publish standard-form sub-contracts for use in sub-contracts and consultancy contracts. They are designed very carefully to fit in with the main contract. It is a very dangerous practice for employer's agents to recommend anything other than back-to-back contracts in such situations. The term 'back-to-back' denotes documents where the sub-contract terms match closely the main contract terms.

3.4   RISK IN GENERAL CONTRACTING

The basic disposition of risk on general contracting is extremely important. It is often misunderstood and poorly explained. For example, a contract in Liverpool for refurbishment work to a church made insurance of the works a responsibility of the employer. The employer did not realize this. During the execution of the work the contractor formed a hole in the roof, which was not made weather-tight overnight. When rain penetrated the roof and caused extensive damage to the organ, it turned out that the employer (client) was uninsured for this kind of damage. The architect was held to have been negligent in not explaining the insurance provision to the employer.[5] This is an excellent example of what can happen when a clause in the building contract allocates the risk, but one or other of

---

[5] *William Tomkinson and Sons Ltd v Parochial Church Council of St Michael and Others* (1990) 6 Const LJ 814.

the parties fails to appreciate it. The most important areas of risk are dealt with below.

3.4.1 Money

The employer is entitled to expect the building to be completed on time. Failure to achieve this will render the contractor liable to pay liquidated damages. By this is meant a fixed rate of money (pounds per time period), which is entered into the appendix to the contract or, as in the JCT 2005 suite of contracts, in the contract particulars. Although there are strict rules in law about the extent and application of liquidated damages (see Section 21.2), these damages should be sufficient to enable the client to continue trading, despite the lack of the new building, or to enable the client to deal with any other form of damages incurred due to the lateness of the building. These sums, then, may be critical to the continuing survival of the client's business. Construction clients should not have to risk their very survival upon engaging in the construction process. In addition to liquidated damages, an employer should ensure that all of the contractors and sub-contractors in the process are backed up by financial guarantees and bonds, otherwise liquidated damages provisions may not be worth the paper they are written on.

The financial effects of the contract are contained within the payment provisions. It is important to understand the way that contractors build up their prices for the purposes of tendering. The primary element of their prices is the cost of the materials and labour, as well as head office overheads. On to these are added premiums, which reflect the market situation and the level of risk associated with the project. The market premium is such that when work is short, contractors' prices will be as low as possible; when there is plenty of work around, the prices will be high. The allocation of various items of risk to the contractor may affect the contract sum. It is a fundamental principle of contracting that the transfer of a risk from one party to another should be accompanied by a financial premium (Wallace 1986). The passing of too many risks to the contractor will result in either inflated tenders, or gross underpricing by short-sighted contractors who would not be able to cope if anything subsequently went wrong. The purpose of standard-form contracts is to allocate risks fairly between the parties and, therefore, standard general contracting should imply a small influence of risk over contractor's pricing policy. However, risks *are* apportioned by this contract, and in some circumstances it may be unwise to allocate them in quite this way. When choosing this procurement method, the criteria for selection at the beginning of this Chapter should be studied very carefully, and the allocation of risk should be made explicit, rather than implicit. This leads to one of the strongest criticisms of standard-form building contracts: the apportionment of risk is rarely questioned and, therefore, becomes implicit. In such situations, the employer is only comparing tenders from contractors competing upon the same risk apportionment. Therefore, the employer is unable to assess the suitability of the form of contract.

3.4.2   Default

The risk of default lies basically with the contractor. All work must comply with the contract documents. The contractor is responsible for the performance of every person on site, whether directly employed, sub-contracted by the contractor or nominated by the employer. In the latter case there are often complex provisions because, as should be expected, the fact that the employer reserves the right to choose who undertakes such work carries with it a certain amount of risk. There is a danger that a contract can result in the employer having no recourse when a nominated sub-contractor fails to perform. An example is where a nominated sub-contractor causes delay to the main contractor. This is usually a valid ground for the contractor to claim an extension of time. If this happens, the contractor would not have to pay liquidated damages for such delay, since the contract period had been extended. Therefore, the contractor will have suffered no loss. In such a situation, the contractor is in no position to sue the sub-contractor for breach of obligations, not having suffered any loss. Unless there is another direct agreement between the employer and the nominated sub-contractor, there is no contractual route for the defaulting sub-contractor to be penalized. This underlines the importance of following strictly the procedures set down for the nomination of sub-contractors. Wallace (1986) feels that this kind of arrangement is a weakness, because it breaks the contractual chain. If the contractor was simply liable for everyone, without exception (as in civil engineering contracts), then liability could be passed down the line (see Chapter 19). Now that JCT contracts do not use nomination, the problem may be less important than it used to be.

A further problem with default is related to the inadequate attention given to matters concerning risk allocation. Since general contracting does not provoke fundamental questions about risk, it is rarely, in practice, priced for. There has often been a propensity among employers to amend several of the contract clauses in the standard forms of main contract and among contractors to amend or replace the standard sub-contracts. The wisdom of these practices is doubtful for the following reasons. When given a batch of contract documents to tender on, many contractors and sub-contractors only price for their costs and the market position. In this way, they fail to add in a premium for the risk. It seems from much of what is written about construction contracts that this area of knowledge is poorly understood by those who are most affected by it. If this is true, then the practice of pricing a risk in a contract must be rare. Therefore, it would be very easy to include onerous contract terms on unsuspecting contractors and sub-contractors. The result can only be a higher incidence of bankruptcies and insolvencies, as the people carrying most of the risk are least able to bear it when things go wrong. This is often construed as the contractors' hard luck for being so gullible, but the question must be asked, 'who pays?' In all cases, whatever the legal rights and obligations between the contractor and the sub-contractor, it is the employer who is left with a half-built facility that must still be completed. The practice of passing the risk arbitrarily down the line until someone finally absorbs it is not, therefore, good business sense.

3.4.3 Completion

This is a complex concept in English building contracts. It rarely means that the building is finished. Under most construction contracts it is taken to refer to the time at which the contractor is entitled to leave the site and hand the building over to the employer. This does not relieve the contractor of liability for what has been built, but rather imposes an obligation upon the contractor to repair any defects that become apparent during the defects liability period. The contractor's obligation to complete the works is modified by the legal doctrine of substantial performance. The employer's obligation is to allow the contractor to complete the works: in other words the contractor has a *right* to complete the works, and this right cannot easily be compromised. The only way in which completion, or substantial completion, can be avoided is if either party determines the contract.

Basically, the parties to the contract are obliged to carry their obligations to their conclusion, unless there are express clauses setting out circumstances relieving them of this obligation. Most general contracts carry such clauses, for example, giving an employer the power to determine the contractor's employment if the contractor fails to make regular progress, suspends work or becomes insolvent. Similarly, a contractor may have the right to determine the contract if an employer obstructs the issue of certificates, continuously suspends the works or becomes insolvent. Such termination can only take place within the express terms of the contract. The absence of such terms would leave the parties reliant upon common law remedies (see Chapter 23).

3.4.4 Time

One of the primary requirements, particularly of commercial clients but also equally important for public sector agencies, is to be able to predict the time for completion with some degree of reliability. Buildings form very large parts of any client's investment in their business, and the use of a building is usually critical to the success of the client's continuing function. Therefore, general contracts are very definite about time. There will be a date by which the contract must be completed or a period within which it must be completed. This is entered in the appendix or, as in the JCT suite of contracts, in the contract particulars. A delay will usually allow the client to deduct liquidated damages (see above, Section 3.4.1). However, certain types of delay are permitted, provided that the specific events are identified in a clause which entitles the contractor to an extension of time. They include some items (such as exceptionally adverse weather), which do not appear in the list entitling a contractor to extra money. The result is that contractors often prefer to claim for events other than exceptionally adverse weather but employer's agents try to include a disproportionately large portion of any extension of time under this heading. This enables them to deal with the claim without committing the employer to extra expenditure. The relationships between claims for time and claims for money thus become very important. The key factor is that the granting of an extension of time does not automatically entitle the contractor to extra money. Therefore time and money claims must be dealt with separately (see Chapters 14 and 16).

3.4.5   Quality

The issue of quality is very important. There are many firms chasing registration with the British Standards Authority to become accredited firms under BS 5750 or ISO 9000. Such an appellation is an excellent aid to marketing. It is unfortunate that the Standards define quality as 'conformance to requirements'. This means that if the employer requires a cheap and nasty installation, and the contractor provides it, then conformance to the standard has been attained, but is it quality? Clearly, there are tremendous benefits in having a documented procedure to ensure that what is being supplied is close to what was requested, but such a system does not necessarily ensure quality. There is an aspect to quality that cannot be measured at all. This is the subjective reaction to stimulation, a perception that is sympathetic with something in the observer's emotional make-up. This has nothing to do with measurement. A system that concentrates upon documentation and measurement as evidence of quality may militate against any qualitative reaction in observers. Perhaps this is not the place for a philosophical discussion about what constitutes quality; but the current emphasis on *documented* quality raises this question in every aspect of construction. The general contracting procurement method and its variants are ideal candidates for documented control systems, because the involvement of the major contributors, those who produce the building, is entirely dependant upon the contract documentation. To the extent that a client wishes to predefine the contractor's targets, and to have legal recourse in the event of their not being achieved, general contracting has what it takes. However, ensuring that a contractor produces what was specified is not the same as ensuring that the contractor produces a good, high quality building.

3.5   STANDARDIZED APPROACHES TO GENERAL CONTRACTING

General contracting has been around for a little over a century and a half. It was ideally suited to the emerging construction industry between the Industrial Revolution and the post-war boom. More recently, it has been criticized for its inflexibility and attachment to outmoded ideas of construction organization. There are many factors that come together to influence the form and content of standard-form contracts. Not all of these factors are in the best interests of the client. Unfortunately, the way the risks are apportioned under a general contract, while fair to all of the parties, is not well understood by many who use the forms. It almost seems that the existence of a drafting committee, representing all of the major interested power groups in the industry, relieves many people of the responsibility for thinking about and understanding the contracts they sign. Every person who engages in the construction process should seek to understand the contracts they are using, especially when they are using amended or unilaterally drafted contracts. The aim is not to become litigious, but to understand better the positions from which adversarial conflicts develop. A good understanding of each other's positions would enable adversaries to resolve their disputes quickly and cheaply, without recourse to the law. As always, in the final analysis, it is not what the contract says that matters, but how the people in the team interact.

4 Design and build

Design and build is a procurement method that has been in use for a long time. The process is found in many industries. When somebody buys something, the usual process seems to be to buy a product that has been designed by its producer. Examples of this are found in fields as diverse as shipbuilding and microelectronics. Indeed, before the emergence of architecture as a profession distinct from fabrication, pre-industrial society used to procure buildings by a process of design and build. It was the separation of responsibility for fabrication from responsibility for design that led to the emergence of so-called traditional general contracting in the nineteenth century. This separation of design from construction in the building industry has, for a long time, been the source of many problems (Murray and Langford 2003), and remains so.

4.1   BACKGROUND

It would seem, then, that design and build is what would happen if the construction industry were to suddenly come into being, without the evolution of professional institutions that pre-define certain roles in the process. It could be seen as the most logical way to procure a building, given a clean slate to start with. However, such a 'clean slate' does not exist, and over the last 150 years the construction professions have carved out their respective niches.

The roles of the professional consultants continue to develop and change. Consequently, the methods of procuring buildings also change. Because the contractual relationships in the construction process are difficult and complex, most people who participate in the process prefer to standardize the contracts. The emergence of standard-form contracts for design and build took a long time.

The principles of design and build (DB) contracts will be illustrated in this chapter by reference to the JCT approach. The publication in 1981 of the JCT Standard-form Contract With Contractor's Design coincided with the publication of the Contractor's Designed Portion Supplement to JCT 80 (CDPS), both of which were revised in 1998. In 2005 major changes took place again. First, CDPS has now been incorporated into the Standard Building Contract, JCT SBC 05, where it is now called Contractor's Design Portion (CDP). References to CDP can be found throughout the contract. The parties can use the CDP provisions but they don't have to use them. Second, the 1998 DB form has been updated and renamed as Design and Build Contract (JCT DB 05). Hence, the JCT suite of contracts provides two approaches. The JCT DB 05 form is for use where the contractor's design *responsibility* extends over the whole of the works, even though significant parts of the design may have already been done before the contract is executed, being embodied within the employer's requirements. SBC 05 with its CDP provisions is appropriate when the contractor's design responsibility is for only a portion of the works, the remainder of which will have been designed by consultants in the usual way. There is, therefore, a significant difference between the two. This Chapter focuses on JCT DB 05.

Examples of successfully completed design and build projects are too numerous to mention. They range from housing, through industrial and commercial projects, to major complexes. It is clear that this procurement method has very wide applicability. This is because there is always some kind of decision about defining the project in order to appoint a DB contractor. The work that takes place prior to the appointment of a DB contractor may involve design to a greater or lesser extent, and the range of options in terms of how much design is carried out by the DB contractor is wide. At one extreme is the type of deal where the client selects the contractor before any consultants have been approached. This may include acquisition of the site as well as procuring the finance and leasing the building to the occupier. Such arrangements involve specifying the performance of the facility, and leaving the means by which it is achieved entirely up to the contractor, hence the label, Performance-Based Contracting. Most DB contracts would involve initial designs being carried out by consultants on behalf of their clients, and these partial designs are then put to the contractor who would then make proposals about how to complete the work. Frequently, in such a situation, the DB contractor would also be asked to contract with the client's designers, so that there is some continuity of the design effort, but the remaining design work is the contractor's responsibility. This involves formally transferring the designers' contracts from the client to the builder, and this legal transfer of responsibility is called 'novation' (see Section 4.2.3). At the other end of the scale is the procurement of a standard building, or system building. While this may appear to match the essential features of design and build, the actual design work in such an arrangement can be minimal, and may have been carried out at some time in the past by someone other than the contractor. In this case, the contractor simply provides a pre-designed building to a client's standard specification.

While on this topic, it is worth mentioning that there are different applications of system building. When an experienced designer uses this process as if it provided a kit of parts, system building lends itself very well to design and build procurement. An architect who knows the system well, and understands its limitations can manipulate ideas very rapidly. This can help to overcome problems caused by poor or inadequate briefing, which is a particularly important aspect in design and build. It is most important that the brief is concise and unambiguous at an early stage. By using system building the pressure is taken away from the early stages. This does not mean that design and build is exclusively for system building; its applicability is extremely wide. System building has been responsible for the supply of some dreadful failures of the construction industry. The alliance of system building with design and build has produced an unfortunate association of ideas. The problems and disadvantages of bad system building are not problems that are caused by utilizing design and build.

4.2   FEATURES OF DESIGN AND BUILD CONTRACTS

There are some essential features unique to design and build contracts. These can best be dealt with in terms of how the employer describes the requirements for the job; how the contractor proposes to achieve them; the pricing mechanism; and the roles and responsibilities within the process.

4.2.1   Employer's requirements and contractor's proposals

The first of the essential features of a design and build contract is that the employer approaches a contractor with a set of requirements defining what the employer wants. The contractor responds with proposals, which will include production as well as design work. As mentioned in Section 4.1, the scale of design work included depends on the extent to which the employer has already commissioned such work from others. The contractor's design input varies from one contract to another, ranging from the mere detailing of a fairly comprehensive design to a full design process including proposals, sketch schemes and production information. There will usually be some negotiation between the employer and the contractor, with the aim being to settle on an agreed set of contractor's proposals. These proposals will include the contract price, as well as the manner in which it has been calculated.

Once the employer's requirements and the contractor's proposals match, the contract can be executed and the contractor can implement the work. The contractor will be totally responsible for undertaking the design work outlined in the contractor's proposals, for fabricating the building, and for co-ordinating and integrating the entire process. This includes the appointment of consultants if the contractor does not have the necessary skills in-house. The employer may also choose to appoint consultants in order to monitor the various aspects of the work, but this is not always the case, and they would not have a contractual role in the way that they would under general contracting.

4.2.2   Price

A feature sometimes present in design and build deals is a guaranteed maximum price (GMP). This helps clients to feel reassured that they are not signing a blank cheque. As an incentive to the contractor, any savings made by completing the project for a price below the GMP may sometimes be shared between the client and the contractor.

The price in JCT DB 05 is governed by means of a contract sum analysis (CSA). There is no need for bills of quantities in JCT DB 05. The nature of a CSA is very different to that of bills of quantities, its form not being prescribed by the contract. A CSA can be presented in any form appropriate to the circumstances of the project, but some of the purposes of a CSA have their parallels in bills of quantities. These are calculation of stage payments (in some circumstances), valuation of employer's change instructions, and exercising the fluctuations clauses (where applicable). According to JCT's guide to JCT DB 05, the CSA should

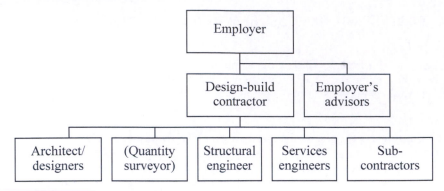

Figure 2: Contractual relationships in design and build

enable the design content of the work to be reflected in any estimation of the value of variations (Joint Contracts Tribunal 2005a: 12).

4.2.3   Roles and responsibilities

One of the most significant features of design and build arrangements is the lack of an independent certification role in the contract. There is no architect or contract administrator to settle differences between the parties and there is no independent quantity surveyor responsible for preparing the basis upon which contractors tender. This changes some of the basic assumptions about the roles required on construction projects, and the consequences of this are discussed later.

Figure 2 shows the contractual relationships encountered in design and build contracts. The role of the quantity surveyor is shown in parentheses because, although cost information and economic advice are essential, there is no need for traditional quantity surveying in this form of procurement. There is no standard method of measurement, no bill of quantities, and no *contractual* role for the quantity surveyor. However, a contractor may choose to employ a quantity surveyor as part of the DB team. Alternatively, or in addition, the employer may choose to employ architects, quantity surveyors, engineers and so on, but they would not have the same role that they would have in general contracting.

Many quantity surveyors have welcomed the departure from traditional bills of quantities and the British approach to cost control which is sometimes seen as counter-productive in serving the best interests of the client. Such quantity surveyors have seized the opportunity of becoming specialist client advisors in all aspects of the economics of building and the procurement process. There are many consultants who are qualified to advise the client during the whole of the process. These consultants come into the category of employer's advisors and would be used by the employer when dealing with the contractor. It is often the case that employers who are more experienced in traditional general contracting will set up their own project team, with a counterpart for each of the professional advisors in the contractor's project team. The contractor's team may or may not be in-house.

The contractual relationships in design and build offer some advantages over other methods of construction procurement. The most important advantage, as clearly shown in Figure 2, is that the contractor is responsible for everything. This 'single-point' responsibility is very attractive to clients, particularly those who may not be interested in trying to distinguish the difference between a design fault and a workmanship fault. (For the design duties of a D&B contractor see Section 13.2.1.) This single point responsibility also means that the contractor is not relying on other firms (for example architects) for the execution of design or for the supply of information. By removing these blocks to effective communication, experience has shown that programmes and budgets are more likely to be adhered to, and the speed of building is likely to be quicker (Pain and Bennett 1988).

The design and build process increases the opportunities for harnessing the benefit of the contractor's experience during the design stages of the project. Many of the developments in procurement processes, and much of the work in the field of study known as 'buildability', have had this purpose in mind. The benefits of the integration of designers and builders are more economic buildings as well as a more economic and effective production process.

One reported disadvantage of design and build is where there is a conflict between aesthetic quality and ease of fabrication, the requirements for fabrication will dominate. A further criticism has been that a design and build contractor will put in the minimum design effort required to win the contract. These two criticisms suggest strongly that quality, particularly architectural quality, will suffer under this procurement process. However, this is not a valid criticism of the process itself, but rather of some of the people who may be exercising it. As such, it is a criticism that can be levelled at any process. Further, like much of the criticism of design and build, it is based upon institutionalized ideas about roles and responsibilities and thus suffers from the same weakness as all stereotyping. It should be noted that the RIBA code of conduct allows architects to take part in construction in different roles, for example as directors of design and build construction companies (Royal Institute of British Architects 2005b: 4). Therefore, it is not necessarily the case that design and build militates against good design practice.

The relationships between the consultants can be of several types. For example, the employer may employ an architect (or designer) to work up some initial proposals. These would form the basis of the employer's requirements, and the employer could subsequently assign the architect's contract to the builder. Alternatively the client may approach directly a design and build contractor who employs in-house consultants. This inherent flexibility is one of the strengths of design and build as a method of procurement. This flexibility happens not just between projects, but also within projects. For example, a commercial client with a prestigious office building may have some very detailed requirements, comprehensively designed by an architectural team, concerning the façade and the entrance lobby of the building. Other less prestigious parts, such as the corridors, staircases, toilets and lifts, may then be left to the contractor to design.

The extent of the contractor's design *role* is sometimes much less than the associated design *responsibility*. A particular variant of design and build involves an initial approach that progresses through the design process in much the same way as general contracting. But, instead of preparing bills of quantities and putting the project out to tender, it is plausible to use the substantially completed design as a set of Employer's Requirements. In addition, by requiring the successful design and build contractor to employ the design team that has been used to prepare the design, the contractor then takes on responsibility for the design, without having had any previous involvement with the design team. The transfer of contractual obligations and rights from the employer to the contractor is an example of a more general legal technique known as novation, which simply refers to the replacement of one of the contracting parties by another. For this reason, the approach is often known as *novated design and build*. Although, on the face of it, this appears to be a variant of design and build, it more closely resembles general contracting with a significant shift of risk apportionment. The widespread use of novated design and build obscures the true impact and frequency of what might be termed *integrated design and build*, the subject of this chapter.

4.3   USE OF THE JCT DESIGN BUILD FORM (JCT DB 05)

On its inside cover, the contract JCT DB 05 gives basic advice about the circumstances under which it should be used. Four aspects are listed. First, there must be a need for detailed contract provisions. Second, the client, called the employer in JCT contracts, has to prepare the employer's requirements and convey them to the contractor. Third, the contractor undertakes to complete the design in addition to the execution of the works. Fourth, an employer's agent is appointed to administer the conditions. Below, the question of appropriateness is analysed in more detail.

Considering the range of uses to which the form has been put, and its inherent flexibility, it would not be relevant to define where it should be used in terms of specific types of building. Rather, the dominant features are the nature of the employer's requirements regarding risk apportionment, the nature of the client's experience, and the availability of suitable contractors to undertake the work.

The characteristics of projects where design and build would be suitable can be considered under the following headings:

- The client's familiarity with construction.
- The relative importance of client priorities (time, cost, function, quality, value for money, etc.).
- The technical complexity of the project.
- The need to make variations to requirements as work proceeds.
- The patterns of responsibility and communication.
- The need for an early start on site.

Each of these is discussed below.

4.3.1   The client's experience

Unlike some forms of procurement, it is not necessary for the design and build client to be an 'expert' client. However, the so-called traditional method of procurement, general contracting, has evolved as a response in many ways to the lack of trust endemic in the construction industry. Clients of general contracting like the feeling of control that the high degree of documentation gives them.

However, if the price of this control is the difference between hitting cost targets and over-running cost targets, then sometimes it is seen as too expensive.

Novice clients, who know nothing of the construction industry, particularly clients requiring small works, will often stumble into design and build without realizing it. For example, an uninsured householder whose chimney blows down in a storm may invite one or more local small contractors to come and look at the damage and to provide an estimate for remedying it. In such a case, a builder might ask the householder whether the replacement chimney needs to match the old one, or even if it can be dispensed with and capped off in the absence of an open fire. This process may take place with a few builders, all of whom will have their own ideas about how best to achieve an economical solution. The client considers the options and eventually appoints one of the builders to do the work. This shows how a complete novice can inform a series of contractors about the employer's requirements and enter into negotiations with each of them about the contractor's proposals. Modifications are made to requirements and proposals until they match, and prices are discussed until the client feels that the most economical and effective job will be done. Therefore, while design and build clients may be very sophisticated, it is equally likely that they may not be. Because of this, unlike some other forms of procurement, the choice about whether or not to use design and build does not depend upon the client's familiarity with construction.

4.3.2   Relative importance of client priorities

Whatever system of procurement is used, the client ought to make clear to the tendering contractors the objectives of the project, and how they relate to each other. It is generally felt in the construction industry that quality is the first thing to suffer in design and build contracts. However, the evidence does not support this (Pain and Bennett 1988). There are good quality buildings from design and build, and there are low quality buildings from general contracting and other procurement methods. There is no correlation between procurement method and perceived quality of the product.

Cost certainty for the employer is one of the advantages of design and build. It is a fixed price contract (though not necessarily a firm price one, see Section 7.3), so the risk associated with pricing is entirely the contractor's, except to the extent that fluctuations clauses apply. However, notwithstanding the cost certainty, design and build is not necessarily cheap. Single point responsibility and fixed price contracts mean that the contractor carries more of the risk than a general contractor would. Risk attracts a premium, so it is to be expected that a design and build contractor would add a premium to a tender to allow for this extra risk. Similarly, design and build encourages economical solutions, and enables value to be considered as well as price. A truly economical approach to the employer's problem may not produce the cheapest building in terms of capital outlay. In this respect, design and build may be more expensive than traditional general contracting. On a high-risk project it may be inappropriate to pass too much risk over to the contractor, so other forms of procurement should be considered. Of course, when the market is weak, or in recession, contractors will not add a premium for risk: in effect it is wiped out by what can be viewed as a negative market premium, i.e. pitching the price below the cost of doing the work. While this may make design and build particularly popular in a weak market, the danger is that contractors will only be able to generate profits from the manipulation of cash flow and from claims.

As a procurement method, design and build offers many different combinations of priorities. A client whose highest priority is aesthetic quality can appoint a design and build contractor with a reputation for architectural merit, perhaps a specialist design and build firm that includes architects as directors. A client whose highest priority is speed or money should appoint a design and build contractor with a record of success in hitting such targets. Clearly, there are numerous ways of configuring the priorities for a project, and the client should make explicit decisions about these before inviting contractors to respond with their proposals. It is essential that the client takes an opportunity to look at previous work by, and to speak to previous clients of, the contractor. Different contractors have different skills, and the selection process should take some account of this in choosing the most appropriate contractor (see Section 9.3.3).

An element of competition is usually seen as advantageous when appointing a contractor. Traditionally, contractors submit tenders based upon a description of the works. In this situation they are competing on price alone, since they are all pricing the same description of the works. One of the strengths of design and build is that the contractor's proposals will include design solutions to problems posed in the employer's requirements. Here contractors are competing not only on price but also on any criteria the client thinks important. This presents opportunities for making the level of accommodation the selection criteria, where a client can put forward a budget and ask the bidders to demonstrate how much building they can supply for the money. Clearly, the element of competition is far more flexible in design and build than in traditional arrangements.

One of the biggest disadvantages of design and build is that employer's requirements must be very clear and unambiguous. Also, they should not be subject to change during the project. This is not such a problem in general contracting because the brief can be developed alongside feasibility studies and sketch schemes.

4.3.3 Technical complexity of the project

When dealing with technically complex problems, the usual solution is to specialize. The range of specialists in construction is extensive and comprehensive. Design and build is one of the few procurement processes that is not conducive to the employer's selection of specialist sub-contractors. In general contracting the process of nomination emerged so that the employer may reserve the right to select particular specialists that the contractor must employ. In construction management the employer appoints all of the trade and specialist contractors directly. This cannot be done easily under a design and build contract. The employer who wishes to specify particular contractors to undertake particular parts of the work must do this in parallel with the design and build agreement but beyond the terms of a standard-form design and build contract. This means that if technical complexity is to be confronted by the use of specified (nominated) specialists, design and build is unsuitable. Even though most design and build contractors will accommodate such specialists if they are named within the employer's requirements, careful drafting will be required to establish a direct relationship between the employer and the named sub-contractors. Such a relationship is not established in the clauses of JCT DB 05.

The technical complexity of the client's process may lead to solutions more akin to engineering than to building. While this does not mean that design and build is unsuitable, it is likely that in such circumstances the use of other forms of procurement will be favoured. This is because of the more comprehensive pre-contractual documentation that may be required by the client with a technically complex project.

4.3.4 Variations to requirements as work proceeds

Variations (i.e. changes to client requirements) are a constant source of problems. They are one of the most frequent causes of claims and often lead to litigious disputes. A client who wishes to reserve the right to alter requirements during the fabrication process should *not* use design and build. The process demands early agreement between employer's requirements and contractor's proposals. A change in either of these documents makes the agreements awkward. The valuation of variations can be difficult without a comprehensive contract sum analysis, and the employer's insistence on time and cost targets is less convincing if the require-ments are altered. Therefore, a client who needs to retain the right to make variations should either consider an alternative procurement method, or should consider allowing the design and build contract to be completed on its original basis, making variations the subject of additional contracts after the conclusion of the project. The limited scope for variations and changes is thus a weakness of the design and build process.

4.3.5 Patterns of responsibility and communication

Single point responsibility is the most obvious advantage offered by design and build. There is no division of responsibility between design and fabrication, so the finished building should reflect the trade-offs made between the design exigencies and the fabrication exigencies. This question of quality has been addressed above, and will be returned to later (in relation to the standard of design, for example, see Section 13.2.1 for the fact that JCT DB 05 excludes the implied term of 'fitness for intended purpose'). In terms of choosing the procurement method, it is essential to appreciate that design and build does *not* mean poor quality. It is a method to be used where the project is one for which it makes sense to combine responsibility for design with responsibility for fabrication. The types of project for which design and build contracts are suitable are those where the contractor's responsibility for design extends over the whole project, whether the design is partially completed by others or not. This is distinct from the situation where the contractor's responsibility for design only covers a particular portion of the works, for which other contractual options should be considered (such as performance specifications or contractor's design clauses).

4.3.6   Early start on site

Since the contractor is undertaking the design work, there are opportunities to overlap the design and construction processes and thus to make an early start on site. This may be attractive to those clients, particularly in the public sector, who need to start spending their budget within a short time of the money being allocated. For such clients the appeal of an early start on site is that they can spend some of the budget before the money is withdrawn in favour of departments with more urgent needs.

In this sense design and build will give the benefits that any form of fast-track construction will give, but with the same penalties. The benefit of fast tracking is that the overall construction process can be speeded up by not delaying constr-uction until the whole of the design is completed. However, too much overlapping will give rise to problems from the need to revise early design decisions, as the design is refined. If the project has already started on site by the time that these revisions are made, work may have to be undone before further progress can be made. In extreme cases, this can lead to fast tracking taking even longer to complete than a traditional method.

4.4   CHARACTERISTICS OF JCT DB 05

The contract itself is divided into articles of agreement, conditions and schedules. In drafting JCT DB 05, the JCT strived for a high grade of conformity with the other forms of their 2005 contract suite. JCT DB 05 is therefore of approximately the same complexity as SBC 05. It is in fact a modified version of JCT SBC 05 rather than a different contract, with about 75% of the text being identical. Rather than describe all of the clauses, those areas where the contract differs from SBC 05 are highlighted below.

4.4.1   Articles of agreement

The articles of agreement contain the recitals, the articles, the contract particulars and attestation provisions.

In the first recital, the employer specifies the scope and location of the works in a few lines in the normal way, and as usual this is defined as 'the works'. This is followed by the statement that the employer has issued to the contractor the 'employer's requirements'. The second recital records the fact that the contractor has submitted contractor's proposals and a tender figure. The figure is presented in the form of an analysis, and this analysis is referred to as the 'contract sum analysis'. The third recital makes it clear that the employer has examined the contractor's proposals and contract sum analysis and is satisfied that they *appear* to meet the employer's requirements. There is a footnote to the effect that where there is any divergence between the employer's requirements and the contractor's proposals, the former shall be revised to ensure that there is no divergence. This procedure ensures that the employer's requirements and the contractor's proposals match. In non-JCT design and build forms of contract there may not be formally documented contractor's proposals at the tendering stage. In this case, the onus is on the contractor to ensure that what is built meets the requirements of the employer. The JCT position, however, is that the employer takes responsibility for ensuring (albeit superficially) that the employer's requirements are met by the contractor's proposals. The difference seems to be that in JCT DB 05 the contractor is obliged to produce whatever is described in the proposals, whereas under other contracts the contractor is obliged to build whatever is described in the requirements.

The contractor's obligations are clarified in JCT DB 05 at Article 1 and clause 2.1.1, which state simultaneously that the contractor must carry out and complete the works in compliance *with the contract documents.* According to clause 1.1, the contract documents comprise, *inter alia*, the employer's requirements. In consequence, the employer's duty to check that the contractor's proposals seem to meet the employer's requirements does not absolve the contractor of liability for any of the work to be done.

The fourth recital deals with the Construction Industry Scheme and the fifth highlights that the works can be divided into sections.

The first article specifically states that the contractor's obligations are to both *complete the design for the works and carry out and complete the construction of the works.* The phrase *complete the design* indicates that some of the design may already have been done. The second article defines the amount of the contract sum, and the third article names the employer's agent for the purposes of acting in the employer's interest in interpreting contract documents, monitoring progress of the work, receiving or issuing information requests or instructions, and generally acting for the employer in exercising any of the employer's rights or duties under the contract. Clearly this person is a key player in helping the employer to discharge the contract effectively but plays a much less important role than, say, a contract administrator under general contracting. The employer's agent may be an architect, building surveyor, quantity surveyor, engineer or any other kind of person. It is not necessary to appoint an employer's agent from one of the construction professions. However, it is advisable to ensure that whoever is appointed is thoroughly conversant with the construction process and its products. The agent should also have a deep understanding of the employer's requirements. The absence of such a person will greatly reduce the reliability of the process.

The fourth article states that the employer's requirements, the contractor's proposals and the contract sum analysis are those referred to in the contract particulars. This removes any possible misunderstanding about the basis of the contract. The need for agreement between the employer's requirements and the contractor's proposals will often produce a series of pre-contract negotiations over their content. This will lead to revisions and refinements to the documents as each is adjusted until they are the same. Consistency is vital. By identifying the precise documents from within the contract, severe misunderstandings can be avoided. The fifth and sixth articles deal with roles under the CDM regulations and the seventh article sets out that the parties may exercise their statutory right of referring their dispute to adjudication. Finally, the eighth and the ninth articles deal with disputes or differences leading to arbitration or legal proceedings.

While both the recitals and the articles contain some space to insert specific data of the project, most of these data are to be completed in the contract particulars. The guide to JCT DB 05 stresses that, by abolishing the appendix and introducing the contract particulars, the new structure brings together all the project-specific data in sequence and at the front of the contract (Joints Contracts Tribunal 2005a: 3).

4.4.2   Conditions of contract

The conditions of contract are divided into nine sections in which related clauses are brought together. The clauses show some very interesting differences from those of JCT SBC 05. The similarities are also interesting, but space precludes their discussion here.

No architect is specified in JCT DB 05, but there is instead an employer's agent. However, the status of the employer's agent is not the same as that of a JCT SBC 05 contract administrator, since all responsibilities for contract administration are ascribed to the employer and not to the agent in JCT DB 05.

In terms of discrepancy between documents, the articles of agreement and conditions will prevail over other documents (clause 1.3). There are provisions about how to deal with discrepancies *within* either the employer's requirements or the contractor's proposals (clauses 2.14.1 and 2.14.2), but no mention is made of how to deal with discrepancies *between* them. This falls back to the third recital, which makes it incumbent on the employer to deal with discrepancies between these two documents before the contract is executed. Discrepancy from within the contractor's proposals will not result in change to the contract sum, whereas discrepancies from within the employer's requirements may be treated as a 'change' (formerly known as a variation).

Clauses 3.5 to 3.14 deal with employer's instructions. These are analogous to architect's instructions under JCT SBC 05 but the employer's power to issue instructions is not as wide under JCT DB 05. For example, the employer cannot issue instructions affecting the design of the works, without the consent of the contractor. The supply of documents is addressed in clauses 2.7 and 2.8. A feature of these clauses is that the contractor must supply to the employer drawings of the works as built, as well as information about the maintenance and operation of the works, in effect a maintenance manual. This provision is clearly necessary because the contractor is responsible for the consultants' detailed design work. As a result, any information that would have been supplied by the consultants to the employer under the general contracting system needs to be catered for under the design and build system.

Clauses 2.1 and 2.2 describe the contractor's obligations. Clause 2.1.1 covers, *inter alia*, statutory obligations and includes arrangements for planning permission. Clearly, planning permission and building regulations approval need to be covered by this contract, to allow for the circumstances where the contractor is appointed at a very early stage in the process. However, there may also be circumstances where the contractor is appointed after these approvals have been sought. Clause 2.1.2 applies except to the extent that the employer's requirements explicitly state that the project complies already with statutory requirements. In other words, if the employer's requirements are silent on this point, the contractor is responsible for obtaining all necessary approvals and permissions.

Variations are called 'changes' in this contract. The employer's right to issue instructions requiring a change is set out in clause 3.9. Apart from that, changes are dealt with in Section 5. This section is very similar to Section 5 of JCT SBC 05 in that both refer to rules for valuing changes only in the absence of agreement between contractor and employer. They differ only regarding the valuation of the new contract price in the absence of such agreement in that JCT DB 05 refers to the contract sum analysis rather than to rates in the bill of quantities.

The situation regarding sub-contractors is very much simpler than used to be the case under more traditional JCT contracts. The DB form does not have the option of listing sub-contractors as provided for in JCT SBC 05. There are supplementary conditions allowing for the use of named sub-contractors. The employer has a right to object to the sub-letting of any work, including the sub-letting of design work.

The provisions for extensions of time, and those for loss and expense, follow very closely the provisions in JCT SBC 05, with the addition of delay in receiving permission or approval from a statutory body (2.26.12) as being relevant for extra time and/or for extra money.

The payment clause is an interesting departure from traditional practice. Clause 4 contains alternatives, in that interim payments may be by means of stage payments (option A) or by periodic payments (option B). The contract particulars must state which alternative applies. In option A, stage payments are made, not at timed intervals, but upon completion of particular items of work. These are entered in the contract particulars by means of brief descriptions, with cumulative values assigned to them. By this means, the contractor is paid as each identifiable part of the work is completed. The alternative is more like JCT SBC 05's provisions, in that the work is valued at timed intervals, usually of one month. There are familiar arrangements covering retention, changes (variations) and fluctuations. Since there is no architect or quantity surveyor, the contractor makes application for the money to be paid. The employer is then obliged to respond to an application, stating how much will be paid (clause 4.10). If there is disagreement about the amount to be paid, the contractor has various recourses available, such as arbitration (if the arbitration option has been chosen), suspension or interest on the outstanding amount.

The processes for the final account and final statement documents are set out in detail in clause 4.12 and the effect of the final statement is limited in terms of its conclusiveness only to certifying payment due.

Most of the other provisions hold few surprises; indeed, the main surprise is that such a large amount of text is directly transferred from SBC 05.

4.5   RISK IN DESIGN AND BUILD

The apportionment of risks in design and build contracts is unique among procurement methods. This uniqueness is brought about by the single point responsibility and by the nature of the relationships between employer's requirements and contractor's proposals.

4.5.1   Money

Whether the contract sum is derived by negotiation or by competition, it is a price for the whole of the contractor's responsibilities, including design, production and any necessary statutory approvals such as planning permission, unless expressly mentioned in the employer's requirements. The contractor must do whatever is necessary to achieve the employer's requirements, whether or not particular items of work or materials have been included in those requirements. This means that the client's financial commitment is very clear, and there is not much risk associated with this for the client. The risk of the cost exceeding the price lies entirely with the contractor. Of course, any changes in the employer's requirements or any hindrance by the employer will result in the employer being liable for extra money. However, of all of the procurement procedures, design and build offers the highest confidence in the contract sum.

4.5.2   Completion of the project

The contractor is committed to complete the project, and the employer is committed to allow the contractor to complete. Nonetheless, the design and build contract may be terminated prematurely in some circumstances. These events, which are listed in clauses 8.4–8.6 and 8.9–8.11, are the same as those in JCT SBC 05 clauses 8.4–8.6 and 8.9–8.11 (see Chapter 23).

4.5.3   Default by contractor

Any fault in the finished building will be the liability of the contractor. It is not necessary for the employer to attempt to distinguish whether a particular problem is a design fault, a manufacturing fault or an assembly fault. The contractor is liable for the performance of the work. This is why single point responsibility is so appealing to clients of construction and is often cited as the greatest strength of design and build.

4.5.4   Time

Again, single point responsibility means that the contractor is responsible for ensuring that the project is completed on time. Any delays beyond the control of the employer would be at the risk of the contractor in the absence of express provisions to the contrary. As in JCT SBC 05, there is a list of relevant events at clause 2.26 of JCT DB 05, dealing with extensions of time. These provisions are substantially the same as those for JCT SBC 05, the difference being that in DB 05 it is the employer who decides what extension is to be granted, there being no contract administrator or architect to exercise this function. As with all JCT forms, many of the provisions relating to extensions of time are there to protect the employer's right to deduct liquidated damages. Extensions of time are *not* solely in the contractor's interest, see Section 14.5.

4.5.5   Quality

Quality is not compromised simply by using the design and build form of procurement. There are design and build contractors who do not understand aesthetics or architecture just as there are clients who do not understand these things. The fact that design and build allows such people to join together and produce buildings may be seen by some as a disadvantage or by others as an advantage. However, if the client requires a high quality building, in the subjective sense of the word, there are no features of this procurement method that would necessarily compromise that requirement. In fact, the nature of the tendering documentation, and the inherent flexibility of the system, may encourage better architecture. On the other hand, if the employer requires conformance with objective criteria that can be measured, there is nothing in design and build to compromise this either. Since the contract is based upon a comprehensive statement of employer's requirements, there is theoretically a better chance of producing a high quality building in the regulated sense than there would be by writing a bill of quantities and putting that out to competitive tender.

The reputation of design and build has suffered from criticisms by some construction professionals of projects involving system building and standardization. This type of project often leads to very poor buildings. It is unfortunate that in the minds of some people, design and build has become synonymous with system building. This is not the case in practice, and criticism of system building should be distinguished from criticism of design and build. In any event, most buildings procured under general contracting are not of substantial architectural merit. It is not the procurement process itself that causes good quality, but rather the people used within that process. The process selected may certainly help, or it may hinder, but it is of secondary importance.

The only reliable way to encourage quality in a building is to obtain a clear statement about what constitutes high quality for a particular client, and to have that statement embodied within the employer's requirements.

4.6   APPROACHES TO DESIGN AND BUILD

Design and build is a logical and clear method for procuring a wide range of buildings. Its increase in popularity has been steady, and there are very few reports of clients who are dissatisfied with it. As a procurement method it is a realistic and worthwhile alternative to general contracting or construction management. There is no real limit on the type or scale of project for which it can be used, but it is inadvisable to use it for high-risk projects, or for adventurous schemes. It offers a high degree of cost certainty.

As with any procurement method, it is the selection of the personnel for the project team that is the most important thing. Having selected the team, the choice of design and build as a procurement method is unlikely to compromise good practice. It has been used for centuries and its use seems likely to continue.

5   Management contracting

Judging by some of the literature on management contracting at the end of the 1980s, it seemed that it was a new direction and a solution to all of the problems inherent in construction projects. Sadly, it was not. Management contracting has actually been in use for a considerable time, although it is only since 1987 that there has been a standard-form contract for this type of procurement. Prior to 1987 there were many instances of management contracts, but they were let using either modified standard forms or contracts drafted by one of the parties (usually the contractor). For example, the project which led to the case of *IBA v EMI and BICC*[1] used the IEE/IMechE Model Conditions. These conditions were used for both the contract between employer and management contractor and that between contractor and sub-contractor, even though they were entirely inappropriate to cover the main contract (Parris 1985). Another project run as a management contract was the construction of the British Library in London. This was entered into before the Joint Contracts Tribunal issued its first standard-form management contract, and was let on the Property Services Agency's own terms. Management contracts have been used for decades, but it seems that their use is strongly linked to buoyant markets, as will be seen in this chapter.

5.1   BACKGROUND

The characteristics of a management contract are that the client engages the management contractor to participate in the project at an early stage, contribute construction expertise to the design and manage the construction (CIRIA 1984). Because of these requirements, it is normal for the management contractor to be an experienced builder or construction company, but this is not a pre-requisite. The management contractor is *not* employed for the purposes of undertaking any of the works, but solely for managing the process. In effect, management contracting is a procurement method consisting of 100% sub-contracting. Every item of building work is sub-contracted to **works contractors**.

This sub-contracting feature is what distinguishes management contracting from the **construction management** approach, discussed in Chapter 6, under which the separate works contracts are all made directly between the employer and the works contractors.

The contractual relationships in management contracting are summarized in Figure 3, which clearly shows that the relationships follow a pattern identical to general contracting (see Figure 1). The opportunity for the contractor to become

---

[1] *Independent Broadcasting Authority v EMI Electronics Ltd and BICC Construction Ltd* (1978) 14 BLR 1, HL.

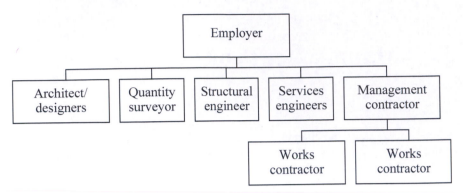

Figure 3: Contractual relationships in management contracting

part of the design team arises not from the contract structure but from early appointment and the pattern of liability and responsibility outlined in the contract. These seek to establish the management contractor's participation on a level that can be equated with that of the professional consultants. It is important to recognize this feature, as it is one of the main reasons management contracting is favoured by contractors. As a form of procurement it offers contractors the opportunity to have the same status in the project team as that of the consultants – to dispense with their labour force, plant and equipment – and reduce their operating costs significantly. Furthermore, as befits a status on a par with the consultants, the management contractor's *risk* in connection with the project is reduced to a level similar to that of the design consultants (to the extent that this can be achieved in such a contractual framework). The aim is to distribute the contractual risk associated with the construction of the building entirely between the client and the works contractors, so that the management contractor is not legally responsible for the defaults of works contractors. This leaves the management contractor with very little contractual risk, although still the contractual channel through which risks are transferred between client and trade contractors.

As we have mentioned above, an important purpose of the management contracting approach is to provide the project team with access to the experience of a contractor at an early stage. It is for this reason that management contracts would distinguish a pre-construction stage from that of construction. Because of this distinction, there is in fact an opportunity for the employer to terminate the relationship after the design stage but before construction work actually commences.

Although such projects involve 100% sub-contracting, management contracts are typified by an array of linked documents. This linking creates some special and complex relationships. We shall be examining the range of documentation that constitute a typical management contract, and considering the types of legal relationship set up by this documentation. It is relatively straightforward to set up a management contract under most forms of documentation, since it is analogous to 100% sub-contracting, provided that the appointment of the main contractor is made in an appropriate way. Indeed, both JCT and the Engineering and Construction Contract have versions for management contracting.

In this Chapter we shall look at the conditions favouring the use of management contracts, before considering the way in which risks are usually apportioned under this type of contract.

5.2 USE OF MANAGEMENT CONTRACTING

The guidance note issued with the first JCT management contract of 1987, which at the time of writing is the only guidance available from JCT, indicates suitable circumstances in which to use this form of contract (Joint Contracts Tribunal 1987a). Several of the suggested criteria are not new; indeed, they are individually important on a variety of different types of standard-form contract. Further, since the guidance note specifically states that not all these conditions are necessary for a successful management contract, it is perhaps only when the majority of these factors act in combination that the management contract should be used. The conditions mentioned are as follows:

1. The employer wishes the design to be carried out by an independent architect and design team.
2. There is a need for early completion.
3. The project is fairly large.
4. The project requirements are complex.
5. The project entails, or might entail, changing the employer's requirements during the building period.
6. The employer, while requiring early completion, wants the maximum possible competition in respect of the price for the building works.

Each of these aspects is dealt with below.

5.2.1 Independent architect and design team

This means that the employer wants the architect and design team to be independent of the contractor. There is nothing new about wishing to have an independent architect and design team; indeed, independence is a normal characteristic of design consultants. The reason for specifically mentioning it in this context is that management contracting is clearly being offered as an alternative to, and perhaps in competition with, 'design and build' packages, where independence of the designers from the builders is not intended.

5.2.2 Early completion

There is also nothing new about wanting early completion, and there are various approaches available to the client who seeks it. Chief amongst these is what is called **fast-track construction**, under which the contract is let as soon as sufficient work is documented and ready to start on site. Although this *encourages* faster overall times, due to the overlapping of design with construction, it causes severe problems of communication. Information must be very carefully co-ordinated if site problems are to be avoided. Indeed, there are many people who claim – with good reason – that this overlapping of design with construction actually slows down the process of construction, because of the concomitant increase in variations, disputes and the escalation of disputes. It is these extra problems of fast tracking that management contracting is possibly designed to avoid, and not just general lack of speed for which traditional contracting is notorious.

5.2.3   Size of the project

It might be thought that the difference between a 'large' project and a 'small' one would be fairly self-evident, but this is a very subjective criterion. It is not at all clear whether size is meant to refer to cost, area, height or level of accommodation. Perhaps it is hoped that clients will simply make a value judgement that their project is 'fairly large'.

5.2.4   Complexity of the project

Complexity is perhaps more easily recognized than size alone. In any event, complexity should be considered together with size. It is the combination of size and complexity that results in the need for management contracts. Take for example a project such as a prestigious modern office development in the heart of London, with a restricted site and high levels of building services required together with extremely high level finishes. The risks for the contractor on such a project are so high as to lead to an inflated price, as the tendering contractors try to cover the risks. It may be in such circumstances that the client is in a stronger position to be able to bear the risks, especially if the client is a property developer who builds frequently. Where a client with large resources builds frequently, the level of uncertainty associated with contractual risk is reduced, according to the principles outlined in Section 1.3.1. If this occurs, it is in the interests of the client to choose a contract form that reduces or even eliminates contractual risk for the contractor, and a management contract is a prime candidate.

5.2.5   Changes to the employer's requirements during construction

The need for variations is another feature which is not particularly novel. As we will see in Chapter 15, there are usually provisions in various standard-form contracts enabling the employer to change the requirements for the project. Exercising the provisions for variations can involve the client in enormous cost and delay, and it is essential that variations are kept to a minimum if claims, prolongation and disruption are to be avoided. It is strange, then, that the guidance notes for the management contract seem to suggest that the management contract is going to facilitate, or somehow make it easier, for the client to alter the requirements for the project. These notes ought really to emphasize that changes should be *avoided at all costs*. That said, however, it would seem that the structure of this procurement path, in which the work is let in packages, means that the freezing of design decisions can be left to a later stage than is possible under the traditional process. For example, the earthworks and substructures packages can actually be let and started on site before much of the detail of the superstructure has been finalized.

5.2.6 Competition on price

Although seeking early completion, the employer may still want a keen price for the works. These two priorities are usually in direct opposition, and consequently have to be traded against each other. However, proponents of management contracting suggest that, because the works packages can be let separately, the lowest price for each package can always be chosen. This will result in a keener price because each package of work will be undertaken by a specialist who will be more competitive for that part of the works. Had that same works contractor tendered for the whole of the works, then they would not have been as competitive.

This argument is spurious. When the practice of front and back loading of bill rates is considered, some parts of the works will naturally show wide variations in pricing if they are tendered for as a whole, but analysed as separate packages. However, once the works contractors are only invited to tender for packages of work, their start-up costs will be multiplied, their rate loadings will not be able to spread over so many items, and the differences and advantages in specific groups of rates will disappear.

More importantly, the depression of prices hoped for by applying this tendering mechanism, if it is successful, will result in a far greater likelihood of cost overruns for each package. Ultimately, there is even the risk that works contractors will screw down their prices so far that they become insolvent, at great expense to the client.

5.3 PRINCIPLES OF MANAGEMENT CONTRACTING

As outlined above, the Joint Contracts Tribunal published their first standard-form of management contract in 1987. The most current available version at the time of writing is from 1998, namely the Standard Form of Management Contract 1998 (JCT MC 98). A new version is planned for 2007. Typically, the structure of a management contract would be broadly in line with the other contracts in the suite. Project-specific data will define the scope of the work as well as certain of the responsibilities under the contract.

The use of project-specific data enables the details to be easily tailored to suit individual projects. It will also provide a succinct summary of what it is the parties are actually agreeing to do. Based on management contracts used in the past, the following data would usually be defined in a management contract.

5.3.1   Description of project

There should be few constraints on how a project is described. This allows a certain amount of flexibility in the use of a mixture of text, diagrams, sketches and calculations; but considerable care is necessary. This description may be referred to recitals of a contract and would thus be fundamental in defining the purpose of the contract.

5.3.2   Prime cost

Unlike general contracts, a management contract is not a lump sum contract. The amount to be paid to the management contractor is the prime cost of all work done under the contract, plus the management contractor's fee. In other words, the management contractor will be paid whatever he or she spends, plus a fee. Excluded from payment should be any costs incurred as a result of the negligence of the management contractor in discharging his or her obligations, as well as any costs already recovered as part of the pre-construction management fee.

Although there may be no specific contract sum in this approach to procurement, there may be instead a contract cost plan. This will have been prepared by a Quantity Surveyor, based upon project drawings and specifications. However, the management contractor would have *no* contractual responsibility if the prime cost of the project exceeded the contract cost plan.

Not all of the work needs to be on a cost reimbursement basis. Certain specific items of the management contractor's direct work (not part of one of the sub-contracts), such as site staff, sundry costs, labour and/or materials provided by the management contractor, may be dealt with on a lump sum basis instead of cost reimbursement.

5.3.3   Services provided by management contractor

It is interesting to note that the roles of a contract administrator and a quantity surveyor probably won't be defined in a management contract. The management contractor simply has to interact with the full range of the professional team. The roles of the professional team remain to be co-ordinated via their own professional terms of engagement. Some approaches to management contracting may come with integrated forms of appointment for the full team.

The services to be provided by a management contractor can be divided between the pre-construction period and the construction period. During the pre-construction period alterations could be made to the services intended to be provided at the subsequent construction stage by mutual agreement (of course, *any* changes can be made to *any* contract by mutual agreement). The pre-construction stage would consist normally of the advisory role to be played by the management contractor, for example in advising on the breakdown of work packages or assisting with negotiations. After the pre-construction stage there are services such as programming and planning, monitoring off-site preparation work, instituting effective cost control techniques, labour relations and site management.

5.3.4 Lists of project drawings and of site facilities

The contract should list the project drawings that help to define the work of the contractors and this list should be referred to from the recitals to be sure of incorporating the drawings as part of the contract documents. It is useful to ensure that all drawings are signed by or on behalf of the employer and the management contractor.

Site facilities and services will be provided by the management contractor for use by the works contractors. These will need to be defined for each project as part of the definition of the management contractor's specific responsibilities.

5.4   RISK IN MANAGEMENT CONTRACTING

To state that one of the main purposes of a management contract is to limit the main contractor's exposure to risk is not as odd as it may sound. The kind of project regarded as most suitable for a management contract approach involves a high degree of commercial risk. And, as we suggested in Section 1.3.2, one of the fundamental principles underlying the apportionment of risk should be that, where a risk is transferred from one party to another, a financial adjustment must be made to balance it. Consequently, if an experienced and competent contractor tendered for a general contract for such a project, there would be a significant premium on the price to absorb the excessive risk. Accordingly, it is not only the contractor who benefits from a low risk contract (i.e. a contract which does not transfer many risks to the contractor), but also the employer.

Many users of management contracts are developers. They are experienced and frequent clients. Following the principles of risk exposure outlined in Section 1.3.1, it is clear that constant exposure to hazards reduces the level of uncertainty and its attendant risk. This type of client is well placed to absorb some of the risks attached to the process. Consequently, the risk that has been taken away from the main contractor should properly be transferred to the employer.

In the early days of management contracting, the intention was indeed to create a 'no-risk' contract for the contractor. This was seen as the best way to encourage the contractor to act as professional consultant. However, as management contracting increased in popularity, some clients sought to shift many of the risks back to the management contractor. The main risks involved were those associated with responsibility for works contractors, time overruns, defects maintenance, preliminaries and design. Effectively, this process resulted in the management contractor carrying as much risk as the main contractor under general contracting, the only difference from a general contract being that 100% of the work had been sub-let. This shifting of the burden of risk led in turn to one of two results: either the management contractor would absorb the risk and put up the price, whereupon relations between the management contractor and the employer became strained as they tended toward the adversarial; or the management contractor would let the works contracts under more onerous conditions, passing the risks to sub-contractors who were less able to bear them.

5.4.1   Money

The most significant difference in risk allocation between a management contract and more traditional forms of contract lies in the price to be paid to the contractor for the work. While the management contractor has an obligation to control costs, the employer has to pay whatever the management contractor spends, plus an amount for the management fee, which may be either a lump sum or a percentage of the prime cost. This means that the contract between the employer and the management contractor is a cost reimbursement contract. The amounts to be paid may be certified by a contract administrator, in the usual way, and payment can be based on interim certificates, stating which parts of the works are covered, so that money can be apportioned correctly between works contractors. The management fee should also be apportioned and included in interim certificates, so that the management contractor is paid in instalments.

Apart from questions of price, it has been common practice to include a clause that, in the event of a works contractor claiming from the management contractor as a result of the default of another works contractor, the management contractor is committed to pursuing recovery from the defaulting works contractor. If this is not possible, for example because the works contractor has become insolvent, then the employer would have to make up the shortfall. This, of course, is a mechanism for effecting a vital transfer of risk from the management contractor to the employer.

5.4.2   Time

Management contracting is designed to encourage overlap between design and construction. For this reason, the chances are that a building will progress from conception to completion more quickly than under a traditional arrangement.

Management contracts often contain special provisions relating to time, most notably an **acceleration clause**. This may well be an optional clause that would only apply if the contract particulars state that it is to apply. At common law, it is not permitted for one party unilaterally to change the terms of a contract once the contract has come into existence, although a variations clause does of course give such a power in respect of what is to be built. Perhaps because an acceleration clause is so unusual, the works contractor may make reasonable objection, although it is not clear what could constitute reasonable or unreasonable in such circumstances (Brown 1988).

The acceleration provision should be echoed in the works contract. It usually operates through a process consisting of a preliminary instruction by the contract administrator, an opportunity for the management contractor and the trade contractors to object, followed by withdrawal or modification of the preliminary instruction. After acceptance of such a preliminary instruction, the management contractor would then need to respond to the contract administrator, in respect of each works contract, with the amount of money as a lump sum required by each works contractor in order to achieve the new completion date. The contract administrator, after receiving this information, would then confirm that the employer is willing to pay the price for acceleration. If the employer is so willing, then the contract administrator can issue the instruction to accelerate.

Such a highly procedural clause is an attempt to compromise between the needs of the employer for flexibility, and the needs of works contractors to be paid for incurring expenditure. A clause like this can be brought into effect to cancel or reduce a previously granted extension of time. The strange thing about this is that anyone should consider it necessary to draft such a clause. Its effect is merely to state what would be the case anyway; that the parties to a contract are free to re-negotiate the terms of their agreement at any time. No contract, standard or otherwise, can disallow the parties from bilaterally agreeing new terms as the project unfolds.

In dealing with the extension of time provisions, an important distinction has to be drawn between qualifying events that entitle a *works contractor* to an extension, and those that entitle the *management contractor* to an extension of time for completion of the entire project. The former group closely would typically resemble those found in JCT SBC 05, but the management contractor's position is very different. The project time may be extended if there are any causes that impede the proper discharge of the management contractor's obligations under the contract (including omissions or defaults by the employer). However, the management contractor would *not* be entitled to any extension where the delay has been caused or contributed to by any default of the management contractor *or of a works contractor* (including in each case their servants, agents or sub-contractors). As a result, while a works contractor may have an extension of time for delay caused by another works contractor or by the management contractor, this will not automatically constitute grounds for extending the management contractor's time for completion.

Rules like this for extensions of time, coupled with a management contractor's liability to pay liquidated damages to the employer if the whole project is not completed by the completion date, appear to impose a very heavy burden of risk on the management contractor. However, there is often a clause in management contracts that effectively means that, where delay is caused by a works contractor, the employer can only recover liquidated damages from the management contractor to the extent that the management contractor is in turn able to recover them from the works contractor concerned. Such a clause would make it a duty of the management contractor to take all necessary steps to ensure that works contractors meet their obligations and would also make it a duty of the management contractor to pursue claims for breach of contract by a works contractor, even to the extent of arbitration and litigation. Even this does not ultimately impose any serious risk on the management contractor, since the employer would be bound to reimburse the management contractor for any money spent in pursuing these obligations. The effect of such a clause is to maintain the chain of contracts, so that losses can be pursued, without attaching independent liability to the management contractor for the performance of the construction work. This is the chief distinguishing characteristic of management contracting.

5.4.3 Quality

To govern the quality of both materials and workmanship, the management contractor can be apportioned full liability to the employer for any breach of the terms of the contract, *including any breach by any of the works contractors* under the terms of their works contracts. But this apparent allocation of a considerable risk to the management contractor is turned on its head by the kind of clause mentioned above, which limits the management contractor's liability to whatever can be recovered in turn from the defaulting works contractor. The philosophy underlying this provision has been expressed by saying that *no independent liability can be attached to the management contractor for any defects in workmanship and materials, on the basis that it is the works contractors who are the parties responsible for the completion of the works* (Brown 1988).

5.5   APPROACHES TO MANAGEMENT CONTRACTING

In conclusion, it is clear that there are many differences between this and other forms of contract, and the complex interlinking documentation may at first sight be bewildering. However, a familiarity with general contracting is an enormous benefit in this respect, as a large number of clauses will be either identical or at least very similar. What must always be remembered is that management contracting does not aim to be an automatic solution to the problems inherent in procuring buildings. It is no more than a mechanism whereby the employer can harness the experience and knowledge of a co-operative and highly motivated building contractor/expert. It will not by itself alter awkward people, nor reorientate contractors who are habitual claimers. Clients need to be very careful in identifying their priorities, and even more careful than usual in selecting personnel to appoint to the project team. And, of course, this team includes the management contractor. What has become clear since standard forms for management contracting emerged is that the use of management contracting has dwindled. This may be a result of bad experiences putting people off. However, it is more likely that recessionary conditions in the construction market altered the basis for management contracting. In a recessionary market no contractor can afford to charge a premium for risk because competition for work is so stiff. In such conditions, the major reason for choosing a management contract disappears. It seems that if contractors are willing to take risks without charging a premium then employers are willing to let them.

6 Construction management

The single most important distinguishing feature of **construction management**, and the one which distinguishes it most clearly from **management contracting**, is that the employer places a direct contract with each of the specialist trade contractors. In order to co-ordinate these contracts, the employer utilizes the expertise of a 'construction manager' who acts in the role of consultant. This technique overcomes many of the problems of the management contracting method, because there is no direct contractual link between the construction manager and the trade contractors.

6.1 BACKGROUND

Construction management has its origins in the USA. There the need for large buildings to be erected quickly and reliably, coupled with increasing technical complexity, led to the involvement of an ever higher number of technical people in the design, programming and construction of a building. The management of these people became less of an architectural issue, and more of a management issue. It seems that the specialist construction manager emerged during the 1930s (Gutman 1988). It was probably the result of a few clients, architects and contractors becoming familiar with each other's working patterns and evolving a method of procurement that was suitable to the task at hand.

In the United Kingdom, evidence that the USA construction industry could perform better, and increasing dissatisfaction with the industry's output, prompted people to examine the management techniques used in the USA. It was assumed that the approach adopted by the Americans (i.e. construction management) could not simply be transferred across the Atlantic. It seems that management contracting was the UK's first attempt at duplicating the American practice. As such it was a compromise between general contracting and construction management. As management contracting grew in popularity and became more generally acceptable, the basic problems of the UK construction industry emerged once again and showed the method to be unsuitable to the majority of British construction projects. It was clear that the 'traditional' forms of procurement were unsuitable for the increased technical complexity of modern construction projects but equally clear that the construction professions in the UK found it very difficult to accept wholeheartedly the philosophy of management contracting. This became evident in two ways. First, management contracting became more a marketing tool for ambitious contractors than a real alternative management strategy. Second, clients modified many clauses in the contract in order to shift contractual risks back to the contractor. The combination of these two factors resulted in the situation where many so-called management contracts were no different from the general contracting. As would be expected under these circumstances, the same old problems emerged and management contracting rapidly fell into disrepute.

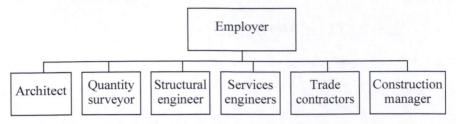

Figure 4: Contractual relationships in construction management

The basic problem with management-based procurement systems lies with the contractual links, information links and authority. As we saw in Chapter 1, the British construction professions have their roots in systems of authority that date back to the middle of the 19th century. Because these professional traditions are so well established, it is difficult to form a team of people who have not worked together in the past and then to expect them readily to modify their traditional roles. Therefore, it seems that novel forms of procurement work best when the members of the team have worked together before. When the situation is unfamiliar or risky, participants feel more vulnerable than usual. Under these circumstances, people tend to entrench themselves contractually, to attempt protection from litigation. Unfortunately, it is under these precise circumstances of unfamiliarity and complexity that the newer forms of procurement are most needed. It is ironic that the increased need for flexibility and dynamism results in rigid hierarchies of outdated management structures (Bresnen 1990).

The apparent failure of management contracting opened the way for attempts to introduce a purer form of construction management into the UK. Figure 4 shows how the contractual links of such a procurement system are formed. It is important to emphasize that this kind of Figure shows only the *contractual* links; it does not hope to convey the implications of construction management for a project's organizational structure.

6.2   USE OF CONSTRUCTION MANAGEMENT CONTRACTS

The Joint Contracts Tribunal published their first and still current standard-form construction management documentation in 2002 (a new edition was announced as forthcoming at the time of writing this book). In this Chapter, we will refer to this documentation (JCT CM 02) as an example of construction management standard-forms.

The construction management method of procurement is most suitable where some or all of the following circumstances are present:

1. The employer is familiar with construction, and knows some or all of the professional team.
2. The risks associated with the project are dominated by timeliness and cost (for example the employer may be a private sector employer requiring a commercial building).

3. The project is technologically complex involving diverse technologies and sub-systems.
4. The employer wants to make minor variations to requirements, as the project proceeds.
5. There is scope for separating responsibility for design from responsibility for management of the project.
6. The employer requires an early start on site.
7. The price needs to be competitive, but 'value for money' is more important than simply securing the least possible cost.

These criteria are somewhat similar to those favouring the use of management contracts. As with management contracting, some or all of these circumstances may combine to make construction management the best option. However, as with any form of procurement, the personnel have the biggest impact. They must understand what is required of them and their position in the team. If this is misunderstood, the chances are that everything else will be and the procurement method will not salvage the situation. There are many similarities between management contracting and construction management; the differences between them are more to do with the application of management principles than with detailed contractual provisions. The circumstances in which it is desirable to use construction management are discussed below.

6.2.1 The employer's experience

If construction management is to work properly, the employer must take an active role in the management of the process. There must be regular and effective feedback from the project team to the employer. The appointment of an employer's project manager may help this process. The need for familiarity applies not only to the *product* of the construction industry but also to the *process* of construction. In order to appreciate this fully, it is desirable if not essential that the employer has some experience of having worked with the construction manager, and some of the other consultants, on previous occasions. This is not a procurement method for the inexperienced.

6.2.2 The importance of time and cost

Risk is very difficult to identify in construction projects. Nonetheless, in order to configure an appropriate management strategy for a project, the employer must make some explicit decisions about the nature of the risks associated with the project. For construction management to succeed, the entire team, including the employer, should understand that the process is intended to be speedier than other procurement methods and the need for speed may compromise decisions about least cost. Fast projects need quick decisions. This requires decisive project team members and a responsive employer. It precludes the investigation of problems from every angle.

6.2.3   Technical complexity

Construction management seems to have been used mainly on 'shell and core' projects. These are buildings, which are not fitted out by the developer until they are let. This reduces the level of technical complexity and also eliminates many site co-ordination problems. It may therefore seem very odd to quote technical complexity as a reason for using construction management. The point is that complex building projects need to be managed with dynamic and flexible working relationships. This implies that the management structure will be adaptable throughout the life of the project and the significance of each participant's contribution will vary from one stage to another. While construction management is suited to technically complex projects, it is equally suitable for technically simple projects.

6.2.4   The need to make minor variations

The need to make variations may not be paramount but minor changes can be accommodated. If frequent changes are envisaged, their administration could snarl up the communication and supervisory processes and make the project more like traditional forms of procurement. If no changes are envisaged, other procurement systems such as design and build may be more appropriate. Changes can be accommodated because they can be made during the process, before individual work packages have been let.

6.2.5   The separation of design from management

Not all projects are suited to construction management. In some cases, design is such an intrinsic part of the project that it cannot be separated from management. In such a situation, design *is* management. However, there are many commercial buildings where design is not the overriding feature of the job. In such cases, the designer's role is reduced to manipulating the relationships between spaces in the early stages and thereafter manipulating the façade. If all other activities related to design have been delegated to consultants, then the problem of co-ordination ceases to be a design problem and becomes a pure management problem. This is the situation most suited to construction management.

6.2.6   The importance of an early start

Many employers are driven by the need to optimize their use of finance. In a competitive business environment an early start on site can be vital to the success of the project. And, in order to get started on site at the earliest opportunity, questions relating to the fundamental design issues must be settled and resolved at an early stage. The comments in the preceding paragraph are pertinent. If the project must generate income at the earliest opportunity, the employer cannot afford the luxury of having a comprehensive design statement translated into a sophisticated architectural concept. The employer's need is simply to house economic operations in a shell that is marketable.

6.2.7   The importance of 'value for money'

Although economic competition is advisable in many cases, it does not mean that least cost is the primary objective. To compete successfully in the market, business premises must be attractive, integrated with their surroundings, and flexible. In addition, their provision must be timely. Thus 'value for money' may be more important than least cost. There must be some mechanism whereby alternatives to competitive tendering may be considered and incorporated. Construction management has the capacity to allow for contracts either to be put out to bid or negotiated. Selection of trade contractors can be based on the most appropriate criteria *for each package*.

6.3   THE CONSTRUCTION MANAGEMENT PHILOSOPHY

Construction management is primarily a management philosophy and secondarily a set of contracts. Therefore, before discussing the details of clauses, the impact of the management philosophy will be considered.

The central feature of a construction management project is that the employer contracts directly with the trade contractors who are doing the work, and the co-ordinator of the construction work has no contractual responsibility for their performance. If the construction manager *does* have contractual liability for the performance of the trade contractors, the arrangement is not really construction management at all, and will probably be found on analysis to be some form of management contracting. That said, however, a construction manager *will* be liable where defective performance by a trade contractor is due to the construction manager's breach of its own duty to exercise the reasonable skill care and diligence of a properly qualified and competent construction manager in managing, administering, planning and co-ordinating the trade contractors' work.[1]

The separate trade contracts are each let directly by the employer, in much the same way that the employer has a direct contract with each of the consultants. In this way, the construction manager is a construction consultant in the same way that the architect is a design consultant. While the construction manager would be expected to manage the overall process in terms of information flow and co-ordination, it is not beyond the bounds of possibility that the architect might manage the process during the design stage, using the construction manager as a consultant for advice on fabrication, assembly and co-ordination. When the time comes to work up the initial proposals into detailed 'bid packages', it is sensible for the construction manager to take a more dominant role while the architect remains a consultant. At every step, even though the construction manager might be responsible for management of the process, design decisions lie squarely with

---

[1] *Great Eastern Hotel Co Ltd v John Laing Construction Ltd* (2005) 99 Con LR 45.

the design team. Design co-ordination remains the responsibility of a design manager.

Depending on the nature of the project, overall responsibility may shift during its different stages, or it may start and finish with the construction manager. This implied flexibility (relating the actual management structure to the needs of the particular job) is essential for successful management of the project. Indeed, it would probably not make sense to have a rigid standard-form contract for construction management since, by standardizing the contractual terms, this flexibility might be lost. For this reason, JCT CM 02 includes a detailed list of obligations of which some must be deleted according to the particular project. Such options help to avoid the danger that once a procurement method becomes established, issues that need to be weighed carefully in the balance at the outset of the project cease to be issues. In a fixed standard approach of any type, attention ceases to be focused on the strategic questions and roles become taken for granted. This could make the whole system prone to the problems that it has been designed to overcome. Therefore, it is good practice to ensure, as JCT CM 02 does, that the standard form cannot be entered into unless the parties make decisions about the apportionment of risks and the management of particular risks, according to the circumstances of the project in question.

The appointment of a construction manager necessarily removes much supervisory and management responsibility from the architect, particularly in the production information and construction stages of the process. This requires that design and management are separate responsibilities. If the two functions are not separated, then the procurement process may well resemble design and build rather than construction management. An inherent problem with such separation is the question of who signs certificates. This is traditionally the role of the design leader as contract administrator. In construction management the contract administrator is not the design leader. It is not obvious who should certify work. Typically, funders and insurers insist that designers sign certificates and therefore, despite the split of design from management, it is almost inevitable that both designer and construction manager have a central role to play in certification and, by implication, other aspects of contract administration. For these reasons, design cannot be fully separated from management.

The intention to split design from management requires the construction manager to be a specialist in management. This calls for skills quite different from those of a general contractor and probably different from those of a management contractor. The wise employer should steer clear of firms who claim to be specialists in all these things. In any event, the employer should always be wary of contractors using the appellation as a marketing tool, rather than as an accurate description of the services offered. Construction management firms in the UK estimate that it can take at least 18 months to acclimatize someone from a general contracting background into the construction management philosophy.

6.4   CONTENTS OF JCT CM 02

As mentioned earlier in this chapter, the Joint Contracts Tribunal published their first and still current construction management documentation in 2002 (JCT CM

02). This documentation is largely based on common practice within the industry and the development of the detail involved intensive industry consultation (Hughes 1997). Here, we will elaborate on the two most important parts of this documentation: First, the construction management agreement between the client and the construction manager (JCT C/CM 02) and second, the construction management trade contract between the trade contractor and the client (JCT TC/C 02)[2]. For a comprehensive list of the different parts contained in JCT's CM documentation see Section 7.5.3.

6.4.1   Construction management agreement (JCT C/CM 02)

The construction management agreement JCT C/CM 02 is a professional services contract similar to those standard forms in use for architectural services. In line with most JCT contracts of the time of its publication, JCT C/CM 02 consists of articles of agreement, conditions, schedules and an appendix. (In contrast, more recent JCT contracts dispense with an appendix and use contract particulars, as part of the articles of agreement, instead.)

Of particular interest are the construction manager's obligations under JCT C/CM 02 since these obligations specify the construction manager's professional role. Within the standard form, these obligations are set out in clause 2 of the conditions which consists of no less than 13 sub-clauses (clause 2.1 to 2.13). Clause 2.1 contains the overarching management obligation stating that the construction manager has to manage the procurement of the project in accordance with the project brief, cost plan, and health and safety plan. The following subsections specify a large number of specific tasks. Above all, clauses 2.1.1 and 2.1.2 state that the construction manager provides services in line with the fourth schedule. This contains a detailed list of the construction manager's duties, subdivided into the pre-construction period (Part A), construction period (Part B) and additional services (Part C). The agreement, in line with the guide to the CM documentation (JCT 2002a), advises that the fourth schedule is a model only which has to be modified according to the particular project. For example, if the construction manager is employed after the design is completed, the default design management duties in the pre-construction period are irrelevant and therefore these duties must be deleted. Furthermore, clause 2.1.3 makes reference to obligations laid down in the sixth schedule (obligations to provide site facilities and to provide services to support trade contractors) and clause 2.1.4 states that the construction manager has to act as an agent for the client towards the trade contractors. The obligations in clauses 2.2 to 2.13 mainly concern four matters: they stipulate the construction manager's duty as certifier regarding the trade contracts; they state that the construction manager must act with that level of care expected of a reasonably competent construction manager in a similar project; they oblige the construction manager to take out professional indemnity insurance (the extent of which has to be described in the appendix); and they deal with the construction manager's duties arising from the CDM Regulations. Interestingly, it is also clear

---

[2] The terms 'trade contract' and 'trade contractor' have been used in the CM documentation in order to avoid confusion with the management contract documentation which uses the terms 'works contract' and 'works contractor'.

that the construction manager is not responsible for a bad design (clause 2.6). This, of course, is due to the fact that the design responsibility lies with the design consultants employed by the client or, as the case may be, with the trade contractor.

6.4.2   Construction management trade contract (JCT TC/C 02)

JCT TC/C 02 consists of articles of agreement, conditions, schedules and an appendix. As with other construction management trade contracts, JCT TC/C 02 is very similar to a traditional construction contract. In other words, this standard form strongly resembles JCT 98, the predecessor of JCT SBC 05. To mention but the most obvious example, both the contractor of JCT 98 (and JCT SBC 05) and of JCT TC/C 02 are under the obligation to carry out and complete the works in compliance with the respective contract documents (clause 2.1 JCT 98, clause 2.1 JCT SBC 05, clause 1.9.1 JCT TC/C). Other similarities concern, for example, the provisions referring to materials, goods and workmanship, opening up for testing and inspection, termination, and payment.

However, some features of JCT TC/C 02 depart from the traditional approach. In terms of differences between JCT TC/C 02 and traditional construction contracts, there are two main issues. The first concerns the problem of who signs certificates (see Section 6.3 above). In this regard, different solutions are chosen by JCT's CM documentation in general and by JCT TC/C 02 in particular: some certificates must be issued by the construction manager (for example practical completion certificate, clause 2.13.1 JCT TC/C 05 – but note that the construction manager is obliged to agree with the architect, clause 2.2 JCT C/CM 02); some certificates require a counter-signature of the architect (for example interim payment certificate, clause 4.11.2 JCT TC/C 02); and some require that the client issues the relevant certificate (for example final project completion certificate, 4.19.1 JCT TC/C 02). The second source of differences relates to the fact that there are several trade contracts rather than only one construction contract in a project. In fact, each trade contract is one of an interrelated series of contemporaneous trade contracts which together accomplish the building project. This situation leads to several problems. For example, while the defects liability period starts when the construction manager certifies practical completion of the works of the trade contractor (clause 2.13 JCT TC/C 02), this period does not end until the end of the *project* defects liability period (which starts with the issue of the project completion certificate and whose length is stated in the appendix of JCT TC/C 02). Especially for a trade contractor who is involved at the beginning of the project, this means a higher risk than in a traditional construction contract. The focus on the *project* defects liability period also means that each trade contractor has to wait until the end of the project, rather than the contract, for the final release of the balance of retention money (guide to the CM documentation, JCT 2002a: 16). In contrast to this, clause 2.18 of JCT TC/C 02 requires that the client protects the completed works and clause 2.18(ii) JCT TC/C 02 releases the trade contractor from an obligation to remedy a defect if the defect is a result of a client's failure to protect the completed works. This is a reasonable solution to the problem of who has to protect the completed works of the trade contractor. Before the publication of JCT's CM documentation it was more typical that trade contractors were obliged to protect their own completed works even though they were not present on site; an onerous condition which has been avoided in JCT CM 02.

6.5 ALLOCATION OF RISK IN CONSTRUCTION MANAGEMENT

The risk apportionments in construction management contracts are substantially the same as those in management contracts, due to the similarity of the occasions upon which they should be used. However, there are subtle differences.

6.5.1 Time

The obligations as to time are entirely related to the construction manager's programme. This means that an employer who wants speedy progress should appoint a construction manager who has proven experience of being able to complete projects quickly. Too ambitious a programme will result in inflated tenders from the trade contractors, and too conservative a programme will result in a slow project. The skill of the construction manager lies in judging the speed of construction.

Construction management has been shown to produce some of the fastest building known in this country. However, this must be weighed against the type of project being undertaken. Shell and core building is very fast anyway, and produces a building which still requires the occupier to fit it out.

The risk of delay lies, in theory, entirely with a defaulting trade contractor. In this regard, our comments relating to time risks in general contracting (Section 3.4.4) apply. However, only trade contractors with financial resources can actually be penalized. Claims for delay could be far more than a small trade contractor's annual turnover and could easily result in insolvency. This is why trade contractors may be required to have sureties and guarantors. Without these, the risks to the employer could be enormous.

In common with management contracts, there is a procedure in JCT's construction management documentation to accelerate the construction process. The relevant acceleration clauses can be found in clauses 2.6-2.10 of JCT TC/C 02.

6.5.2 Money

The direct contract between the client and the trade contractor, without the intervention of a main contractor, ensures prompt payment of certificates. This should improve performance and minimize costs of finance to the trade contractor. These costs would be passed on to the employer anyway, so it is in the client's interest to help the contractor to keep them down. Experienced developer-clients report savings of between 5% and 30% when trade contractors realize the full implications of being in direct contract with the employer.

Since each contract is with the client, there is a high degree of confidence in each of the contract sums. The contract sum in each case cannot be altered except by express provisions of the contract. The client's choice between tendering and negotiating should encourage the use of the correct technique according to the particular circumstances of each trade contract. Provided that the client restricts the use of variations, the final price of the project should be predictable.

There is a choice about how the price is defined. Either there is a trade contract price, subject to negotiated variations where appropriate, or there is a tender sum based on a bill of quantities. It seems that the quantities in such a bill will be remeasured on completion of the work, so unlike a 'with quantities' contract, the quantities in a trade contract are not contractual. This avoids the traditional hold that a contractor can have over a client as information becomes more definite. The risk of actual quantities being different from those anticipated should now lie with the trade contractor; this is as it should be, because such issues are well within the skill and experience of a trade contractor.

6.5.3   Default

The inclusion of design responsibility in trade contracts increases the possibility of default. This design liability is essentially no different to that undertaken by sub-contractors in traditional procurement methods. Since there is no intermediate contractor between the employer and the contractor there is no need for collateral design warranties, except those intended to benefit third parties, such as funding institutions.

6.5.4   Completion

The circumstances under which either party can determine the contract are fairly straightforward and familiar. There is usually no blanket provision for the employer to be able to determine the contract at will, as there is in management contracting. This means that, as usual, each trade contractor is not only obliged to complete the works but has a right to do so.

The need for the employer arbitrarily to terminate the project is not compromised. Since the work is divided into packages, this risk is substantially reduced. No one contractor has the right to continue working for the whole of the project, or, if they have, then perhaps the packages have been divided up incorrectly. Although the employer is obliged to conclude each active work package, there is no obligation to finish the entire project. The employer's risk is thus substantially reduced, without a corresponding increase in the risk to individual contractors.

6.5.5   Quality

The extent to which the employer is protected from having to accept inferior work depends on the adequacy of the architect's description and specification of the work to be carried out. Trade contractors are obliged to ensure that all work conforms to the descriptions in the contract documents, and that the relevant testing and inspection have been carried out. In fact, this is no different to the approaches in general contracts, except that performance specifications are used frequently in construction management and therefore there is more chance that trade contractors are motivated to focus upon the performance of their contribution, rather than the letter of a detailed schedule of items of work.

6.6   APPROACHES TO CONSTRUCTION MANAGEMENT

This procurement method is fairly unusual and only used in certain, very specific circumstances. As well as the JCT documentation, there is also a version of the Engineering and Construction Contract for use with this procurement method. The most important aspect of construction management is that it involves a philosophy quite different from other procurement methods. The danger of a standard form is that it will make it easier for inexperienced parties to get involved with this form of procurement that is inherently riskier than others.

7 Procurement methods and risk allocation

The purpose of this Chapter is to outline the major areas of risk in construction projects. We shall also examine some of the methods of dealing with these risks. The apportionment of risk leads to a consideration of contract structure and procurement strategy including the important issue of deciding which form of contract to use.

7.1 TYPES OF RISK IN CONSTRUCTION CONTRACTS

In Chapter 1 we considered the nature of risk. Here we detail the particular types of risk that are of interest in construction projects. Any construction project by its very nature involves certain unavoidable risks. These have occasionally been analysed and classified, but such analyses are all too infrequent. The following examples summarize many of the risks (Abrahamson 1984, Bunni 1985).

- Physical works – ground conditions; artificial obstructions; defective materials or workmanship; tests and samples; weather; site preparation; inadequacy of staff, labour, plant, materials, time or finance.
- Delay and disputes – possession of site; late supply of information; inefficient execution of work; delay outside both parties' control; layout disputes.
- Direction and supervision – greed; incompetence; inefficiency; unreasonableness; partiality; poor communication; mistakes in documents; defective designs; compliance with requirements; unclear requirements; inappropriate consultants or contractors; changes in requirements.
- Damage and injury to persons and property – negligence or breach of warranty; uninsurable matters; accidents; uninsurable risks; consequential losses; exclusions, gaps and time limits in insurance cover.
- External factors – government policy on taxes, labour, safety or other laws; planning approvals; financial constraints; energy or pay restraints; cost of war or civil commotion; malicious damage; intimidation; industrial disputes.
- Payment – delay in settling claims and certifying; delay in payment; legal limits on recovery of interest; insolvency; funding constraints; shortcomings in the measure and value process; exchange rates; inflation.
- Law and arbitration – delay in resolving disputes; injustice; uncertainty due to poor records or ambiguous contract; cost of obtaining decision; enforcing decisions; changes in statutes; new interpretations of common law.

In examining this list of items, it is important to consider the extent to which they can be priced for at tender stage, and indeed the extent to which they can be predicted at all. Some things may be extremely unlikely to occur; others may be likely to occur but difficult to specify accurately. These factors should have a significant effect on responses to different risks. Extensive research has been undertaken in this area and it has been shown that the development of an adequate contract strategy must be based upon choices about how to respond to risks (Smith *et al.* 2006). Each risk can be analysed regarding its likely range of outcomes. Responses to the risks can then be chosen.

7.2   DEALING WITH RISK

There is a common misconception that everyone wishes to avoid risk. Uncertainty is indeed a problem but it must be remembered that the aim of engaging in construction is to take *calculated* risks. The life-blood of any business is to make money by dealing with the risks that others do not want to bear. Therefore, rather than shy away from them, we should make risks explicit so that rational commercial decisions can be taken about who should bear them.

Understanding this point helps construction consultants to advise their clients about risk allocation. The aim of contract choice should always be to distribute risk clearly and unambiguously. Unfortunately, there is a general failure within the construction industry to appreciate this and the results of this failure are seen in excessive claims and litigious disputes.

The process of dealing with contractual risk falls into three stages:

1. **Identify the risks:** The list given above exemplifies the types of risk that may occur. This is a useful preliminary checklist for promoting discussion about those risks that may be important on a project. Identification of risk must be linked to a clear statement of the client's priorities for a project. For example, if the timing of the project is critical, the severity of time-related risks is automatically increased.

2. **Analyse the risks:** The second step in the process is to analyse each of the risks, in terms of likely frequency of occurrence, likely severity of impact and the range of possible values in terms of minima, maxima and medians for each of these aspects. This may be fairly subjective so that it can be done quickly but is an important step in raising awareness about risk exposure. Some risks may be deemed to be so critical that they need detailed quantitative analysis but most risks are dealt with more subjectively because they have lower priority. A note of caution should be sounded in relation to analysing risks. There is a routine and habitual tendency among those who appraise projects to be over-optimistic about risks and costs (HM Treasury 2003). Indeed, were it not for this tendency to optimism, many projects would never get off the ground. The effect of this 'optimism bias', though, is that risks, costs and programmes are often underestimated.

3. **Respond to the risk:** The range of possible responses is discussed below. In terms of identifying a contract strategy, the previous steps will provide an insight into the priorities of the client and the major risks involved. The final step is to decide who is best placed to manage a risk. The choices lie

between employer, consultants, contractor or insurers. Any decision about laying off risks to others must involve weighing up the frequency of occurrence against the level of premium being paid for the transfer (see Section 1.3.1). It is also important to consider the extent to which the risk can be controlled by certain parties. For example, risks connected with the design of the project are best controlled by the designers; hence liability for defective design is usually allocated to them. For this reason, different procurement options allocate the risks associated with sub-contractors in different ways.

This discussion should make it clear that any arguments about whether one standard-form contract, or one procurement system, is 'better' than another are specious. Each has a role to play in certain circumstances and the consultant who habitually recommends one over another, without identifying and analysing the attendant risks is hardly acting with that degree of skill and care which the client is entitled to expect.

We may now consider in more detail the range of possible responses to contractual risks. The choice lies between transfer, acceptance, avoidance or insurance of risk or even, perhaps, doing nothing.

7.2.1   Transfer of risk

Risks are inevitable and cannot be eliminated. They can, however, be transferred. In accordance with the basic general principle given in Section 1.3.2, the transfer of a risk should usually involve a premium. Because of this, it is unwise to try to burden other parties with risks that are difficult to manage. However, Shash (1993) has discovered that although the risk profile of a project may influence a contractor's mark-up it does not seem to affect a contractor's willingness to bid.

The transfer of risk is achieved through appropriate wording in the clauses of a contract. It is absolutely fundamental to any study of building contracts to understand that contractual clauses are intended to allocate risks. To appreciate the extent to which they transfer risks, it is necessary to understand what the legal situation would be with and without the relevant clauses. This is a basic aim of this book and it is intended that the reader should develop an understanding of the transfer of risk by studying it.

A good example of the transfer of risk is provided by the way in which building contracts deal with bad weather. JCT SBC 05 clauses 2.28 and 2.29.8 (JCT IC 05: 2.19, 2.20) make provision for the contractor to be entitled to an extension of time in the event of exceptionally adverse weather. However, clauses 4.23 and 4.24 (JCT IC 05: 4.17, 4.18) do not give the contractor any entitlement to *financial* compensation in this event. The overall result, therefore, is that this particular risk is shared by the employer and the contractor. By contrast, GC/Works/1, clause 36(2)(e) specifically excludes an extension of time for bad weather (Property Advisors to the Civil Estate 1998c). Under this form of contract, therefore, the effects of the weather are entirely at the risk of the contractor, although the guidance notes point out that the employer's basic contract period should be calculated with an allowance for weather conditions (Property Advisors to the Civil Estate 1998b: 86, 87).

This particular apportionment of risk raises a very interesting issue. Clearly, if the contractor carries the risk of delay due to possible bad weather, a premium can be included in the tender for this. However, what happens if the employer delays the contractor by supplying information late? The contract period can be extended for this eventuality, but the delay may mean that the work is subject to bad weather, which would not otherwise have affected it. It is not exactly clear what would happen in this eventuality because of the complex interaction between two different causes of delay. This example shows how a simple allocation of risk can become very complex when the effects of other risks are taken into account.

As a general principle, it is unwise for the employer to try to pass to the contractor a risk that is difficult to assess. Conscientious and skilled contractors will increase their prices to deal with them or they may insert qualifications in their bids to avoid them. Unscrupulous or careless contractors will disregard these risks when preparing their bid and may then find themselves in difficulty at a later stage. Once this happens, they may try to pass the cost back to the employer. If this fails, they may even be forced into liquidation, which will not help the employer at all.

7.2.2   Acceptance of risk

Clients should avoid imposing undue or unbalanced risks on to contractors. It is not good business practice to try to steal an advantage over the other party in this way. It may be possible in the short term to lay off a risk on to someone else without having to pay a premium for it. In the long term, however, someone will have to pay. Therefore, if one continues to indulge in unbalanced contracts, eventually the worst will happen and the unwilling bearers of the risk will probably be sent out of business. When this happens, it is their creditors who pay the price and the client will be one of those creditors. In the long term, this process of shifting too much risk on to contractors who must accept it in order to stay in business will gradually reduce the available number of contractors tendering for work. All their competitors will have been put out of business.

There are many risks that should lie with the client. If none of the project participants can control or mitigate a particular risk, it is unwise to assign that risk to them. Any explicit transfer will carry a huge premium. Similarly, a client who repeatedly engages in building procurement is wasting money by paying a premium for people to take on risks that need not be passed on (see Section 1.3.1). The client should also carry highly unpredictable and poorly defined risks because the alternative will be tenders that are so inflated as to be unacceptable. Examples of such risks are those associated with war, earthquakes and invasions, which would be impossible to quantify or predict.

7.2.3   Avoidance of risk

The third option to consider is avoidance. Once the risks have been identified and considered, it may be decided that some risks are simply unacceptable. Their explicit definition may persuade the employer that the building project should be redefined or even abandoned. By examining the funding limits of a project and the range of possible outcomes of some of the more likely risks, the feasibility of a project may be compromised. Redefining the project is a method of avoiding a risk altogether. For example, if the financial viability of the project is entirely dependent on the existence of a particular government subsidy, and if there are legislative moves afoot to end such subsidies, it may be prudent to redefine the project so that it is not dependent on such ephemeral support.

In addition to risks between contractor and employer, each consultant takes on risks. The principles of risk identification and avoidance apply here too. Indeed, the most important message from a RIBA report on risk avoidance for architects is to *ensure that the commission is clear and unambiguous* (Cecil 1988). The clarification of responsibilities, remuneration and expenditure at the beginning of the consultant's appointment will help to avoid many problems for consultants.

7.2.4 Insuring against risk

In dealing with 'acceptance of risk' above, we have remarked on the similarity between insurance and laying-off risks. Insurance is an option in some situations. Several risks on the list can be insured, and most of the standard-form contracts insist on certain types of insurance. Standard insurable risks are items such as indemnity against third party claims for injury and insurance against fire. There are also opportunities to insure against loss of liquidated damages and other forms of consequential loss. The actual insurance needed for any particular project should be considered very carefully. Consultants will usually take out professional indemnity insurance, to cover themselves and their clients against the risk of failure to perform the duties with the requisite level of skill and care. These issues are discussed more fully in Chapter 16.

7.2.5 Doing nothing about risk

It frequently happens that none of the project team considers the risks at the outset of a project. If a client is poorly advised and advisors fail to appreciate the level and disposition of risks, the eventual occurrence of disasters takes everyone by surprise. On the other hand, consultants who carefully consider the balance of risks decide that they already lie with those parties who can best control them, may also choose to do nothing. Both courses of action may look the same to the casual observer. In the latter case, perhaps, the people concerned should not keep quiet about the fact that they have considered the apportionment of risk and decided to do nothing but should make their decisions explicit so that they can be challenged.

A further example of 'doing nothing' lies within the main standard-form contracts. Certain events are not envisaged by the contracts, which therefore make no mention of them. In such a case, it would be a mistake to believe that the contract does not apportion the risk. Even by remaining silent, the contract still allocates the risk to one or other of the parties. The problem with allocating risks in this way is that it may lead to misinterpretation and ambiguity, which is a risk in itself. Such a situation can rapidly lead to disputes and claims when problems arise.

7.3   ALLOCATING RISK THROUGH METHODS OF PAYMENT

One of the most fundamental issues in apportioning risk is the way that prices are calculated. The issue is about who takes the risk of the actual price differing from what was estimated. The contractor's work is generally dealt with as a list of quantified items (bill of quantities), to be priced by the bidding contractor. It is convenient to analyse prices in two categories; either 'fixed price' or 'cost reimbursement'.   All items of work fall into one of these types. The difference between them is as follows (Hackett *et al.* 2007):

- Fixed price items are paid for on the *basis* of a contractor's predetermined estimate, including risk and market premiums (Section 3.4.1). The estimated price is paid by the employer, regardless of what the contractor spends.
- Cost reimbursement items are those paid for on the *basis* of what the contractor spends in executing the work.

These two types are not independent. It would be unusual to find a contract exclusively discharged by one method alone. Usually both methods will be used, to varying degrees, in combination. The contract will be described in terms of the dominant method. Thus, most of the items in a bill of quantities for a job let under JCT SBC 05 (i.e. JCT's 'with quantities version' of their standard building contract) will be fixed price. The employer will pay the rate in the bill multiplied by the relevant quantity in the bill. Similarly, under an ICE 7 contract, which is a remeasurement contract, the rate in the bill will be multiplied by the quantity of that item fixed. In both cases, the basis of payment is the contractor's estimate and the amount of money paid has no relationship with the contractor's costs. This type of contract is known as a **fixed price contract**, even though it may have some cost reimbursable elements. Such elements could include **fluctuations**, which are often related to actual changes in market prices.

By contrast, a fixed fee **prime cost contract** will give detailed provisions for payment based upon the contractor's actual expenditure. Even though the contract is based upon cost reimbursement, the contractor's attendance and profit margin is based on a pre-determined proportion of the prime cost. As such, the calculation of that portion bears no relation to actual costs and is therefore of a fixed price nature. This demonstrates how each type of contract contains elements of the other.

The important point in terms of the distribution of risk is that, under fixed price arrangements, the contractor undertakes to submit an estimate for the work and agrees to be bound by that estimate. Thus, any saving over the original estimate will be to the contractor's benefit, and any overspending will be the contractor's loss. Under cost reimbursement arrangements, the employer takes the risk of the final price being different from the estimate, keeping any savings and paying for any increases.

One final point in defining types of contract is that the phrase **firm price contract** should not be confused with **fixed price contract**. Firm price usually refers not to the method of calculating the price but to the *absence* of a fluctuations clause, increasing the chance of the final price being the same as the tender sum. Chapter 14 deals with contractual provisions relating to payment.

7.4   PROCUREMENT CRITERIA

Now that the main areas of risk have been discussed and the range of responses to risks has been identified, we can turn to the selection of procurement method before looking at some of the typical standard-form contracts available.

It is important to understand where each form of procurement should be used. If this is understood, then the procurement method can be chosen in relation to project type. Many clients are reliant upon the advice they get from their consultants. That is, after all, why they appoint them. The power struggles between the professional institutions in construction shed light on the relative importance of each of the procurement methods. General contracting (GC) represents the traditional triumvirate of architecture, quantity surveying and building. It is particularly useful for those traditional quantity surveyors who rely on the large amount of documentation for their staple workload. The decline of GC is one of the pressures behind the diversification of the whole profession of quantity surveying. It is increasingly common to find quantity surveyors practising as *building economists* or *procurement advisors*. Builders are also fairly comfortable with GC because they only have an obligation to build what is documented; therefore they have little liability, other than for materials and workmanship. Typically, the architect's role in GC provides the greatest scope for design leadership, especially given the architect's certification role during the construction process (see Chapters 12 and 17).

Clearly, there have been many problems with the traditional system and one possible way to overcome the traditional problems is simply to turn the tables and have the contractor employ the architect (a typical design and build pattern). On the face of it, design and build (DB) can make GC seem odd by comparison. To appreciate this, consider repairing a car; what customer would specify every item of labour and material and ask the garage to price this schedule? In DB practice, the fabrication exigencies prevail over design ideas, and therefore the ideal of equal status between builder and architect may not be achieved. Worse, the healthy debate between the opposing views may be quietly resolved within the DB organization, and the conflicts are thus hidden. One of the driving forces behind the development of construction management (CM) is to engender an open and participative process, where decision-making is overt. Participation is not possible in a hidden process.

Construction management is intended to balance the power struggle, and to give the design exigencies equal weight with the fabrication exigencies. In a sense, management contracting (MC) went some way toward this by introducing the management contractor at an early stage. But the management contractor remained separated from, and often subservient to, the architect.

This discussion demonstrates that it is basically unrealistic, if not impossible, to develop an ideal procurement system. Just as there is no one best way to organize a firm (Woodward 1965), so there is no one best way to organize a project (Hughes 1989). Many projects suffer from inadequate or inappropriate procurement decisions. The industry lacks a sensible and systematic policy for choosing appropriate procurement systems. The most useful protection that can be offered to a client is a sensible policy for choosing a procurement strategy for each building project.

Least   **Most**

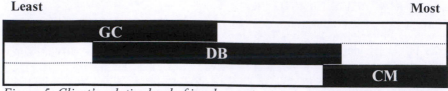

Figure 5: Client's relative level of involvement

To provide a framework for guiding clients through the necessary discussions on procurement method, this section introduces a potential solution to this problem. What follows is based upon the descriptions and assumptions underlying each method in Chapters 3–6, which have highlighted the most suitable circumstances for each of the procurement methods. This will be useful in developing a procurement decision policy for a client, and as a precursor to choosing or drafting an appropriate contract. The most important criteria for choosing procurement methods are:

- Involvement of the client with the construction process.
- Separation of design from management.
- Reserving the client's right to alter the specification.
- Clarity of client's contractual remedies.
- Complexity of the project.
- Speed from inception to completion.
- Certainty of price.

The next seven Sections discuss each of these criteria in turn, showing how each procurement method fares. For clarity, MC is not included because it is rarely used and only in the same circumstances as those appropriate for CM.

7.4.1 Involvement of the client

The first question to consider is the extent to which the client wishes to be involved with the process. Some clients would wish to be centrally involved on a day-to-day basis, whereas others might prefer simply to let the project team get on with it and pay for it when it is satisfactorily completed. There are many points between these extremes. The decision will depend, at least partly, on the client's previous experience with the industry and on the responsiveness of the client's organization.

GC demands the least from clients because it involves delegating most of the management functions to an architect (or civil engineer). While it is not necessarily advisable, it is certainly possible for the project to be left in the hands of the team for its full duration. By comparison, DB does not involve a contract administrator in quite the same role. This places extra demand on the employer. CM removes the role of a main contractor completely and thus the client takes on the most active role. These relative levels of client involvement are summarized in Figure 5. The bars on the Figure are an indication of the range of levels for which each procurement method might prove suitable.

Figure 6: Separation of design from management

7.4.2   Separation of design from management

The relationship between design and management was mentioned in Section 6.2.5. The principle that applies is the extent to which design philosophy should form the basic unifying discipline of the project, or whether more quantifiable aspects should prevail. In the former case, it would be inadvisable to remove management from the traditional purview of the design leader. Such a distinction may emasculate the architectural processes and reduce it to mere ornamentation. But this is not always the case: not all projects are architectural. There is a marked difference between the procurement methods in this respect, as shown in Figure 6.

GC unites management with design by virtue of the position of the architect (or civil engineer) in the process. As design leader and contract administrator, the architect is in control of most of the major decisions in a project. Moreover, general contracts usually contain mechanisms for the contract administrator to retain the final word on what constitutes satisfactory work. In building contracts, such items have to be described as such in the bills of quantities (for example JCT SBC 05: 2.3.3, JCT IC 05: 2.2.1) whereas civil engineering contracts tend to reserve much wider powers of approval for the engineer (for example ICE 7: 13.1). GC therefore displays the least separation of design from management.

On the face of it, DB contracts should also exhibit no separation of design from management since both functions lie within the same organization. While this is indeed true, it also means that where design issues must be weighed against simple cost or time exigencies, issues are resolved within the DB firm. This excludes the involvement of the client from such debates. Further, DB contractors are typically contractors first and designers second (though not always). The fact that DB projects are let on a lump sum will motivate the DB contractor to focus on time and cost parameters over other considerations. Therefore design becomes separated.

In CM there is a clear and deliberate separation of design from management. The design leader has a role in co-ordinating and integrating design work, but the construction manager must ensure that design information is available at the right time and that trade contractors' design is properly integrated. Additionally, the construction manager is the leader of the construction site.

7.4.3   Reserving the client's right to alter the specification

There are three reasons for altering the specification. First, the client may wish to change what is being built. Second, the design team may need to revise or refine the design because of previously incomplete information. And third, changes may

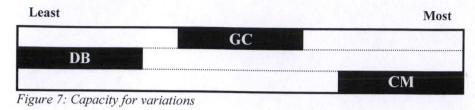

Figure 7: Capacity for variations

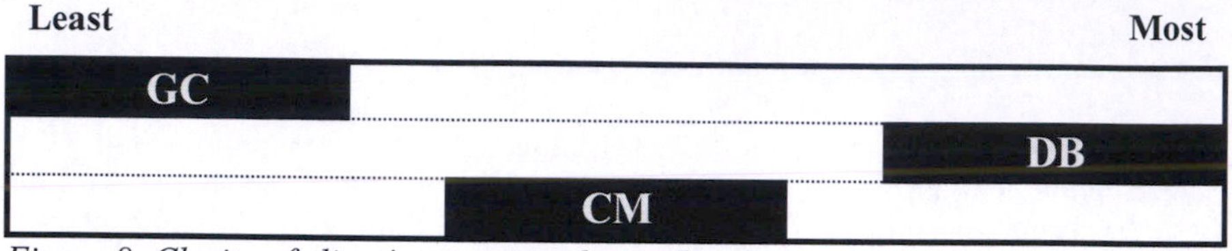

Figure 8: Clarity of client's contractual remedies

be needed as a response to external factors. Although it is quite clear that a construction contract imposes obligations on the contractor to execute the work, it is often overlooked that this also gives the contractor a right to do the work and that right cannot lightly be taken away. If a client wishes to make changes to the specification as the work proceeds, or wishes to allow the design to be refined for whatever reason, then clauses will be needed to ensure this. The procurement decision affects the extent to which the contract structure (rather than clause content) facilitates changes, as shown in Figure 7.

General contracts typically contain detailed clauses defining what would be permitted as a variation, but common law restricts the real scope of variations clauses to those that could have been within the contemplation of the parties at the time the contract was formed.[1] Therefore, despite extensive provisions for instructing and valuing variations, their true scope is limited. DB contracts usually lack the detailed contractual machinery and bills of quantities for valuing variations. Further, as a lump sum contract for an integrated package, variations to the specification are awkward and best avoided. CM involves a series of separate packages, each of which can be finally specified quite late in relation to the project's overall start date, but before the individual package is put out to tender. Therefore, this procurement method has the highest flexibility.

7.4.4   Clarity of client's contractual remedies for defects

An important part of the contract structure is the degree to which the client can pursue remedies in the event of dissatisfaction with the process. Some contract structures are simple, enabling clear allocation of blame for default, whereas others are intrinsically more complex, regardless of the text of actual clauses. One of the fundamental aims behind a contract is to enable people to sue each other in the event of non-performance or mis-performance (defects). The relative ease with which contractual remedies can be traced is shown in Figure 8.

---

[1] *Blue Circle Industries plc v Holland Dredging Co (UK) Ltd* (1987) 37 BLR 40, CA.

Figure 9: Complexity of projects

The least clear contractual remedies arise from GC because the contractor is employed to build what the client's design team has documented. Therefore, any potential dispute about some aspect of the work first has to be resolved into a design or workmanship issue before it can be pursued. The involvement of nominated sub-contractors typically makes the issue much more complicated and difficult. By contrast, DB with its single point responsibility carries the clearest contractual remedies for the client because the DB contractor will be responsible for all of the work on the project, regardless of the nature of the fault. However, this clarity would be compromised if the client had a large amount of preparatory design work done before the contractor was appointed. Under CM, the direct contracts between client and trade contractor also help with clear lines of responsibility, but the involvement of a design team and a variety of separate trades, some with design responsibility, make the situation more complex than DB, although not as complex as GC. Although it is not shown here, MC is unlike CM in this respect. In terms of the clarity of remedies it is like an extreme version of GC and is the worst procurement method for clear remedies.

7.4.5 Complexity of the project

Complexity cannot be considered in isolation because it is inextricably bound up with speed and with the experience of those involved with the project. It is also a very difficult variable to measure. Although technological complexity is a significant variable, it can be mitigated by using highly skilled people. Of more concern in the procurement decision is *organizational* complexity. Roughly speaking, this can be translated into the number of different organizations needed for the project. Because of fragmentation, this means that a project with a large number of diverse skills is more organizationally complex than a project with fewer skills, even though the few may be more technologically sophisticated. Each of the procurement methods can be used for a wide range of complexities, as shown in Figure 9, but the limits differ.

GC involves an organizational structure that is probably too complex for simple projects. The idea of fully designing or specifying all of the work to be done before a contractor bids for it assumes that the design is beyond the skill of the contractor. In simple projects this is not so and it would be better to rely on the skill and judgement of a contractor by utilizing DB. Complex projects may require nominated sub-contracts because of the need to incorporate the design skills of specialist trade contractors. When this forms a large part of the contract work, GC breaks down as it is based on the assumption that the main contractor will be *doing* most of the work, not simply managing work of others. With its single point

Figure 10: Speed from inception to completion

responsibility, DB is ideal for simple jobs but this does not preclude its use for more sophisticated work. However, if the project is too complex, DB contractors may lack the experience and skill needed for the high levels of co-ordination and integration required. For very complex projects, CM is most suitable. Perhaps construction managers tend to be more experienced, but the contract structure of CM means that the construction manager should have no vested interest in conflicts between design and fabrication exigencies or between trade contractors. This independence, coupled with a professional management role, renders CM ideal for dealing with organizational complexity. However, there is no reason to avoid CM just because a project is simple. It has been used with great success on small projects where the participants are familiar with CM and with each other.

7.4.6   Speed from inception to completion

One of the most distinctive features of construction projects is the overall duration of the process. Since a single construction project typically constitutes a large proportion of a client's annual expenditure *and* a large proportion of a contractor's annual turnover, each project is individually very important. Many developments and refinements to procurement methods have been connected with a desire to reduce the duration of projects. Much of the process of construction is essentially linear. Briefing, designing, specifying and constructing must follow one from the other. If these steps can be overlapped, then the overall time can be reduced significantly, provided that there is no need for re-work due to changes and wrong assumptions, in which case too much overlapping can slow the process and cancel any gains due to overlapping.

Because of the need to design and specify the whole of the works before inviting tenders, Figure 10 shows that GC is generally the slowest method. This overall slowness often leads to techniques for starting on site early such as the letting of early *enabling* contracts, like demolition or earthworks. Another technique to speed progress is to leave much of the detailed design until after the contract has been let by including large **provisional sums** in the bills of quantities, a bad practice that should generally be avoided. Other procurement methods are inherently quicker simply because they enable an early site start. Since a DB contractor will be undertaking design, early assumptions are fairly safe. Further, DB is generally used for projects that are straightforward. CM can be very quick because the relationships are conducive to quick working and overlapping.

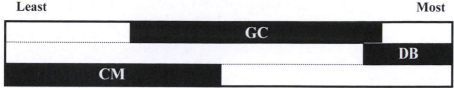

Figure 11: Certainty of price

7.4.7 Certainty of price

Price certainty is not the same as economy. It is doubtful that there is any correlation between economy and procurement method. The reliability of initial budgets is highly significant for most clients. However, this must be weighed against the financial benefits of accepting some of the risks and with them, less certainty of price.

GC comes in a wide variety of forms. One of the most wide ranging variables is the pricing document, the bill of quantities. This may be a fully measured bill (known as 'with quantities'), a bill of approximate rates ('approximate quantities') or there may be no quantities at all ('without quantities'). However, even with a schedule of rates, contractors are paid according to their own pre-priced estimates in line with the principles of fixed price contracts explained in Section 7.3. Therefore, Figure 11 shows that GC tends toward the more certain but there is a huge range of variability in this because the final price depends on many other factors such as the adequacy of the initial budget, the quality of the design and so on. By contrast, DB is a contract for a lump sum for all of the work required. Although the contractor may add contingencies in to the price to deal with the unexpected, they remain the responsibility of the contractor. This may, in theory at least, result in a higher price, but the benefit is that the final price is agreed at the outset. Finally, CM consists of a series of contracts, which are let as the work proceeds. Therefore, it is impossible to be confident about the final price until the project is nearly completed. Moreover, there is no fundamental reason to prevent some of the contracts being let as cost reimbursement packages.

7.4.8 Balancing the requirements

The seven individual criteria provide a basic framework that can never be comprehensive: it is merely a technique for recording observations and views. The aim of the Figures is to act as a vehicle for guiding the client through the decision. Clearly, there may be other, more important factors for particular clients. Similarly, some of the factors chosen here may not apply to each project. However, the principle remains. Before entering into contracts, a client should be advised about the procurement options and taken through a series of decisions so that the most appropriate procurement method can be identified.

7.5   IDENTIFYING AND CHOOSING PROCUREMENT METHODS

It is clear from the discussion so far that there are many different variables to take into account when selecting procurement methods. Moreover, it can be difficult, in practice, to be clear about which procurement method is actually being used, since procurement methods differ from each other in unique ways. For example, GC differs from DB in terms of where the responsibility for design lies, whereas it differs from CM in terms of how site-co-ordination is achieved. Recent developments in procurement practice have made the situation more complex. For example, the use of private sector finance for public sector projects.

7.5.1   Private finance initiative

In the UK, this is called the private finance initiative (PFI). There are also public-private partnerships (PPP) where companies from the private sector enter into some kind of partnership with the public sector for supporting or providing a public service. PFI involves the public sector paying a regular service charge to a private sector consortium for the operation of a particular facility that is funded, designed, built and operated by the consortium for a specified number of years, with the capital asset transferred to the public sector agency at the end of the period. PPP involves more diverse approaches and might not include transferring the ownership of the asset at the end of the period. One key aspect of these approaches is that the public sector seeks to transfer much of the development risk to the private sector partners, involving major financial support by the banks that enter into consortia with contractors, designers and facilities managers. The operation of these consortia is discussed in more detail in Gruneberg and Hughes (2005) who suggest that consortia are formed only for the purposes of winning work and that members of consortia tend not to be quite as open and co-operative as clients would like to believe. Apart from the use of consortia to procure construction work, the key difference between PFI and, say, GC is in the funding for the project. But PFI differs from everything else in this way, and may, in theory, be associated with GC, DB or CM contractual arrangements.

In other parts of the world, this approach to construction has been called design, build, finance and operate (DBFO), and sometimes build, own, operate, transfer (BOOT). The differences between PFI, DBFO and BOOT are vague and inconsistent, and differences between individual examples of each are just as marked. These are not precise terms.

One key feature of PFI is that the client contracts with a consortium for the provision of the service over a period of time. In its turn, the consortium contracts with designers, builders and facilities managers. Even though any or all of these companies may be part owners of the consortium, and separate parts of the same group, they are, in law, separate contracting entities, so the contract between the consortium and the contractor will be comparable to any other building contract. It is important not to confuse or conflate this with the PFI contract between the client and the consortium. Thus, builders of PFI projects are usually paid for their work as they do it, rather than at the end of the construction process.

7.5.2 Partnering

Another recent development in construction procurement practice is referred to as partnering. This term means different things to different people, but it is primarily about team-working. The idea of partnering has become so popular in the UK contracting scene that there are now several standard-form partnering arrangements available, for single projects (project partnering) as well as for multiple projects over time (strategic partnering). An example of the former type is the ACA Standard Form of Project Partnering PPC2000 (Mosey 2003). Here, the key stakeholders of the project, i.e. the client, contractor, consultants and key specialists, sign only one single, integrated contract (no other contracts are needed, such as appointments of professionals or building contracts). Other standard forms for partnering, such as option X12 of NEC 3, do not create a multi-party contract (Institution of Civil Engineers 2005c: 108). Strategic partnering, in contrast, involves developing long-term commitments from both parties to a contract. The aim is to move the focus of attention away from getting the cheapest or quickest solution for a particular job, and towards a longer term understanding of the purposes of the project, and understanding from both parties about what each wants to get out of the project (for a more detailed exposition of the concept, see Bennett and Jayes 1998 and also Bennett and Peace 2006). These policies have their roots in widespread business practice where a long-term relationship enables buyers and sellers to avoid litigious disputes because the relationship becomes an important part of the process. The essential feature of strategic partnering is that it provides a method for selecting the contractor (or other supplier) other than the more traditional approaches of competitive tendering (see Table 2 in Section 7.5.5). Examples of a standard-form arrangement for strategic partnering are the JCT Framework Agreement (JCT FA 05), NEC 3 Framework Contract and the ICE Partnering Addendum.

7.5.3 Performance-based contracts

One final variant in procurement methods is a kind of contract that is quite normal outside the construction sector. This is where things are paid for based upon their function, rather than on their material and labour content. Construction output tends to be paid for on the basis of the resources that the contractors will use, plus overheads and profit. Moreover, it tends to be paid for on a monthly basis, while it is being built, so that the contractor does not have to finance the development. Even in PFI projects, the contractor is paid monthly for work in progress, but by the funders of the development, rather than by the client. (The contractor may have put 10% equity into the consortium, but draws on the finance from the moment that work starts on site.)

In private sector property development, the developer would fund the work, and sell the completed building to the occupier. Housing and some speculative commercial developments are paid for like this. Since the payment is for a completed facility, based upon its market value, this transaction is based on the performance of the facility, rather than based on work in progress. This is what is meant by performance-based contracting in construction. It is radically different for the construction sector (as opposed to the property development sector) because it changes fundamentally the financial structure of the firms who take part. Under GC, for example, when a contractor is paid monthly for work in progress, payments by the contractor to suppliers and sub-contractors may be made after payments are received from the client. Therefore, the contractor needs to invest nothing in the development, and can survive purely on cash flow. As long as the cash balance is positive, the contractor can invest this positive balance and earn money from the interest, or use the money to build something else for subsequent sale, acting as a developer. If the cash flow becomes negative, it is likely that such a contractor would quickly become insolvent.

In order for a contractor to be able to take part in a performance-based contract, the company would need assets that are bigger than the potential losses that might happen in the event that the building failed to perform. This is what happens in other sectors of the economy, where it is typical for the individual transaction to be so small in relation to the size of the supplier, that the occasional failure or dispute does not render the supplier insolvent. In construction, individual contracts are frequently larger than the annual turnover of the contractor, and sometimes of the client, too. This is why performance-based contracting is not widespread and would only develop where the financial circumstances underpinning the solvency of suppliers is reliable and large. PFI is an example of a performance-based contract in that the client pays a service charge based upon performance of the consortia. But for the building contractor, it is not a performance-based contract, because contractor's payments are based on work-in-progress. Thus, the most obvious example of a performance-based contract involves an intermediary (the consortium) between the client and the contractors. This is not the place for a detailed analysis of this approach to procurement. It is mentioned here for the sake of completeness.

7.5.4   Identifying procurement methods

The discussion so far reveals that each procurement option differs from the others on its own basis. In other words, in many cases they are not simply alternatives to each other and the procurement decision is not, therefore, simply a case of choosing between this bewildering array of options, since they can be combined in different ways. For example, a general contract may or may not be partnered, it may be financed by the private or public sectors, and it may or may not involve much sub-contracting. To help distinguish the various options, the flowchart in Figure 12 should help.

The key thing about identifying procurement methods in practice is that each differs from the other on its own basis. So it only takes a few key questions to decide which method is being used. The flowchart shows the major differences between the procurement methods at the most general level. This is the level at which procurement methods are discussed in the literature on construction procurement, but it does not provide much help or guidance in choosing from among all the procurement variables and combining them in the best way for a particular project. In order to specify a procurement system in detail at the outset of a project, more details are needed.

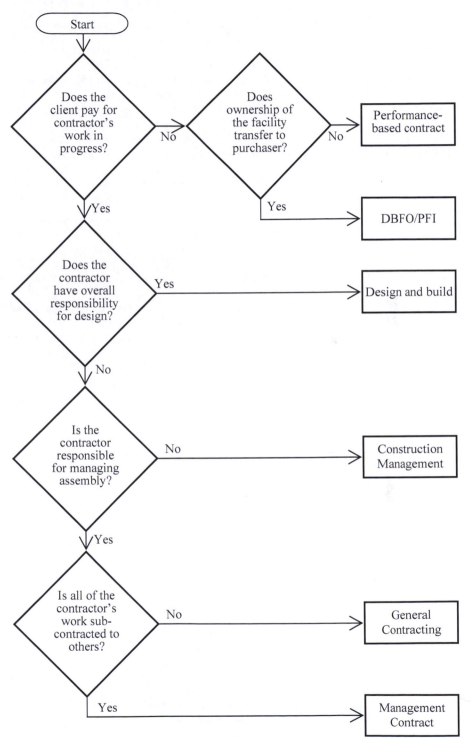

Figure 12: Identifying procurement methods

Table 2: Conceptual definitions in procurement

Category	Examples
Source of funding	Owner-financed, public sector-financed, developer-financed, PFI, PPP.
Selection method	Negotiation, partnering, frameworks, selective competition, open competition.
Price basis	Work and materials as defined by bills of quantity, cost reimbursement, whole building, a fully-maintained facility, performance.
Responsibility for design	Architect, engineer, contractor, in-house design teams, supplier.
Responsibility for management	Client, lead designer, principal contractor, joint venture, construction manager
Amount of sub-contracting	0–100%

7.5.5   Choosing procurement methods

The range of procurement options and the different ways of combining them reveal a confusing array of methods, as each is distinguished from the other on a different basis. What is needed is a more conceptual way of distinguishing the various methods from each other, because contracting, funding, selection and payment methods can all be combined with different modes of ownership in many different ways. Simple labels such as PFI, partnering, general contracting, and so on, communicate little about the way that a project is structured. In fact, these different methods are not mutually exclusive categories of doing work. By grouping the procurement variables, it ought to be possible to describe any construction procurement method. These aspects of procurement are variously represented in every procurement method, even though a procurement method is usually characterized by only one factor. For example, PFI involves private sector investment in major infrastructure projects, whereas in DB the contractor's responsibility for design is the defining characteristic. To bring more clarity to the distinctions between procurement methods, the categories in Table 2 are suggested.

Using Table 2, it is clear that describing a project as PFI does not indicate where the responsibility for design lies. Describing a project as general contracting does not indicate how much sub-contracting there is, or how the contractor was selected. To adequately describe how something is procured, all six categories are needed.

In setting up a construction project, discussion and decisions are needed under each of these six areas. The source of funding may not be an option, as it may simply be a matter of policy or even regulation for any given client. But it needs to be explicit. The methods of selection open to the client may be a matter of choice or preference, but in many cases, particularly in the public sector, there are regulations about how suppliers may be selected. The choice about how the price is calculated is a mechanism for apportioning risk, as discussed in Section 7.3. Responsibility for design and responsibility for management define where the liability lies in the event that the project does not meet expectations. The amount of sub-contracting is an indication of the scale of the co-ordination and management task for a project, but is also a response to the technical complexity of the project, as discussed in Chapter 1.

By working through each of the categories and making a clear decision or statement about the approach used in the project, the detail of the procurement method will be a lot clearer than would be the case defining it at the broad level of Figure 12.

One interesting and somewhat surprising point about Table 2 is the number of combinations of the six variables. Given that there are at least five options in each of the six categories, there are at least 15,625 different ways of procuring a building. In fact, there are many more, because the options given here are only examples, and the amount of sub-contracting and subsequent consequences for the supply chain provides many more different ways of structuring a procurement method.

8 Contract choice

Following the procurement decision, which identifies the contract structure, the final decision relates to the particular contracts to be used. First, there is a choice between standard or non-standard contracts. If standard contracts are to be chosen there are many to choose from and some of them are introduced briefly here.

8.1 USE OF STANDARD CONTRACTS

Although not essential, it is common practice when procuring a building to appoint the contractor on a standard-form building contract. These forms, as will be seen, have emanated from various parts of the construction industry and for various reasons. Each of them has a role to play in the choice of strategy. In order for the process to work effectively, it is essential that the standard forms be understood in terms of the way they distribute risk. There are many aspects to the arguments for and against standard forms (Hughes and Greenwood 1996), discussed below.

8.1.1 Legislating for the industry

Drafting committees often express the sentiment that they are legislating for the whole industry. Whatever their reasoning, such an aim is way beyond their remit. It must be remembered that the UK is a constitutional free market. It is well-established contract doctrine that the parties to a contract are free to choose the terms of their contract. There are leading cases where this sentiment has been expressed specifically.[1]

8.1.2 Similar projects demand similar contracts

It is interesting to note that the Banwell Report (1964) recommended that the building industry should develop and use a single standard-form contract for all construction projects. This idea was given a further push in the 1990s with a recommendation of the Latham Report that steps should be taken to apply the *New Engineering Contract* (now called *Engineering and Construction Contract*, see Section 7.6.4) as the universal standard for the entire industry (Latham 1994). The pursuit of a universal standard contract has fuelled much of the development of contracting in Great Britain. The aim of standardizing building contracts to this extent is simply unrealistic. When the complexity of a modern airport or hospital

---

[1] *Eurico SpA v Philipp Bros., The Epaphus* [1987] 2 Lloyd's Rep 215, CA.

building is compared with that of a housing development, it is clear that the contractual issues are completely different. For example, delays due to bad weather, a contractor's inability to obtain materials, delays in the supply of information, onerous site conditions, insolvency, etc. differ in their likelihood, severity and magnitude. If these issues differ, then the apportionment of the risks surrounding them must also differ. Therefore, different contracts are needed. Unfortunately, the differences in technological complexity and types of client (to name but two of the variables) are so large that the aim of developing a universal standard form is not realistic. Clearly, there is a need among clients and contractors for different contracts to suit different situations but this contradicts the equal need for standards to be applicable to as wide a range of projects as possible.

8.1.3   Equitable distribution of risk

One purpose behind using *standard-form contracts* is to allocate risks fairly between the parties. The use of a standard form ought to imply that contractors need not include an allowance in their prices for risk. However, risks *are* apportioned by any contract, standard or not, and in some circumstances it would be unwise to allocate them in this way. When choosing a procurement method the criteria for selection should be studied very carefully and the allocation of risk should be made explicit, rather than implicit. This leads to one of the strongest criticisms of standard-form building contracts: the apportionment of risk is rarely questioned and, therefore, becomes implicit. In such situations, the employer is only comparing tenders from contractors competing upon the same pattern of risk apportionment. Therefore, the client can make no assessment of the suitability of the form of contract.

8.1.4   Difficulty of understanding contract complexities

One of the main reasons for standardizing the contracts used in building is because the contractual complexities can be difficult to appreciate. It is widely held that the use of a standard form will help to increase familiarity with all the contractual provisions. Unfortunately, this ideal is rarely achieved for at least two reasons. First, the standard forms are rarely used as printed. It is common in the industry for people to amend the printed form, by striking out clauses they do not like and adding in their own preferred clauses (Greenwood 1993). The wisdom of this practice is questionable because an amended contract may fall into the category of an 'employer's standard terms of business' for the purposes of the Unfair Contract Terms Act (see Chapter 9). Secondly, the structure of the construction industry encourages firms to concentrate on particular types of work. Civil engineers and builders rarely interact. The various groups of consultants can thus fall into the trap of only knowing about one standard form. If this happens, their understanding of the contractual issues involved is very narrow, and there is often a failure to appreciate the wider issues in the context of English law. This can lead to misunderstandings which are perpetuated simply because, once such consultants realize how complex the standard form is, they do not wish to add to that complexity by using a different one! The only real answer to this problem is to understand the principles of contract law first and then to apply these principles to standard forms, so that the allocation of risk within each form can be appreciated.

8.1.5   The purpose of contract documentation

There seems to be a wide range of uses to which contract documents are put. The basic defining aim of a contract is to record accurately the terms of a business agreement. While the vast majority of commerce and commercial law is based upon this assumption, the circumstances of the construction industry are such that the concept of contract often becomes severely distorted. For example, some approach the drafting of contracts as if the primary aim was to set down the management procedures and administrative processes for the work; others take a much more confrontational approach and seek to use the contract primarily as an agenda for litigation and a vehicle for taking an obstructive and uncompromising stance; others see the standard forms as a nationally negotiated prescription of what can or cannot be done by the various participants. On the face of it, perhaps these distinctions are not too important. But ultimately, construction contracts are subject to the same rule of contract law as any other type of contract. Therefore, it is extremely important to ensure that there is some consistency in the way that different industrial sectors draft their agreements and contracts. Contracts are drawn up with the intention of relying upon them in a court of law at some point in the future. However, the nature of business is such that a potential trading partner can be put off by too much emphasis on negotiating contractual issues while the bargain is being struck. There is a fine balance between under-emphasis and over-emphasis on contract terms during the negotiating process. Both are *equally* damaging and it may be more useful to use approaches that enable business deals to be recorded more accurately than is currently the case. But given that language is imprecise and constantly evolving, this is a problem that will never go away.

8.1.6   Appropriateness of contract form

Standardizing contract terms enables parties to reduce the emphasis on specific contractual terms during the bargaining process. Thus, they are very appealing. On the other hand, they tend to be drafted by committees representing powerful interest groups in the industry. The danger is that standardization goes hand in hand with an adherence to outdated methods of organization and professional patterns of responsibility, regardless of whether these are appropriate to the needs of a particular project or client. It is the failure to identify appropriate roles and a suitable organizational structure that lies at the root of the question of standardization.

8.2   CONTRACT DRAFTING

The drafting of non-standard forms demands great skill and knowledge. There is extensive experience of contract drafting generally, much of which influences the modern standard forms used in the construction industry. Although standard-form contracts were heavily criticized in the past,[2] the drafting bodies have responded positively to such criticism and have developed the style and presentation of their contracts. Also, over the last few decades, the Technology and Construction Court, a specialist court of the Queen's Bench Division, has developed a much more informed approach to construction contracts and disputes, and there is now a much stronger connection between construction practice and construction law.

The prospect of drafting a contract from first principles for every project is too daunting to most in the industry. Many are tempted to undertake their own revisions and amendments to the standard forms and frequently make the situation worse by producing conflicting or contradictory clauses. Another solution is to develop a library of standard approaches to the main issues of risk apportionment and choose the most appropriate for each project. One advantage of such an approach would be the discussion that would be precipitated during the negotiating stages of a project. The use of options in contracts is not widespread, but is gradually emerging as a method of designing a more generally acceptable standard form. For example, JCT contracts of the 2005 suite contain a section called 'contract particulars' where the parties should specify the project-specific data. Often, the parties can choose from several options which are elaborated on in the conditions and, in greater detail, in the schedules. In order to complete and execute such a contract, the employer and contractor must agree each of the optional terms. In other words, the explicit negotiation and agreement of particular terms is far more important than the ability to choose one of a number of pre-printed contracts. If drafting a new contract is to be avoided, the only practical alternative (currently) is to use or adapt one of the many standard forms already available.

8.3   JCT CONTRACTS

Despite its name, the Joint Contracts Tribunal does not sit in judgment on others. It is an affiliation of interest groups within the construction industry, which operates as a forum for discussing and determining the content of the clauses of the standard-form building contracts. It does not appear that this body was ever meant to be democratic, or even representative. Its origins lie in 1870, when the Builders' Society (now the CIOB) and the RIBA together produced the first 'standard form' of contract. This was for use only in London and seems to have been an attempt to overcome some of the problems of the day, caused by the use of *ad hoc* building contracts (Spiers 1983).

Over the years the meetings of the various interest groups continued and, after much argument, negotiation and debate, the first national standard form became available in 1909. This was drafted by the RIBA and the NFBTE (now the Construction Confederation) with help in mediation by the Institute of Builders. It

---

[2] *Bickerton v North West Metropolitan Regional Hospital Board* [1977] 1 All ER 977; *English Industrial Estates Corporation v George Wimpey & Co Ltd* (1972) 7 BLR 122, 126.

was largely based on the 1870 form, and it is said that much of the original 1870 style, if not the actual content, could still be detected in 1980 range of JCT forms (Spiers 1983).

Since being roundly criticized by Latham (1994), the JCT has been completely reorganized to involve more appropriate approaches to the negotiation and drafting of standard-form contracts: The current membership of the Joint Contracts Tribunal is as follows:

- Association of Consulting Engineers.
- British Property Federation.
- Construction Confederation.
- Local Government Association.
- National Specialist Contractors' Council.
- Royal Institute of British Architects.
- Royal Institution of Chartered Surveyors.
- Scottish Building Contract Committee.

In May 2005, the JCT launched its new suite of contracts. Most of the standard forms were published between May and December 2005, others in 2006, but some were still announced as forthcoming at the time of writing, for example the forms for management contracting and construction management. The forms published in 2006 have this year in their titles. For example, the Prime Cost Building Contract is called JCT PCC 06. In its new suite of contracts, the JCT not only updated many of its existing forms but also expanded the range of forms. One important new form is the Framework Agreement which is for use in collaborative working relationships (partnering). Although this is not the place for a detailed analysis of the changes implemented in the 2005 overhaul,[3] the major new features are worth highlighting.

- A more modern and consistent layout has been used.
- The structure is consistent: contract specific data is now in 'Contract Particulars' instead of in an Appendix, and now appears towards the front of the contract.
- Clauses are now grouped together in a consistently numbered set of headings.
- The language of the contracts is clearer.
- New definitions have been introduced; several provisions have been shortened, for example termination, insurance.
- Some terms have been re-named, for example specific editions for local authorities are no longer provided, so 'architect' is now 'architect/contract administrator' throughout, defects liability period is now the rectification period, determination is now termination, extension of time is now adjustment of time, because the duration of the contract may be reduced.
- What were separate supplements are now integral parts of the main forms, for example fluctuations, sectional completion, collateral warranties and contractor's designed portion are not now separate documents.
- Removal of provisions for nomination of sub-contractors or suppliers.
- Removal of Performance Specified Work provisions in SBC 05.
- The default level of retention has been reduced from 5% to 3%.

---

[3] For an overview see Lupton (2006). For an in-depth analysis see Davison (2006).

- Litigation has replaced arbitration as the default method of dispute resolution (subject, of course, to the right to adjudication).

Lupton (2006: 100) summarizes these changes as having brought 'a vast increase in user-friendliness ... improvement in layout and drafting ... result in a more manageable document, which it is likely to be more popular than its predecessor'. To illustrate the complexity of the range of forms needed to deal with the network of contracts in a construction project, a far-reaching example of the forms produced by the JCT is as follows.

8.3.1   JCT SBC 05: Standard Building Contract

This form of contract, issued in 2005, is regarded as the 'industry standard' against which all others are measured. It is a long and complex document, which consolidates all the revisions to the earlier forms of contract (JCT 63, JCT 80, JCT 98), as well as making fundamental changes to accommodate the new features of the 2005 suite of contracts as outlined above. It is intended for use in general contracting and is published in three versions:

- SBC/Q: Standard Building Contract with Quantities.[4]
- SBC/AQ: Standard Building Contract with Approximate Quantities.
- SBC/XQ: Standard Building Contract without Quantities.

There are also certain sub-contract forms designed for use with JCT SBC 05 and two guides:

- SBCSub/A: Standard Building Sub-Contract – Agreement.
- SBCSub/C: Standard Building Sub-Contract – Conditions.
- SBCSub/D/A: Standard Building Sub-Contract with sub-contractor's design – Agreement.
- SBCSub/D/C: Standard Building Sub-Contract with sub-contractor's design – Conditions.
- SBC/G: Standard Building Contract Guide.
- SBCSub/G: Standard Building Sub-Contract Guide.

In addition to the forms produced by the JCT, there are forms for domestic sub-contracts, produced by the Construction Confederation, including a domestic sub-contractor's designed portion supplement.[5]

Among the major characteristics of JCT SBC 05 is the requirement that an architect/contract administrator and a quantity surveyor must be appointed. The architect/contract administrator has wide but strictly defined powers. These include the issuing of certain instructions on behalf of the employer, and also the independent certification of payments, the quality of work, the contractor's performance and so on. The quantity surveyor is responsible for the valuation of work done and the valuation of variations.

---

[4] If the version is not specified, the acronym JCT SBC 05 usually refers to this version.
[5] These forms are published by the Construction Industry Publications Ltd. (CIP), a wholly owned subsidiary of the Construction Confederation.

8.3.2   JCT IC 05 Intermediate Building Contract

This general contract form is the successor of the Intermediate Form of Building Contract (IFC 98) which was first published in 1984. There is also a version with contractor's design for projects where the contractor is required to design discrete parts of the building (JCT ICD 05: Intermediate Building Contract with contractor's design), but note that this form is not a design and build contract. The forms are published with the following associated documents:

- ICSub/A: Intermediate Sub-Contract Agreement.
- ICSub/C: Intermediate Sub-Contract Conditions.
- ICSub/D/A: Intermediate Sub-Contract with sub-contractor's design Agreement.
- ICSub/D/C: Intermediate Sub-Contract with sub-contractor's design Conditions.
- ICSub/NAM: Intermediate Named Sub-Contract Tender & Agreement.
- ICSub/NAM/C: Intermediate Named Sub-Contract Conditions.
- ICSub/NAM/E: Intermediate Named Sub-Contractor/Employer Agreement.
- IC/G: Intermediate Building Contract Guide.
- ICSub/G: Intermediate Sub-Contract Guide.

It is expected that JCT IC 05 will be as widely used as the predecessors of this form. When first published in 1984, its aim was to bridge the gap between the lengthy provisions of the standard building contract (now JCT SBC 05) and the simplicity of the Minor Works Form, which was being used in contracts for which it was totally inappropriate. JCT IC 05 is shorter than JCT SBC 05 but is still fairly complex. It was published in response to demand for a contract that bore a closer resemblance to JCT 63. At the time, JCT 80 was not finding the general level of acceptance that had been hoped for by the Tribunal. JCT IC 05 is designed for works that use the basic trades and skills of the industry, adequately specified in some detail, and which do not require complex building services engineering packages.

The main characteristic of JCT IC 05 is its flexibility. It can be used for either local authorities or private sector employers and either with or without bills of quantities as the cost control document. Interestingly, while nomination has disappeared from the other forms, the simpler mechanism of naming that was used in IFC 98 has been retained in JCT IC 05, in the third recital, clause 3.7, and schedule 2 (see Chapter 20).

8.3.3   JCT MW 05 Minor Works Building Contract

This form is the successor of the Minor Works Agreement 1998 (MW 98). No ancillary forms are published regarding JCT MW 05. Where the contractor is to design discrete part(s) of the building, the form JCT MW/D 05 should be used (Minor Works Building Contract with contractor's design).

The contract JCT MW 05 is designed for use only on simple, straightforward works. The conditions give a bare outline of the parties' duties and responsibilities, and allocate risks too crudely for more complex situations. While it is very useful for small projects, it can also be used for larger projects as long as the works are of a simple nature. Generally, the form should not be used on contracts exceeding £150,000 (Chappell 2006: 4). Its chief advantage is its simplicity, but this becomes a major disadvantage on larger projects, where its lack of detailed provisions will leave the employer dangerously exposed to many different kinds of risk. While it is tempting to err on the side of simplicity, it should be remembered that long and comprehensive contracts contain clauses that keep the contract alive in the event of what would otherwise constitute a breach of contract. Therefore, simple contracts do not necessarily achieve the aims of the parties.

8.3.4   JCT PCC 06 Prime Cost Building Contract

According to the introductory note within the Prime Cost Building Contract 2006, this form is intended for situations where an early start on site is required and where it is not possible to obtain a precise definition of the nature and the extent of the works until the work is already underway. In such a case, it will not be possible to obtain a firm price for work in advance. As examples for the use of JCT PCC 06, the practice note 'Deciding on the appropriate JCT contract' (Joint Contracts Tribunal 2001) mentions alteration work and urgent repair work, for example after fire damage. However, the form does assume that a certain amount of design has been done, because the conditions assume that the employer has appointed an architect/contract administrator, a quantity surveyor and such other professional *advisors* as necessary for design and administration of the works. The contract also requires an outline description of the works in specifications and drawings, a description of the nature and scope of the works and an estimate of the prime cost of the works prepared by the quantity surveyor. The contractor will use this information to quote a fixed fee or percentage fee for overheads and profit. In addition to this fee, the contractor will be paid the 'prime cost', i.e. the full cost reasonably incurred in order to discharge the contract obligations. A detailed definition of the costs which fall under the category prime cost is given in the first schedule of JCT PCC 06. Since a precise definition of the work is not available, the contractor does not carry out any work at all without an instruction of the architect. The contract contains detailed provisions for control of the works, payment, statutory obligations, insurance, termination and dispute resolution, much along similar lines to other JCT contracts but modified to suit the particular circumstances of such projects.

8.3.5   JCT DB 05 Design and Build Contract

This design and build form is published for use in situations where the contractor undertakes the design of a building as well as its construction. It is described in Chapter 4. There is no provision in JCT DB 05 for the appointment of either a contract administrator or a quantity surveyor. However, there must be an 'employer's agent' who acts on behalf of the employer.

The contract is let on the basis of a document called the 'employer's requirements', which specifies precisely what it is that the contractor is being asked to do.

This will often take the form of a performance specification. 'Contractor's proposals' to comply with these requirements are submitted, and the winning contractor's proposal document is then used as the major control document on the project. In many respects, apart from re-wording to eliminate the contract administrator's responsibilities and to create a situation in which the contractor is responsible for self-certification of work, the style and content of much of this form is very similar to JCT SBC 05. Particular issues of design liability under this and other forms are covered in more detail in Chapter 13, and the principles of design and build contracts are examined in Chapter 4.

8.3.6   JCT MP 05 Major Project Construction Contract

The Major Project form is also a design and build contract. It is intended for use in major projects by experienced parties where the contractor is able and willing to take on a particularly high degree of risk. The contract is deliberately short (considerably shorter than JCT DB 05, i.e. 37 pages compared with 105 pages) because the parties are assumed to have their own in-house procedures to administer the contract. For use with this form, the JCT has also published a guide, a sub-contract (JCT MPSub 05), and a guide to the sub-contract.

8.3.7   JCT MC 98 Management Contract

This form of contract of 1998 is the current edition at the time of writing, although its successor has been announced as forthcoming. It has the following associated documents:

- WKS/1: Section 1: Management contractor's invitation to tender.
- WKS/1: Section 2: Tender by works contractor.
- WKS/1: Section 3: Articles of agreement of works contract.
- WKS/2: Standard conditions of works contract.
- WKS/3: Employer/Works contractor agreement.
- Phased completion supplement for management contract.
- Phased completion supplement for works contract.
- Formula rules for the works contract.

The first management contract was issued in 1987 as a response to demand for a standard form that reflected practice in commercial contracts at the time. There was nothing new about the idea, which in essence had emerged on many occasions since the 1930s. The basic idea is that all work is sub-contracted and the main contractor acts in a managing and co-ordinating role. Before the management contract was issued, a similar pattern of involvement could be achieved under predecessors to JCT SBC 05 by using sub-contractors. The form JCT MC 98 and the principles underlying management contracting are discussed in detail in Chapter 5.

8.3.8   JCT MTC 06 Measured Term Contract

This form of contract is designed for the situation where an employer requires regular maintenance or some other kind of minor works to be carried out, and wishes to engage only one contractor for a specific period. The single contract will then cover a number of separate jobs, since the contractor carries out work from time to time on receipt of instructions from the employer. As already indicated by the name, this form is a remeasurement contract; the work is measured and valued after completion on the basis of an agreed schedule of rates.

8.3.9   JCT CM 02 Construction Management documentation

In 2002, the JCT published its construction management documentation for use in construction management projects. It should be noted that a new edition has been announced as forthcoming at the time of writing this book. The 'main contract' for this procurement method is the CM agreement between the client and the construction manager (C/CM 02). This form is published with the following associated documents:

- TC/C: Trade contract between client and trade contractor(s).
- TCWa/P&T: Warranty given by a trade contractor to a purchaser or tenant.
- TCWa/F: Warranty given by a trade contractor to the project's funder.
- TC/T: Part 1: Invitation to tender for a trade contractor
- TC/T: Part 2: Tender by a trade contractor
- Fluctuations: Fluctuation clauses for use with TC/C
- Guide to construction management documentation.

The CM documentation has been based upon extensive industry consultation and reflects much of the common practice in CM projects. Both the procurement method CM and JCT's documentation are described in Chapter 6.

8.3.10 JCT Consumer Contracts 2005 Home Owner/Occupier

The family of contracts available from JCT covers an increasingly broad spectrum of building projects. The minor works form was unsuitable for domestic customers, so JCT has produced a consumer contract to meet the needs of the domestic market, where individual owner/occupiers are contracting with small building firms for home improvements, extensions or repairs for a lump sum. Actually, a residential occupier of a dwelling can choose between two packages. If a professional consultant (architect, quantity surveyor or engineer) is required to oversee the work, the relevant package is JCT HO/C 05 which is called 'Building contract and consultancy agreement for a home owner/occupier'. This set contains the following documents:

- Building contract.
- Consultancy agreement.
- Guidance notes.

The alternative package, JCT HO/B 05, is appropriate if no consultant has been appointed by the householder. It consists of the following documents:

- Building contract.
- Enquiry letter (a sample for sending to potential contractors).
- Guidance notes.

All of these forms are written in plain English, without technical or legal jargon. Indeed, the Plain English Campaign awarded them the 'Crystal Mark' for clarity.

8.3.11 JCT FA 05 Framework Agreement

In its efforts to respond to the trend towards collaborative working relationships (partnering), the JCT has published the following documents:

- FA 05: Framework agreement (binding).
- FA/N 05: Framework agreement (non-binding).
- FA/G 05: Framework agreement guide.

8.3.12 JCT CE 06 Construction Excellence Contract

The moves towards reform of procurement practice in the UK construction sector that stemmed from the reports of Latham (1994) and Egan (1998) resulted in a suite of documents for use in the public sector known as Achieving Excellence in Construction (Office of Government Commerce 2007). Part of this reform movement involved the establishment by the Department of Trade and Industry of an organization called *Construction Excellence* which is now solely funded by its membership. Being at the forefront of developments in collaborative approaches to construction procurement (i.e. non-adversarial), they have worked with JCT to produce the Constructing Excellence Contract.

8.4 OTHER STANDARD-FORM CONTRACTS

There are bodies other than the JCT who also produce building contract forms. Of these other forms, the most important are as follows:

8.4.1 ICE 7 Conditions of Contract Measurement Version

This is a negotiated form produced by Conditions of Contract Standing Joint Committee (CCSJC), an organization which consists of the Institution of Civil Engineers (ICE), the Association of Consulting Engineers (ACE) and the Civil Engineering Contractors Association (CECA). It was first published in 1945; the current 7th edition appeared in September 1999. This version has been amended several times since 1999. There is only one version, for use with public or private clients, and it is intended for use on major civil engineering projects. These include a wide range of works, such as navigable canals, irrigation schemes, roads, railways, docks, harbours, dams, bridges and tunnels.

The essential feature of the ICE 7 conditions is that they provide a contract between a promoter and a contractor. The promoter becomes the employer under the conditions, and the engineer provides the technical aspects of design and specifications and effectively translates the contract (Haswell and de Silva 1989). The engineer is not a party to the contract and thus has no legal rights or obligations under the contract. However, engineers have their own contract (conditions of engagement) with their employers. This situation is closely analogous to the position of the architect under general building contracts.

The conditions create a 'remeasurement' contract, formerly known as 'admeasurement' or 'measure and value'. This means that the contractor is paid at the contract rates (which may be subject to variation) for the actual quantities of work carried out. This is recognized by the fact that the ICE 7 conditions, like other contracts where quantities are approximate and subject to remeasurement after the work is carried out, make no reference to a 'contract sum'; they refer instead to a 'tender total' and to a 'contract price', which *will* be ascertained in accordance with the conditions (clause 1(1)(j)). Although previous editions of ICE conditions were remeasurement contracts, on the publication of the 7th edition, the drafters renamed this *ICE Conditions of Contract, Measurement Version*.

Work under an ICE 7 contract is to be carried out to the satisfaction of the engineer, whose powers of control and direction are both extensive and apparently arbitrary. The conditions contain wide provisions for the adjustment of the contract price, usually in favour of the contractor. The detailed provisions of ICE 7 are referred to throughout this book, usually by contrast with JCT SBC 05.

8.4.2   Other ICE standard form of contracts

As well as the Measurement Version, the ICE Conditions of Contract are published in the following versions:

- Design & Construct 2nd edition 2001.
- Term Version 1st edition 2002.
- Minor Works 3rd edition 2001.
- Partnering Addendum 2003.
- Tendering for Civil Engineering Contracts 2000.
- Archaeological Investigation 1st edition 2004.
- Target Cost 1st edition 2006.
- Ground Investigation 2nd edition 2003.

Most of these forms, which together form a family of standard conditions of contract for civil engineering works, are published together with guidance notes.

8.4.3   FIDIC conditions of contract

This form is produced by the International Federation of Consulting Engineers (FIDIC) in association with the European International Federation of Construction

(FEIC). It is substantially based on the 4th edition ICE form, modified to enable it to be used anywhere in the world. It contains special provisions to enable the parties to decide under which nation's legal system the contract will be executed. This, traditionally, was the country of the engineer, but recently it has become more usual for it to be the country of the employer.

The basic characteristics of the contract are that an engineer is employed as the client's agent and representative, as well as having certification powers. Legally this role is similar to that of a contract administrator under JCT SBC 05. There are provisions for nomination of sub-contractors, settlement of disputes, extensions of contract period, liquidated damages, and all the other complexities usually encountered in major standard-form building contracts. The provisions are not considered further in this book. Interested readers are referred to the guide issued by FIDIC (2001). The most interesting feature of FIDIC is the way it is split into two parts: Part I of the conditions is of general applicability and Part II contains optional clauses specific to each project.

8.4.4 GC/Works/1 Contract for building and civil engineering major works

The GC/Works/1 contract (Property Advisors to the Civil Estate 1998c) is a re-issue of the third edition, incorporating recent changes in legislation. The documentation is intended primarily for government contracts, but has been designed to be equally suitable for private sector work. The documentation comes in four volumes:

- With quantities general conditions.
- Without quantities general conditions.
- Single stage design and build general conditions.
- Model forms and commentary.

The volume of model forms (Property Advisors to the Civil Estate 1998c) contains a wide range of associated forms, for example performance bond, parent company guarantee and retention bond, as well as administrative documents such as notices and certificates. Contract-specific forms, such as invitations to tender, forms of tender and abstracts of particulars are contained within the three main contracts (listed above), as they vary slightly.

The form was issued after consultation but without negotiation. Its precursors have been in use for many years. A significant characteristic is that many of the contract conditions give binding force to decisions of the employer.

The contract has some unusual aspects, such as payment provisions related to cash flow 'S-curves' (also an option in JCT DB 05), instead of measured work done, and acceleration provisions. Research on past projects has shown that for characteristic types of project, the amount of work done by the contractor accumulates slowly at the beginning, rapidly in the middle, and slowly at the end of the contract. A graph showing cumulative value will have an 'S' shape, hence the expression S-curve. If the particular shape of this graph can be agreed at the outset of the project, then the contractor's monthly payments can be looked up on the graph, rather than assessed by visiting the site.

8.4.5   NEC 3 Engineering and Construction Contract

The first version of this form, the New Engineering Contract (NEC) was published in 1993 (Institution of Civil Engineers 1993). It was as a response to growing discontent with contractual procedures and the prevailing adversarial attitudes in the construction industry. The first objective for the NEC was to be generally applicable to all projects, both building and engineering projects. The second objective was to accommodate all varieties of design responsibilities (giving the contractor either full design responsibility, some design responsibility or no design responsibility). To achieve these two objectives, various optional clauses are offered so that it can be used in different procurement methods. The third objective in developing this form was to avoid legalistic language. In pursuit of this, all legalistic jargon has been avoided. Only common words are used and the clauses are written in the present tense. The fourth objective behind this contract was that it should form a stimulus to good management. The idea is that procedures in the contract should contribute to better management practice and a forward-looking co-operative team spirit.

Latham (1994) identified some basic principles for contracts in the industry and found that the New Engineering Contract, more than any other, embodied these principles. Therefore, he recommended that it be slightly revised and adopted for use throughout the industry. Thus, in its second edition of 1995 (Institution of Civil Engineers 1995), the main contract was renamed *NEC 2 Engineering and Construction Contract*, after the appropriate modifications were made. The title emphasizes the fact that the use of this form is not limited to engineering projects. However, the suite of documents as a whole has retained the overall title of *New Engineering Contract*. In 2005 the third suite of contracts was published and the main contract is called NEC 3 Engineering and Construction Contract (Institution of Civil Engineers 2005b).

The NEC 3 form and its predecessors have become increasingly popular during the last few years. For example, the Channel Tunnel Rail Link was procured on an amended version of NEC 2 and also the English National Health Service bases its contract on an NEC form (Bridgewater and Hemsley 2006). Although the JCT has reacted to this trend with the already described overhaul of its contracts, the news that the Olympics 2012 will be procured on the basis of the NEC form shows that this contract will remain in widespread use.

The new version has not changed the previous NEC approach. The form is published as a set of core clauses common to all projects along with a variety of other clauses. To enable its use under a variety of procurement methods, the form contains in its Main Option clauses a set of pre-configured (merged) versions:

- Priced contract with activity schedule.
- Priced contract with bill of quantities.
- Target contract with activity schedule.
- Target contract with bill of quantities.
- Cost reimbursable contract.
- Management contract.

Altogether, the NEC 3 family comprises the following contracts, all of them published with guidance notes:

- NEC 3 Engineering and Construction Contract.
- NEC 3 Engineering and Construction Short Contract.
- NEC 3 Engineering and Construction Subcontract.
- NEC 3 Engineering and Construction Short Subcontract.
- NEC 3 Professional Services Contract (see Section 2.5).
- NEC 3 Term Service Contract.
- NEC 3 Adjudicator's Contract.

While the aims behind the Engineering and Construction Contract are entirely laudable, the approach that has been taken has been subject to some very strong criticism (Uff 1997) and this criticism has not stopped with the publication of the 3rd edition in 2005 (Bridgewater and Hemsley 2006). The language used is certainly simple. The danger of this is that the contract may have sacrificed clear litigation routes in favour of simple English. It remains to be seen how true this is. More importantly, the use of a contract as a management manual may preclude its use as an effective means of applying sanctions to non-performers. On the other hand, if litigation is impractical because of the way that the contract is drafted, then the only option open to disputants is to negotiate a settlement without recourse to law; an outcome that may be exactly what was wanted. Clearly, those who consider using this contract should look beyond the statement of objectives and carefully consider the extent to which the contract enables them to implement their preferred contract policy.

9 Tendering and contract formation

Construction contracts are governed by the general law of contract, with which most students will be familiar. There are in addition some statutory rules, on payment and the settlement of disputes, which apply only to construction contracts. In this Chapter we consider the meaning of 'construction contracts' for the statutory purposes and examine the way in which general principles of contract law apply in the construction context. We also look in a little more detail at the formation of contracts by a process of tendering, since it is very common for construction contracts to be made in this way.

9.1 THE MEANING OF CONSTRUCTION CONTRACTS

Until recently, the law did not treat construction contracts as a special class, but merely as part of a larger category known as contracts for work and materials. However, this has been changed by Part II of the Housing Grants, Construction and Regeneration Act 1996, which lays down certain rules applicable only to construction contracts. These rules, which concern the system of payment and the right to have all disputes settled through a process called adjudication, are described in later chapters. For the moment, we are concerned only with how the 1996 Act defines 'construction contract' for these purposes.

The first point to note is that the statutory definition is extremely wide. It includes any agreement in writing, or evidenced in writing, under which a party does any of the following:

- Carries out construction operations.
- Arranges for others to carry out construction operations (for example through sub-contracts).
- Provides labour for the carrying out of construction operations.

More surprisingly, it covers many of the professional services relating to construction operations, mentioning in particular architectural, design or surveying work and advice on building, engineering, interior or exterior decorating and landscaping. However (although this does not affect the definition itself), the statutory provisions do not apply to work carried out for a private client, that is to say work relating to a house or flat which one of the parties occupies or intends to occupy as their residence.

As to what is meant by the crucial term 'construction operations', the Act is again very wide-ranging. It extends beyond construction itself to works of alteration, repair, maintenance, decoration and demolition, and even to cleaning carried out in the course of such works. It covers not only structures but also the installation of services and ancillary works such as site clearance, excavation, scaffolding, site restoration and landscaping. However, the Act specifically excludes mining works and the installation or demolition of plant or machinery for the process industry. More surprisingly, perhaps, it also excludes the manufacture or supply of materials or components, except where the relevant contract includes installation as well as supply.

The definition of 'construction operations' has come before the courts on a number of occasions since the 1996 Act came into force, with some rather unexpected results. It has been held, for example, that the work of servicing heating systems in domestic premises constitutes 'construction operations'.[1] However, the supply and installation of shop fittings is not 'construction operations', since the fittings once installed do not form part of the land, as required by the statutory definition.[2]

9.2   THE FORMATION OF CONTRACTS BY AGREEMENT

A contract is a legally enforceable agreement. An agreement is usually defined in terms of an 'offer' made by one party and an 'acceptance' of that offer by the other. This need to identify an offer and acceptance causes few problems where the parties have made it clear throughout their negotiations that their contractual relationship is to begin when and only when a formal document is signed. However, the construction industry does not always operate in this simple way and, even in those cases where a formal document *is* signed by both parties, they quite often regard this document merely as the formal record of a contract which already exists. Indeed, it is by no means uncommon to find a project running for months or even years before 'the contract' is finally executed.

Where no single 'contract document' can be identified, identifying the offer and acceptance may prove rather more difficult, partly because either or both of these may consist of written communications, the spoken word or even mere conduct.

9.2.1   Offer

The traditional distinction drawn by the law of contract is between an 'offer' and an 'invitation to treat'. The significance of this distinction is that, whereas an offer will turn into a contract immediately on its 'acceptance' by the person to whom it is addressed, an 'invitation to treat' has no such status – it is merely a stage in negotiations, inviting the other party to make an offer.

The acid test is whether the other party can bring about a contract by merely replying 'yes'. In the construction context, two particular aspects of the problem call for comment.

---

[1] *Nottingham Community Housing Association v Powerminster Ltd* (2000) 75 Con LR 65.
[2] *Gibson Lea Retail Interiors Ltd v Makro Self-Service Wholesalers Ltd* [2001] BLR 407.

Letters of intent

As mentioned above, it is by no means unusual to find work on a construction project started before (even long before) a formal contract is drawn up and signed. Indeed, it is not unknown for the execution of the formal contract to take place after the job is finished. In such circumstances the employer, or the architect on behalf the employer, may write a 'letter of intent' to the contractor, indicating a firm intention to award the contract in question to them. The effect of such a document depends entirely upon its wording, but the following principles will generally apply.

First, notwithstanding its reference to a future contract, such a letter may be interpreted by a court as evidence that the parties regard themselves as already bound. This is perhaps most likely to occur where the contractor has tendered on the basis of standard terms and where, despite the lack of a signed contractual document, the project goes ahead and the parties operate on exactly those procedures which would apply under the standard form contract. In such circumstances a court is quite entitled to conclude that a contract on those terms is already in existence; all that is lacking is the formal record.[3]

Notwithstanding the possibility described above, it must be said that a letter of intent in itself does not usually give rise to any contractual rights or obligations. This is because a court may well take the view that the letter, by stating that there *will* or *may* be a contract in the future, is an indication that there is no such contract at present. A contractor who carries out work on the basis of such a letter may be entitled, under a legal doctrine called restitution, to be paid the reasonable value of the work carried out, but there are no further legal consequences.

An example of this kind of letter of intent is provided by the case of *British Steel Corporation v Cleveland Bridge & Engineering Co Ltd.*[4] The defendants there, who had successfully tendered for the steelwork in a bank in Saudi Arabia, approached the claimants for the manufacture of a variety of steel nodes, and sent them a 'letter of intent' which proposed that the defendants' own standard-form sub-contract should be used. The claimants did not agree to this, and there were further disagreements over price and delivery dates (plus a telex from the defendants stating that the nodes must be manufactured in a particular order). Nonetheless, the claimants manufactured and delivered the nodes. When the claimants sued for payment, the defendants counterclaimed damages on the basis that the nodes had been delivered late and in the wrong sequence. They argued that a contract had come into existence when the claimants started manufacture.

It was held that, since vital terms of the parties' arrangement had never been resolved, it was impossible to say that the delivery and receipt of the steel nodes created a contract through the parties' conduct. As a result, the claimants were entitled under the doctrine of restitution to be paid on a *quantum meruit* basis for the nodes. However, the defendants' counterclaim failed for, without a contract, there could be no obligation to deliver at any particular time or in any particular sequence. It is also worth noting that, had the point arisen, the absence of a contract

---

[3] See *Stent Foundations Ltd v Carillion Construction (Contracts) Ltd* (2002) 78 Con LR 188; *Harvey Shopfitters Ltd v ADI Ltd* [2004] All ER 982; *Bryan & Langley Ltd v Boston* [2005] BLR 508.
[4] [1984] 1 All ER 504.

would have meant that the claimants were under no legal obligation to complete the job, even though they had started it.

Although the judge in the *British Steel* case refused to hold that a contract had been brought into existence by the commencement of work, he recognized that this *could* happen in appropriate circumstances. Indeed, he cited with approval the earlier case of *Turriff Construction Ltd v Regalia Knitting Mills Ltd.*[5] In that case the claimants, having tendered for the design and construction of a factory for the defendants, were told that they were successful, whereupon they asked for 'an early letter of intent ... to cover us for the work we will now be undertaking'. Such a letter was sent, stating: 'the whole to be subject to agreement on an acceptable contract'. The claimants then carried out the detailed design work necessary to seek planning permission and obtain estimates. When, six months later, the defendants abandoned the project, it was held that the claimants had made it sufficiently clear that they wanted an assurance of payment for their preparatory work in any event, and that the letter of intent constituted that assurance. There was therefore a contract under which the claimants were entitled to be paid.

Where a court, as in the *Turriff* case, is prepared to interpret a letter of intent as creating a contract, it will then have to decide on the precise scope of that contract. It is possible, although very unlikely, that a bilateral contract for the whole of the project work might exist. If so, the employer would not only have to pay for whatever work the contractor carried out, but would also be unable to abandon the project without incurring liability for depriving the contractor of the opportunity to make a profit on that contract.

An interesting dispute over the scope of a letter of intent arose in the case of *Monk Construction Ltd v Norwich Union Life Assurance Society.*[6] The claimant contractors were then sent a letter of intent which authorized them to proceed with mobilization and ordering of materials up to a maximum of £100,000; it also stated that, if no contract materialized, they would be entitled only to their 'proven costs'. As things turned out no formal contract was ever signed, but the claimants in fact carried out the whole of the projected works and claimed to be entitled to a 'reasonable sum' for this. The defendants argued that the letter of intent covered this work as well, so that the 'proven costs' ceiling applied, but this argument (which would have meant that Monk were required to carry out a £4 million project for no profit at all!) was rejected by the Court of Appeal. The court's view was that the letter was intended only to govern the preliminary works, in the event that no formal contract was concluded *and no further work was carried out* by the claimants.

Estimates and quotations

It is sometimes said that, in the context of construction, the law draws a sharp distinction between an 'estimate', which is a mere invitation to treat, and a 'quotation', which is an offer in the legal sense. While there may be at least a grain of truth in this, in that a court will tend to assume that this is what the parties intend, the labels cannot be regarded as conclusive. This point is clearly illustrated by the

---

[5] (1971) 9 BLR 20.
[6] (1992) 62 BLR 107.

old case of *Croshaw v Pritchard.*[7] The defendants there, in response to an architect's invitation to tender, wrote a letter headed 'Estimate' which stated: 'Our estimate to carry out the ... alterations to the above premises according to the drawings and specifications amounts to £1230.' The claimant employer replied 'accepting' this figure, but the defendants thereupon purported to withdraw their 'estimate'. When the claimant sued for the extra cost involved in having the work done by another contractor, it was held that the defendants were liable. Notwithstanding its heading, their letter was an offer.

Where an estimate is *not* treated as an offer, it is obvious that the employer cannot, by purporting to 'accept' it, compel the contractor to do the work at the price stated, or indeed to do any work at all. Further, even if the contractor does begin work after giving an estimate, neither party can insist that payment be measured by the amount of that estimate. If there is no further agreement, the employer will be liable to pay what the court regards as a 'reasonable sum', which may be higher or lower than the quoted figure.

Nevertheless, the fact that the contractor may incur no *contractual* liability in such circumstances is not the end of the matter, a point graphically illustrated by the New Zealand case of *J & JC Abrams Ltd v Ancliffe.*[8] The defendant, who wished to develop a building plot behind his house, was given an estimate by the claimants, who then began work. Despite frequent requests, the claimants repeatedly refused to give the defendant a firm price (on the ground that not all the subcontracts had been finalized), although they rapidly realized that the work would cost far more than originally envisaged. Eventually, the claimants submitted a price of almost double the estimate, which meant that the defendant made a very substantial loss on the project.

When the claimants brought an action for the price of the work, the defendant made a counterclaim for his losses. It was held that the claimants were guilty of negligence in not informing the defendant of changed circumstances and escalating prices until it was too late for him to back out of the project and cut his losses (which, given the opportunity, he would have done). They were accordingly liable in the tort of negligence for all the extra losses which he had incurred.

9.2.2   Acceptance

Assuming that an offer in the full legal sense has been made, a binding contract will come into existence when this is accepted by the other party. However, it is important to note the requirements for a valid 'acceptance' in this context.

Acceptance must be certain and unambiguous

In order to create a contract, a party's acceptance must unequivocally relate to the other party's offer. Further, the resulting agreement must be certain in all its essential terms. If these requirements are not met, there will be no contract.

---

[7] (1899) 16 TLR 45.
[8] [1978] 2 NZLR 420.

In one case,[9] the claimant contractors submitted two alternative tenders for the construction of a freight terminal. One of these was at a fixed price, the other was on a 'cost plus' basis. When the defendants purported to accept 'your tender', but did not specify which one, it was held that there was no contract, since it was impossible to say to which of the offers the 'acceptance' related.

A second case,[10] although rather less straightforward, again demonstrates the need for certainty in contractual dealings. The claimant builders agreed that they would introduce a source of finance for the defendants' proposed motel development, provided that they were then employed on the project and that the defendants instructed their quantity surveyor 'to negotiate fair and reasonable sums in respect of ... the projects ... based upon agreed estimates of net cost and general overheads with a margin for profit of 5%'. The defendants agreed to these terms, and the claimants duly arranged the finance, but they were not then given the work. On these facts it was held by the Court of Appeal that, in the absence of agreement on the sum payable, or the provision of a method of ascertaining it, there was no contract between the parties.

The case described above demonstrates that, before a contract can come into existence, the parties must have reached agreement on all those terms that the law regards as essential (such as the price). However, there is nothing to prevent the parties from agreeing that they will be contractually bound immediately, while continuing to negotiate over one or more 'non-essential' terms.[11]

Acceptance must be unconditional

Where what looks like an acceptance (or, for that matter, an offer) is made conditional upon the happening of some event, no contract is created at that point. For example, the use of the phrase 'subject to contract' in negotiations will in normal circumstances prevent the document in which it is contained from being treated as a binding offer or acceptance. However, it is perfectly possible for parties to make an informal agreement at the outset and for this to be legally binding, even though they expressly contemplate that their agreement will be recorded in formal documentation at a later stage.

In one case,[12] a defendant's tender was accepted by the claimant's architect (with the claimant's authority) in a letter which stated: 'The contract ... will be ready for signature in the course of a few days'. Before this formal document was signed, the defendant discovered a mistake in his tender and sought to withdraw it. However, it was held that he was too late; there was already a binding contract.

Apart from communications that are explicitly made conditional, difficulties may arise in the case of negotiations that take place over a period of time. A purported acceptance that does not exactly reflect the offer will take effect as a counter-offer, destroying the original offer and leading to a contract only if it is itself accepted by the other party. When faced with a long series of letters, each of

---

[9]*Peter Lind & Co Ltd v Mersey Docks & Harbour Board* [1972] 2 Lloyd's Rep 234.
[10]*Courtney & Fairbairn Ltd v Tolaini Brothers (Hotels) Ltd* [1975] 1 All ER 716.
[11]*Pagnan SpA v Feed Products Ltd* [1987] 2 Lloyd's Rep 601, applied in the construction context in *Mitsui Babcock Energy Ltd v John Brown Engineering Ltd* (1996) 51 Con LR 129.
[12]*Lewis v Brass* (1877) 3 QBD 667.

which accepts some of the other party's terms and proposes alterations to others, a court must try to see whether, at any given point, the parties are in complete agreement on everything which is at that time regarded as an essential term. If this is so, then a contract may be held to have come into existence, and the effect of later negotiations can be ignored.

One aspect of this matter that has caused particular problems in the context of construction is what is known as the 'battle of forms'. This refers to the situation where the parties communicate with each other through documents that attempt to incorporate their respective standard terms into the contract. In such cases, where work has actually been carried out, a court may well be able to rule that the last letter prevails, on the ground that its terms have been accepted by conduct. However, every case depends upon a careful examination of the entire correspondence, seen in context, and this may show an earlier contract or even no contract at all.

In practice, battles of forms have been fought most frequently between a main contractor on one side and a sub-contractor or supplier on the other. In *Sauter Automation Ltd v Goodman (Mechanical Services) Ltd*,[13] for example, a quotation submitted by sub-contractors for the supply and installation of boiler equipment was expressed to be subject to their standard conditions, which included a retention of title clause (a type of contract term described in Section 20.6.2). The main contractors, having decided to accept this quotation, sent an order stating: 'Terms and conditions in accordance with the main contract' (which did not contain a retention of title clause). When the sub-contractors, without further communication, delivered the equipment to the site, it was held that this amounted to an acceptance by them of the main contractor's counter-offer.

The *Sauter* case may be contrasted with *Chichester Joinery Ltd v John Mowlem & Co plc*,[14] in which a quotation for joinery submitted by sub-contractors was accompanied by their standard conditions. The main contractors sent a purchase order containing their own standard terms, which stated: 'Any delivery made will constitute an acceptance of this order'. The sub-contractors duly delivered their joinery to the site, but not until after they had sent the main contractors a printed form headed 'Acknowledgement of order', which stated that the order was accepted 'subject to the conditions overleaf'. It was held that, by accepting the joinery, the main contractors had accepted the terms on which it was delivered, which meant that the sub-contractors' conditions prevailed.

Acceptance by conduct

The 'battle of forms' cases cited above demonstrate an important principle. The contractual problems arising out of a lengthy exchange of correspondence may sometimes be resolved by examining the conduct of the parties. It sometimes happens that the conduct of one party can be treated as an acceptance of the terms as they exist at that time. This principle is of great importance in construction cases. Work is frequently started (and may indeed reach a fairly advanced stage) before a formal contract is drawn up and signed. If a dispute arises before this happens, a court may well be able to find sufficient 'acceptance' of terms contained in earlier

---

[13] (1986) 34 BLR 81.
[14] (1987) 42 BLR 100.

documents, either by the contractor in starting work or by the employer in giving possession of the site.

The case of *A Davies & Co (Shopfitters) Ltd v William Old Ltd*[15] is a good illustration of acceptance by conduct. The employer's architect there 'accepted' a tender from the claimants as nominated sub-contractors under a main contract based on JCT 63. The architect's letter stated that the defendant main contractors would 'place an order' with them. When placed, this 'order' in fact introduced a new term. The claimants did not protest about this term, but started work. Because of this it was held that they had accepted the offer which was contained in the main contractor's 'order'.

Retrospective acceptance

Where a formal contract is not signed until after work has commenced, an important question which may arise is whether the contract governs work which is done in the interim period (assuming that the contract itself does not make this point clear). If the contract does *not* apply, then that work will be paid for on a *quantum meruit* basis. The other rights and duties of the parties will have to be governed by whatever terms the court is able to imply in the circumstances. These are unlikely to reflect with any degree of accuracy what the parties would have agreed had they dealt expressly with the matter. Therefore, a court may be fairly ready to imply a term for retroactive effect of the contract when finally formed.

The leading case, *Trollope & Colls Ltd v Atomic Power Constructions*,[16] concerned a sub-contract for which the claimants tendered in February 1959. In June of that year, while negotiations were still continuing, the defendants asked the claimants to commence work and gave them a letter of intent. A contract was not finally agreed until April 1960, and the question that then arose was whether a term of that contract (a variation clause) applied to work done before April 1960. It was held that, in these circumstances, a 'retroactivity' term was to be implied. The judge was quite satisfied that, if the parties had been asked in April 1960 whether the contract was to be retrospective, they would immediately have replied that this was what they intended.

9.2.3   Mistake

For present purposes, the term 'mistake' is used to describe the situation in which an 'offer' made by one party and its 'acceptance' by the other do not truly correspond. When this occurs, it might be expected that the law would treat the resulting 'contract' as invalid, but this is not necessarily the case. The law looks, not at the secret thoughts of the parties, but at what they have said and written. If, on this objective basis, there appears to be agreement, then their contract will be held binding and enforceable.

---

[15](1969) 67 LGR 395.
[16][1962] 3 All ER 1035.

A good example of this 'objective' view of contractual agreement is provided by the case of *W Higgins Ltd v Northampton Corporation*,[17] where the claimants, in tendering for a contract for the construction of 58 semi-detached houses, made errors in completing their tender. As a result of these errors, the claimants' tender price totalled £1613 per pair of houses instead of £1670 per pair. On discovering this, the claimants sought to be released from the contract, but it was held that they were bound by the terms on which they had tendered. Their mistake in formulating those terms was legally quite irrelevant.

There are exceptions to this general principle. In particular, it has been accepted by the courts that, if one party is actually aware of the other's mistake, that party may not be allowed to snap up an 'offer' which clearly was not really intended. The extent of the protection thus given is somewhat uncertain, but some guidance is given by a comparison of two cases. In *A Roberts & Co Ltd v Leicestershire CC*,[18] the claimants submitted a tender specifying a completion period of 18 months. The defendants altered this to 30 months in the formal contract, which the claimants signed without noticing the alteration. Before the defendants signed the contract they became aware, as a result of preliminary meetings between the parties, that the claimants believed the contract period to be 18 months. When, despite this, the defendants simply signed the contract without comment, it was held that the claimants were entitled to have the contract period altered to 18 months.

The *Roberts* case may be contrasted with *Royston UDC v Royston Builders Ltd*,[19] in which bills of quantities listed certain materials and prices as subject to a fluctuations clause in the contract. A director of the defendants told the claimants' surveyor that they would require the clause to cover all materials. The surveyor neither agreed nor disagreed, and the contract was signed without alteration. In these circumstances it was held that there was no ground on which to alter the written contract, even though the claimants had inadvertently paid certain interim certificates which had been based on the assumption that the clause applied.

9.2.4 Consideration

As a general rule, a promise is only enforceable where:

- it is made in the form of a deed (as many construction contracts are); or
- the party seeking to enforce it has given some 'consideration' (something of legal value) in return for the promise.

As to what amounts to 'legal value' for this purpose, the courts have often said that a party who merely carries out an existing duty (whether statutory or contractual) is not providing anything of legal value. Hence, a promise by the other party to pay something in return for having that duty carried out is not legally enforceable.

However, this principle was not strictly applied in *Williams v Roffey Bros and Nicholls (Contractors) Ltd*,[20] a case which has some important practical implica-

---

[17][1927] 1 Ch 128.
[18][1961] 1 Ch 555.
[19](1961) 177 EG 589.
[20][1990] 1 All ER 512.

tions for the construction industry. The defendants there, the main contractors on a refurbishment of a block of 27 flats, sub-contracted the carpentry work to the claimant for a fixed price of £20,000. When the project was partly completed, the claimant informed the defendants that he had seriously underpriced the job and was in imminent danger of becoming insolvent (in which case he would have to abandon the work). The defendants, who were liable to incur liquidated damages if the project were not completed on time, agreed to pay the claimant an extra £10,300, at the rate of £575 per flat, if he carried out the remaining work. The claimant duly completed his work on another eight flats, but the defendants failed to pay what had been promised and, when sued for the money, argued that the claimant had provided no consideration (since he had done no more than he was already contractually bound to do).

It was held by the Court of Appeal that, taking a pragmatic rather than a legalistic view, the defendants had received a genuine benefit from the claimant's work, since it had enabled them to avoid a liability for liquidated damages under the main contract. This therefore amounted to 'consideration' from the claimant, and the defendants were liable to pay what they had promised.

9.2.5   Privity of contract

From the middle of the 19th century, it was a fundamental principle of English contract law that rights and obligations created by a contract were only enforceable by and against the parties to that contract. Thus, for example, if A and B made a contract under which B promised to pay a sum of money to a named charity, that promise would be enforceable by A but not by the charity. This principle, called privity of contract, was subjected over the years to heavy criticism, culminating in a Law Commission report, which proposed that it should be substantially overturned (Arden 1996). That has now been done, in relation to most types of contract (including construction contracts), by the Contracts (Rights of Third Parties) Act 1999.

Section 1 of the Act provides that, in respect of contracts made after 11 May 2000, a third party may enforce a term of the contract if **either**

- the contract expressly states that the third party may enforce the term in question; **or**
- the term 'purports to confer a benefit on him' and the contract does not make clear that the parties did not intend the third party to have a legal right to enforce it.

A third party is not entitled to enforce a contract term unless he or she is expressly identified in the contract; identification can be by name, by description or by reference to a class to which the third party belongs (such as 'any future owner of the building'). Moreover, enforcement by the third party is subject to all other relevant terms of the contract in question.

Where a third party brings proceedings to enforce a contract term, Section 3 of the Act provides that the defendant ('the promisor') may raise any defence or set-off which would have been available had proceedings been brought by the other contracting party. The promisor may also raise any defence or set-off which would have been available against the third party personally, had the latter been a party to the contract. The Act further provides (in Section 8) that, if the contract term that the third party seeks to enforce is covered by an arbitration agreement, the third party too is bound by that arbitration agreement. In effect, the third party cannot deprive the promisor of an existing right to have disputes settled by arbitration.

Section 2 of the Act provides that, once the third party has notified the promisor of his or her assent to the term in question, or has relied on the term in circumstances where the promisor knows or ought reasonably to know that he or she has done so, the original contracting parties are no longer allowed either to rescind the contract, or to vary it so as to affect the third party's right, without first obtaining the third party's consent. This is potentially one of the most far-reaching aspects of the new law, but it should be pointed out that it is subject to all relevant terms of the contract. Hence, if the parties wish to remain free to rescind or alter their contract without the need for the third party's consent, they merely have to insert a term to that effect into their original agreement.

It is important to appreciate that the statutory creation of third party rights does not affect the rights of the original contracting parties to enforce each other's obligations. In principle, this could lead to the promisor being sued by both the promisee and the third party. In order to ensure that this does not result in double liability, Section 5 provides that, where the promisee recovers damages from the promisor to cover loss suffered by the third party, the damages awarded in any subsequent action by the third party against the promisor will be reduced accordingly.

Theoretically at least, the 1999 Act could have an enormous impact on the construction industry, which has traditionally operated on the basis of a chain of contracts and sub-contracts on the assumption that, because of the privity of contract doctrine, each contract in the chain will be enforceable only by the immediate parties. At its widest, it could mean that a sub-contractor's obligations as to the quality of work are to be treated as conferring a benefit on the employer, so as to enable the latter to claim directly against the sub-contractor for any defects. This would be very different from the traditional position of the employer claiming against the main contractor, who then in turn claims against the sub-contractor. Again, it is conceivable that an employer's payment obligations might be treated as conferring a directly enforceable right on a sub-contractor. More likely, perhaps, is the possibility that the contractual obligations which each contractor or designer owes to the employer would be seen as conferring a benefit on purchasers or tenants of the completed building.

The initial reaction of the construction industry and its legal advisers to the Act was an understandably cautious one, with a number of contracts being amended to avoid the creation of any third party rights (as in JCT 98 clause 1.12). However, time appears to have brought about a more balanced response, under which the contract, while still excluding third party rights in general, identifies certain persons such as purchasers/tenants and funders who may acquire third party rights, and also specifies precisely the extent of those rights. This approach has been adopted in clause 1.6 of JCT SBC 05.

9.2.6 Form

As a general rule, English law does not require a contract to be made in any particular form. A contract will be equally valid and binding whether it is contained in a formal document known as a deed, made in writing, by word of mouth or by implication from the conduct of the parties. There are exceptions to this general principle (for example, a valid contract for the sale of land can only be made in writing), but these do not apply to building or civil engineering contracts.

Although no particular form is required for construction contracts, the form in which they are made may have important legal consequences. In practice, many construction contracts are made by deed, which nowadays merely means a document, signed and witnessed, that states clearly that it is intended to take effect as a deed; it is no longer required to be 'signed, sealed and delivered'.[21] The most important consequence of making the contract in this form is that the limitation period for bringing a claim for breach of contract is twelve years from the date of breach instead of the normal six years.[22]

As mentioned above, a construction contract may validly be made by word of mouth or conduct. However, if it is not made or at least evidenced in writing (which includes a deed), it will not be subject to certain important statutory provisions. These are Part II of the Housing Grants, Construction and Regeneration Act 1996 (which deals with payment and the right to adjudication under construction contracts) and the Arbitration Act 1996.[23]

9.3 CONTRACTS MADE BY TENDER

As far as the need for an agreement based on offer and acceptance is concerned, building contracts are no different from any other kind of contract. In principle, the same is true when one examines the *way* in which offer and acceptance come about; but here the point should be made that many contracts for building projects, at least those of any appreciable size, are created by the process of tender. The first questions to be answered are why this should be so and just what it is that the tendering process seeks to achieve.

9.3.1 Purpose of tendering

There are two purposes in tendering. First, a suitable contractor should be selected at a suitable time. Second, the offer of a price is required from the contractor at an appropriate time. This offer (tender) will be the basis for the ensuing contract.

The wise contractor will take account of the conditions of contract when calculating the contract price. The way that risks are distributed in a building contract may have a significant effect upon the contractor's pricing strategy, although in many cases a contractor's need for work may be stronger than the

---

[21] Law of Property (Miscellaneous Provisions) Act 1989, Section 1.
[22] Limitation Act 1980, Sections 5 and 8.
[23] See the Arbitration Act 1996, Section 5 and the Housing Grants, Construction and Regeneration Act 1996, Section 107.

desire to add a premium for a risky project. Along with the conditions, the contract drawings and the bills of quantities (if used) make up the contractual arrangements. The contractual arrangements are dictated by the procurement strategy of the employer. This is often not an explicit decision on the part of the employer, especially if the employer is unfamiliar with the construction industry. The contractual conditions give a legal basis to the rights, liabilities and duties of the parties to the contract. They theoretically form the basis of the organizational strategy adopted for the project. But there are very definite differences between contractual arrangements, procurement strategies and tendering procedures. It is important to remember that there is no direct relationship between the type of tendering procedure and the form of contractual arrangement. Further, there is often little relationship between the form of contract and the procurement strategy, even though there ought to be.

The tendering procedure, then, consists of two strands. In the first instance, the contractor has to be chosen, and in the second the price for the contract works has to be estimated so that it can form a basis for the letting of the contract. Although the two events are often simultaneous, they do not necessarily need to be. The tendering process marks the beginning of a contractual relationship. Therefore, the tender stage of a construction project is as much a beginning as an end. There is sometimes a damaging tendency for some participants to view the tendering procedure as the end of their involvement with the project.

9.3.2 Types of tender

Legal analysis of tendering procedures recognizes that there are two different kinds of tender. The first, which is comparatively rare, is a 'standing offer', under which a contractor tenders for, say, such maintenance work as may be required by the employer over a specified period. The 'acceptance' of such a tender by the employer does not in itself create a binding contract. The employer is not bound to order any work, nor is the contractor prevented from withdrawing before the period is over. However, any orders placed during the period must, if the contractor's offer has not *already* been revoked, be carried out. Standing offers are associated with *term contracts*.

The second, and more usual, type of tender is simply an offer by the contractor to carry out the work specified in the invitation to tender. Once the employer accepts this, then (assuming that the acceptance is not conditional), it forms a legally binding contract.

From a practical point of view, tendering procedures ought to vary according to the kind of contract which they are intended to bring about. However, it is often the case that tendering procedures and forms of contract are not related to each other.

9.3.3 Tendering procedures

A variety of tendering procedures have evolved in the construction industry. The main distinguishing feature is the extent of competition. There is a very strong tradition that the best price can be gained by making contractors bid for work, so that the lowest price gets the job. However, there is also an increasing level of dissatisfaction with competition, because all that it really guarantees is the lowest tender price. This is not the same as the actual price that will be paid for the works once they are finished. For example, a contractor who bids too low, in order to be sure of winning the job, may find that the job actually loses money. Contractors who bid as low as this tend to be those who are desperate for work because they are on the brink of insolvency. While a low tender may seem appealing at tender stage, it becomes very sour when the contractor ceases trading and the employer has to appoint others to complete the works. This can cost the employer a great deal more than a higher initial bid would have done, although there are several ways of incorporating financial protection into the contracts (Hughes *et al.* 1998).

As tendering procedures have evolved to confront this kind of problem, they became codified by the National Joint Consultative Committee for Building (NJCC), which consisted of some of the major professional institutions and trade associations involved with construction (RIBA, RICS, ACE, BEC, FBSC, SECG). Although the NJCC was disbanded in 1996, the guidance notes remain in use. The Construction Industry Board (CIB) published guidance about tendering in 1997 (Construction Industry Board 1997), but that organization, too, has since ceased to exist. JCT have also produced a Practice Note about main contract tendering (Joint Contracts Tribunal 2002). As Hackett *et al.* (2007: 205) point out, the relatively new JCT Practice Note requires no new or excessive procedures by comparison with those of the NJCC codes upon which the practice note is clearly based. We briefly describe below some of the more common tendering procedures, including some of those covered by NJCC Codes of Procedure or Guidance Notes, before looking at more novel forms of selecting a contractor.

Open tendering

This method was probably the 'traditional' method until more sophisticated techniques were accepted. The process begins by the placing of an advertisement in the technical press. The advertisement will carry brief details of the location, type, scale and scope of the proposed works. Contractors who are interested in bidding for the work can apply for the documentation. There will usually be a refundable charge for this documentation to prevent people from applying out of idle curiosity. Either the advertisement, or the documentation, will explicitly state that the employer is not bound to accept the lowest tender, or indeed any tender (see Section 9.3.4).

Open tendering is an indiscriminate request for tenders. This approach may be inadvisable because there is no reliable method of ensuring high quality building. Research has shown that, with open tendering, only about one in twenty contractors' bids are successful. The preparation of such tenders places upon the industry an unnecessary burden of time, effort, and expense. This expense ultimately is passed back to the clients of the industry. Because of the indiscriminate nature of open tendering, contractors can be awarded work for which they are not properly equipped, in terms of either resources or experience. Although an employer is not bound to accept the lowest bid, a committee in charge of public expenditure is under a lot of pressure to accept the lowest. When the lowest bid is accepted, this can easily result in the employer awarding the contract to the builder who has the least appreciation of the complexities of the projects, or the greatest willingness to take risks, or the lowest current workload of all the bidders. It would be unusual, or even lucky, if these factors resulted in the best value for money for the employer.

Because of the problems associated with open tendering, its use has been declining in recent years. Indeed, the NJCC was established specifically as a response to the fact that public attention had been focused on the importance of selective, rather than indiscriminate tendering. However, it is still occasionally used to obtain tenders for building work. Indeed, the indications are that, because of the effect of certain European Union legislation discussed in Section 9.3.4, its use in public sector work is increasing. This is to some extent unfortunate in that, having recognized the large number of problems associated with open tendering and having successfully moved away from it as the main form of tendering in the UK, the European rules meant that we almost reverted back to a far from satisfactory situation. Similar pressures apply in the international market, especially where organizations such as the World Bank are involved. Interestingly, JCT Practice Note 6 does not mention open tendering at all.

Single stage selective tendering

The first step away from the problems of open tendering is to restrict the number of tenderers being invited to submit bids. This is the purpose of **single stage selective tendering**, which is also known simply as selective tendering. Basically, it consists of pre-selecting a limited number of contractors to tender for the work. An employer who builds regularly will usually have an approved list of contractors, from which a short list can be drawn up. In local authority work, for example, specific procedures are typically set up for a contractor to be entered on the approved list. This is quite distinct from the process of being selected for a particular project. JCT Practice Note 6 provides a model form of preliminary enquiry for using in establishing whether particular contractors are willing to bid.

If the employer does not build regularly, then an *ad hoc* list of approved contractors may be drawn up. According to the NJCC, this will consist *of contractors of established skill, integrity, responsibility and proven competence for work of the character and size contemplated* and there should be no more than six on the list (NJCC 1994a). The code of procedure gives the points that should be considered when considering a contractor for inclusion on the list. Clearly, once the pre-selection process has been done properly, any of the contractors is satisfactory to the employer. Therefore, tenders may be considered on price alone, and the lowest one may be selected. Provided that tenders are checked carefully for mistakes by the bidders, and that the list of approved tenderers is regularly revised, some of the worst problems associated with inappropriate selection can be avoided.

However, other problems *may* still occur. Some of the major problems on construction projects happen because the design team do not have the benefit of the contractor's experience at a stage in the design when it would be most useful. Issues such as continuity of work, production cost savings, buildability and subletting specialist items can have a significant impact on the final price of a project.

The successful management of some types of construction project can be totally dependent on getting this kind of knowledge into the design team at an early stage in the process. A variety of techniques have emerged to deal with this, and their purpose is to separate the processes involved with selection of the contractor from the processes for determining the pricing mechanisms to be employed when paying the contractor.

Two stage selective tendering

One of the processes used to involve a contractor at an early stage was known by the NJCC as two stage selective tendering, sometimes called 'negotiated tendering', but now called simply two-stage tendering. It must be emphasized that the purpose is not to involve the contractor with *responsibility* for design. It is to get the main contractor involved, in an advisory capacity, before the scheme has been fully designed. This may lead to complaints from contractors, or more usually, specialist sub-contractors, that they are brought in early to help with design without being paid for design services. However, the Code of Procedure specifically states that this is not the intention (NJCC 1994b).

In this system, the process is split into two stages. The first stage is a process for the selection of a contractor, and for the establishment of a level of pricing for subsequent negotiation. It is at this stage that the competitive selection takes place, based upon a minimum amount of information, which indicates simply the basis of the layout and design of the works. The detail needs only to be sufficient to provide a basis for competition, and a basis for the negotiation of the price during the second stage. In the second stage the pre-contract process is completed. The employer's professional team collaborates with the selected contractor in the design and development of production drawings for the whole project. Also bills of quantities for the works are prepared and priced on the basis of the first stage tender. The result of the second stage is an acceptable sum for inclusion in a contract and complete contract documentation is prepared in conjunction with the contractor.

This form of tendering is used where the building works are of a very complicated nature; where the magnitude of the work may be unknown at the time for selection of the contractor; where an early completion date is paramount or where the professional team wish to make use of the contractor's expertise when finalizing the design (Joint Contracts Tribunal 2002).

Selective tendering for design and build

A further process, for which there was a separate NJCC Code of Procedure, was for use where a **design and build** contract is being undertaken (NJCC 1995). This method is used where the contractor will design as well as build the project. It was envisaged by the NJCC that the form of contract associated with this tender process would be the JCT form of contract with contractor's design (CD 98), the predecessor of JCT DB 05. JCT Practice Note 6 simply deals with the procedure for design an build by making it clear that design *may* be one of the non-price criteria for selection.

As shown in Chapter 4, under a design and build approach there is no architect, consulting engineer or quantity surveyor named in the contract. This form obviates the need for the employer to appoint any consultants. The contractor provides all professional services in connection with the building. The employer's interests are looked after by the person named in the contract as the Employer's Agent: this person may be drawn from any of the professions. The requirements of the employer are set down in a document called the Employer's Requirements; and the tendering contractors put forward their solutions and prices in a document called the Contractor's Proposals.

In this system of tendering, the contractor's tender is the price for which the contractor offers to carry out and complete, in accordance with the conditions of contract, the works referred to in the Employer's Requirements. The word 'tender' in this context is taken to cover the whole of the Contractor's Proposals including both price and design. It is envisaged by the NJCC that any tendering contractor will have been pre-selected to a certain extent, as in two-stage tendering (NJCC 1995)

Negotiation

A more radical approach to the selection of the contractor is offered via negotiation. This approach is more suited to procurement strategies that are markedly different from traditional methods. Construction Management, for example, as described in Chapter 6, involves the use of direct contracts with different specialists. Management contracting (Chapter 5) involves the early appointment of the main contractor whose responsibilities are not the same as they would be under JCT SBC 05 or other, more traditional forms of contract. These newer forms of procurement demand a less adversarial approach at all levels and all stages in the building process. The inherent flexibility demanded by such approaches means that there is no standard method for negotiating a contract.

A variant of negotiation occurs with **serial contracting**, where contractors are asked to bid for a project (perhaps in competition) on the basis that, if they build this one satisfactorily, others of a similar type will follow and the same bill rates will be used. In the light of some of the recommendations in the Latham Report (1994) about partnering, and also in the light of the experience of many seasoned clients of the industry, who refer to **strategic partnering** (Bennett and Jayes 1998), this approach brings many benefits. Not least among them is that the desire to preserve continuing business relationships tends to be stronger than the need to fight particular claims or disputes. Thus, the participants are often focused on co-operation rather than conflict. Another advantage of serial contracting is that contractors have an opportunity to learn more about the client organization and there are opportunities to repeat details from one project to another. This can potentially enable significant cost savings brought about through repetition of tasks. Of course, an alternative, more cynical view is that the threat of withdrawing a sequence of projects is much stronger than threat of withdrawing only one, creating considerable enthusiasm among clients for serial arrangements.

Experience has shown that negotiation is one of the most effective ways of selecting a contractor for 'non-traditional' approaches. In these cases, the deal is negotiated as the relationship develops. It seems that the single most important factor of such a relationship between the employer and the contractor is familiarity. They have worked together before, and they expect to work together again in the future. As with serial contracting, the preservation of a continuing commercial relationship is more important than simply securing the lowest price for the employer or the highest profit for the contractor. Because of this, it is essential that the employer is familiar with some or all of the building process, the professional team, the contractor and the specialists.

Joint ventures

Certain construction projects are so complex that the distribution of liability between the consultants becomes a problem. One solution is to approach the design and implementation of the project as a joint venture between all the consultants and some or all of the specialist contractors. This is an even more radical departure from the so-called 'traditional' methods than the 'management-based' methods. It involves all of the parties to the Joint Venture agreement taking on joint and several liability for the design and/or execution of the project. It is not necessary to set up a special joint venture company to do this, but it is possible. The agreements between the parties to the joint venture will have to be very carefully examined, and probably backed up by performance bonds and/or parent company guarantees (Hughes *et al.* 1998). The NJCC has issued guidance on tendering under this approach (*not* a Code of Procedure), and further details can be found there (NJCC 1994c). The use of consortia for procuring construction work has grown a lot in recent years in the UK, largely as a result of public sector procurement policy. But research has shown that consortia may come together because that is the only way of winning certain types of work, rather than because they feel that it is the best way to operate (Gruneberg and Hughes 2005).

9.3.4   Legal analysis of tenders

The adoption of a suitable tendering procedure will no doubt help to reduce if not to eliminate the potential waste of time, effort and money. Nevertheless, there remain certain difficulties inherent in the way that the law analyses tenders, and these are discussed below.

The parties' obligations

Conventional legal analysis regards an employer's request for tenders as an invitation to treat, and the tenders themselves as offers. It follows that, as a general principle, the employer is under no legal obligation to accept the lowest (or, indeed, any) tender submitted. In practice, this protection of the employer's position is frequently (though unnecessarily) made explicit when tenders are invited, by a statement that the employer does not undertake to accept the lowest tender.

The consequence of this contractual framework is that the costs of tendering (which may be considerable, especially where substantial design work is required) are to be borne by the contractor (for a more detailed discussion about the costs of tendering, see Hughes *et al.* 2006). Of course, these costs can be reflected in the tender price and thus recouped by the successful bidder, but the unsuccessful competitors must in normal circumstances bear their own costs, unless a promise to pay on the part of the employer can be implied. Such a promise might be implied where the preliminary work goes beyond what would normally be expected, or where the employer can make some profitable use of it.

An example of what will suffice to justify departing from the basic rule is provided by *William Lacey (Hounslow) Ltd v Davis*,[24] in which the claimants, who had tendered for the job of reconstructing certain war-damaged premises owned by the defendant, were led by the defendant to believe that they would be given the contract. At the defendant's request, the claimants prepared various calculations, schedules and estimates, which the defendant used to negotiate a claim for compensation from the War Damage Commission (a government body). The defendant then sold the property to a third party, at a price which reflected the benefit of the agreed War Damage claim, but without ever having concluded a formal contract with the claimants. It was held that a promise by the defendant to pay a reasonable sum for these services could be implied.

The claim of the contractors in the *Lacey* case was a strong one, since the client had specifically asked for the work in question to be done and, moreover, had actually made a substantial profit from it. However, a similar principle has been applied in respect of work that the client had only requested by implication, and where the client's profit from that work was still only a potential one.[25]

The employer's basic unfettered discretion in respect of tenders is subject to the following qualifications:

1. By analogy with cases concerning the advertisement of auctions, a person who invites another to tender with no intention whatsoever of accepting that tender will be liable for any expenses that the latter incurs.
2. An employer who expressly promises to accept the lowest tender will be bound by that promise, once a tender that complies with any conditions set has been submitted.[26]
3. In at least some situations (possibly only where public bodies are concerned), the employer may be under an implied obligation to give proper consideration to any tender submitted in accordance with published conditions (as to time, form, etc.).[27] It should be noted that, where this applies, it only requires that the tender be *considered*, not necessarily that it be *accepted*.[28]
4. Local authorities are required by the Local Government Act 1972 to have and to publicize formal contracting procedures, normally involving

[24] [1957] 2 All ER 712.
[25] *Marston Construction Co Ltd v Kigass Ltd* (1989) 46 BLR 109.
[26] *Harvela Investments Ltd v Royal Trust Co of Canada* [1986] AC 207.
[27] *Blackpool and Fylde Aero Club Ltd v Blackpool BC* [1990] 3 All ER 25.
[28] *Fairclough Building Ltd v Port Talbot BC* (1992) 62 BLR 82.

competitive tendering. They are further required by the Local Government Act 1988 to give reasons for their procurement decisions and are generally prohibited from taking into account non-commercial considerations in reaching those decisions.

5. In relation to the procurement of public sector work, legislative rules of the European Union seek to place considerable restrictions upon an employer's discretion. The UK regulations that give effect to relevant EU Directives are discussed below.

European Union control on procurement in the public sector

European Union legislation impacts upon the procurement of public works, supply and services in two main ways. First, the general principles of the European Treaty (which apply to all public sector contracts, regardless of value) prohibit discrimination within the European Union on grounds of nationality (Article 6) and restrictions on the free movement of goods (Article 30) or services (Article 59). In accordance with these provisions, the European Court of Justice has ruled that it was a breach of Article 30 to include, in a contract for a new water main in Ireland, a clause requiring all pipes to comply with a relevant Irish standard, without considering whether materials which complied with other national or international standards might offer equivalent guarantees of safety or suitability. The result of the clause was that a tendering contractor could only obtain the contract by using Irish materials.

Secondly, a complex array of European Union Directives, implemented in the United Kingdom by various statutory regulations, lay down rules governing the procedure for the awarding of any public sector contracts with a value greater than certain thresholds. Similar rules, though somewhat less strict, apply to contracts let by utilities, both public and privatized.

The content of the regulations is extremely detailed, but their general principles are as follows:

- The contract awarding body must publish notice of the intended contract in accordance with specified procedures and time limits.
- Specifications must wherever possible be based on European standards, or on national standards which embody European standards.
- Contracts must normally be let by either an open or a restricted tendering procedure. Open tendering means that anyone is entitled to tender, although bids may be excluded on specified grounds, including technical capacity and financial standing. Restricted tendering means that the employer is entitled to draw up a list of tenderers from all those who register their interest; however, this selection must be made on the basis of a procedure laid down in the regulations.
- In certain exceptional cases (notably where work is required in circumstances of extreme urgency, or where for some legitimate reason the contract can only be carried out by a particular contractor), the tendering procedure may be dispensed with and the contract let by direct negotiation.
- In all cases of competitive tendering, the contract can only be awarded to either the 'lowest' or the 'most economically advantageous' tenderer. The

call for tenders must make clear if the second of these bases is to be used and, if so, the criteria upon which 'economic advantage' will be judged. If this is not done, then the contract must be awarded to the lowest tenderer.

- Once bids have been opened, the employer must not conduct negotiations with individual tenderers in a way which is likely to distort open competition, for example by agreeing any variation in the works to be carried out.

A contractor who has been excluded from tendering or not awarded a contract, and who wishes to complain about a breach of the regulations, must both notify the employer and commence proceedings in court within three months. If this is done, the court can order the employer to correct the infringement, if necessary by suspending the tendering procedure, or may set aside any unlawful decision (for example by removing a discriminatory specification from the tender documentation). The court is also empowered to award damages to compensate the contractor for any loss suffered, and this will be the only available remedy in cases where the offending contract has already been concluded. Thus in *Harmon CFEM Facades (UK) Ltd v Corporate Officer of the House of Commons*,[29] a US-owned company was held entitled to damages covering its cost of tendering where a contract for fenestration work at the House of Commons was let to another company on the basis of 'value for money'. The court held that this was not the same as 'most economically advantageous', which meant that the contract should have been let to the claimants as the lowest tenderer.

Withdrawal of tender

The basic contractual position is that, since a contractor's tender is merely an offer, it may be validly revoked at any time before it has been accepted. If this happens in respect of a main contract, the client may well be disappointed (especially if the tender in question was the lowest), but is unlikely to suffer any great financial loss as a result. By contrast, the withdrawal of a *sub-contractor's* tender may have a disastrous effect on the main contractor, as is demonstrated by the New Zealand case of *Cook Islands Shipping Co Ltd v Colson Builders Ltd*.[30] There the defendant contractors, having obtained a written estimate of transport charges from the claimants, relied on this in pricing their own tender for a contract. The defendants' tender was accepted by the client, but the claimants then announced that they were increasing their prices (which in effect meant that they were withdrawing their original offer). It was held that, in the absence of a binding contractual obligation, the claimants were quite entitled to withdraw, which of course left the defendants with a contractual obligation priced on an unrealistic basis.

As the law currently stands, an English court would probably come to exactly the same conclusion on this matter as its New Zealand counterpart. However, it is of interest to note that the Canadian courts have prevented both sub-contractors and main contractors from withdrawing tenders in similar circumstances; at least once the other party has relied on the tender by making some sort of commitment.

---

[29] (2000) 67 Con LR 1.
[30] [1975] 1 NZLR 422.

Whether English law will develop along similar lines must, for the moment, be a matter of speculation, but there is much to be said for the idea, at least from the point of view of business convenience. Failing this, the practical solution would appear to lie in requiring contractors to supply a 'bid bond', that is, a promise by deed not to withdraw the bid, backed by a financial guarantee (Hughes *et al.* 1998).

9.3.5   Problems in the formulation of bids

Conventional textbook wisdom describes contractor's tendering procedures in a very logical and objective manner. For example, it is always said that in pricing a bill of quantities a contractor will ascertain such things as the price of materials, the cost of labour and plant and the availability of resources. This information is used to **build up** rates that can be entered in the bills. So, for example, if a certain quantity of brickwork is described in the bills, the contractor will determine the cost of the raw materials (based upon the relevant quantity discounts), the rates of pay for the skilled and unskilled elements of the labour, the number of hours actually needed to put together a square metre of brickwork, the number of hours needed to mix the mortar, an amount of money for supervision, plant and so on. These amounts are added together in such a way as to identify a rate for a square metre of brickwork. This rate can then be inserted into the bill and multiplied by the specific quantity required.

This process, known as **estimating**, takes place for every item in the bill. The first problem is that it may take little account of the way in which contractors' costs are incurred; especially if it is based solely on historical information that does not relate the location of the work to its cost. An easy example is that brickwork on the second floor is cheaper than brickwork on the seventeenth floor, because of the distance that the materials have to be moved, but this difference might not be reflected in the built-up rates. It is for this reason that complex rules have developed for measuring work, usually called standard methods of measurement.

The second and more important problem is the 'myth of tendering'. Although contractors have very detailed information related to costs, and can work out very accurate rates for the bills, it is not usually reported that these are frequently *not* the rates that actually appear in the bills. Contractors know about costs (what they pay for their resources); they know about prices (what they charge for their product); and they know about value (what the client is willing to pay for a building). The difference between cost, price and value is rarely appreciated outside quantity surveying (and even quantity surveyors frequently use the word 'cost' when they mean 'price'). It is this distinction that accounts for the difference between the textbook method of tendering, and the additional process that is not usually reported.

Upon being invited to tender, a contractor will first of all make a decision about whether or not the job is wanted. A contractor will almost never decline to tender, for fear of not being asked to tender again in the future. There is a certain amount of stigma attached to a contractor who declines to tender. If the job is not wanted the contractor will submit an inflated tender, called a **cover bid**. While a contractor will know what the rates in the tender ought to be, these will all be altered to make the final figure clearly too high to be acceptable (Hughes *et al.* 1998).

If the contractor wants the job, then a further decision is taken about the state of the market and an assessment of what this type of building is selling for at the moment. This assessment is then modified by the level of risk associated with the project, particularly in terms of the contractual conditions put forward by the employer. If the contract is risky, a substantial premium should be added to the contractor's bid, so that the risks are covered. The calculation is, in theory at least, similar to the actuarial calculations made by insurers when examining and pricing risks.

Finally, the contractor's own cash flow has a significant impact on the way that the tender is put together. Having a detailed knowledge of costs and finance, the contractor can predict the monthly net **cash flow** in or out of the project. If the project takes place near the end of a tax year, the contractor may want to reduce the level of profit appearing on the balance sheets for the purpose of reducing tax liabilities. This can be done by artificially reducing the rates for work at the beginning of the project and adding a corresponding proportion on to the rates for work later on in the project. Alternatively, the contractor may need to get cash in quickly to meet liabilities, or to show a good return for shareholders' dividends. In this case the rates at the beginning of the project can be increased, with a corresponding decrease in the rates at the end. The former process is known as **back-loading** the bill, the latter is known as **front-loading** the bill. Neither process makes any difference to the contract sum, although both can have a significant impact on the continued survival of the contractor's business.

This brief analysis shows that, notwithstanding the contractor's detailed knowledge of costs and prices, the skill of a good contractor is in pitching the contract sum at a level which will maximize the chances of winning jobs that are wanted, while ensuring that profits are adequate. The bill rates are then manipulated to add up to the desired contract sum, and internally adjusted to regulate the cash flow according to the financial position of the contractor.

This shows how dangerous it can be to attach too much credence to individual bill rates when analysing tendering procedures. It also illustrates one of the major pitfalls in using bill rates for valuing variations, as examined in Chapter 15.

10   Liability in contract and tort

Where a contract has been properly made, in terms of the criteria laid out in the previous Chapter, it then becomes necessary to identify precisely what obligations it imposes on the parties – that is to say, what *terms* it includes. This is because where any party fails without lawful excuse to perform fully and exactly a contractual obligation, that party is guilty of a breach of contract. In considering this question, what emerges is that there are three types of contractual term: those *express terms* which are contained in the main contractual document itself; those (also *express terms*) contained in other documents to which the main contract document refers; and those terms which are *implied* by law.

In addition to those obligations created by a contract, a party may incur legal liability under the law of tort. The most important example of this type of liability is that based on negligence.

10.1   EXPRESS TERMS

10.1.1 Terms and representations

Since English law does not as a general rule require a contract to be made in any particular form (see Section 9.2.6), it is in principle possible for the parties' agreement to be spread over a number of documents, and even to include oral statements. Even where the parties have eventually signed a document described as 'the contract', it is open to a court to decide that this document does not paint a complete picture, and that the true contract consists of that document plus other terms. However, a court will not be quick to do this, for there is a presumption that a written document that appears to contain all the terms of a contract in fact does so.

The result of these principles of contract law is the creation of some uncertainty regarding the status of a statement or promise made by one party to the other during the course of negotiations leading to the signing of a formal contract. Such a statement (which might be oral, or might equally be contained in, say, a letter) can be treated in law as a term of the contract. In this case, if it is untrue, the innocent party may bring an action for breach of contract. Alternatively, if it has operated as an inducement to the other party to enter into the contract, the statement may be treated as a misrepresentation, in which case the legal remedies available will depend on whether it was made fraudulently, negligently or innocently. This is not the place to discuss the remedies for misrepresentation, which will be found in any textbook on the law of contract. Suffice it to say that they are generally less beneficial to the innocent party than a right to sue for breach of contract.

In seeking to decide whether a particular statement is a term or a mere representation, the courts make use of several guidelines. Some of these may point in opposite directions. It is for example presumed that a statement made at a very early stage of negotiations is probably a representation, while a statement made by an 'expert' is more likely to be treated as a term. However, what is very often decisive in the context of construction contracts is the strong presumption that, where the parties have signed what purports to be a complete contract document, this represents the whole of their contract. As a result, any other statement, promise or assurance is likely to take effect, if at all, as a mere representation.

10.1.2 Contract documents

As mentioned above, a contract made in writing frequently consists of more than one document. Certain categories of document are routinely found in construction contracts and the main ones are described below.

Contract documentation is the means by which a designer's intentions are conveyed to the client, the statutory authorities, the quantity surveyor, the contractor and the sub-contractors. In addition, the other design consultants each add their own specialized information to the increasing body of documentation as the project progresses. From a management and administrative point of view, every item in this 'contract documentation' is important, because it describes and records all aspects of the project. However, by no means every item comes within the more limited legal definition of 'contract documents', a special term used in construction contracts to describe those documents which are legally binding. In one case, for example, it was held that a preliminary site survey report was not incorporated into the contract, even though it was specifically referred to in the contract drawings.[1]

As to what the term 'contract documents' includes, nearly all the standard-form contracts identify the articles of agreement, conditions of contract, drawings and bills as key contract documents. However, there are differences between forms when it comes to such items as programmes, specifications, project-specific data, etc. The elements of documentation will be examined briefly in turn, before looking at how they relate to each other and how the conditions of contract integrate them.

Articles of agreement

It should be noted at the outset that many English standard forms of contract contain a section called 'definitions' and it is also useful to know that each capitalized term in JCT contracts is defined in such a definition section. The definitions clause of JCT SBC 05 (clause 1.1), states that the 'contract documents' contain, among other things, a document called the 'agreement' which in itself is defined in the contract as the 'articles of agreement'. The articles of agreement record in general terms what the parties have agreed to do. They identify the parties to the contract, what is to be built (the contractor's obligation) and what is

---

[1] Co-operative Insurance Society Ltd v Henry Boot (Scotland) Ltd (2003) 84 Con LR 164.

to be paid (the employer's obligation). They tie these obligations to the conditions and to the other contract documents.

In JCT contracts, the articles of agreement consist of three sections, namely the recitals, articles and contract particulars (see the definition of the term 'agreement' in clause 1.1 JCT SBC 05):

- **Recitals**: In defining the contractor's obligation, the articles of agreement do not descend to any great detail. They merely state, in a section starting with the word 'whereas', that 'the employer wishes to have' the briefly described works carried out. Some standard forms also record that tenders have been accepted, that drawings have been prepared, and so on. This section is called the 'recitals' (although only JCT contracts specifically give it this name). The purpose of the recitals is to describe the background to and the purpose of the contract. They help to provide a basis for interpretation of the detailed provisions contained in the contract clauses.
- **Articles**: The articles substantiate the obligation of the contractor by stating that the works shall be carried out and completed in accordance with the contract documents. Moreover, the articles provide the place in which the parties should specify essential information, such as the amount of the contract sum, the name of the architect/contract administrator and the desired method of dispute resolution.
- **Contract particulars**: This section contains the project-specific data. Older JCT forms contain an 'appendix' as a specific part of the contract documents instead of contract particulars.

Although they may use different terminology and present them in a different part of the form, all standard-form contracts will need to contain these three types of information.

Conditions of contract

The conditions of contract are the very detailed clauses that follow on from the articles of agreement. The purpose of the conditions is to amplify and explain the basic obligations the parties undertake by signing the articles of agreement. The conditions also provide administrative mechanisms for ensuring that the correct procedures are observed. Effective contract clauses of this kind deal efficiently with what would otherwise be breaches of contract, and therefore ensure that the contract is kept 'alive'.

Project-specific data

Certain facts relating to the execution of a building contract will differ from one project to another. To enable standard-form contracts to be used in spite of these differences, the facts concerned can be summarized in a part of the contract variously referred to as appendix, contract particulars, contract data, abstract of particulars and so on, which should be filled in when executing the contract. Thus, JCT 98 contains an appendix of four pages, ICE 7 three pages. JCT SBC 05 contains eight pages of 'contract particulars' as a subsection of the articles of agreement, NEC 3 has a 13-page section called contract data and GC/Works/1 achieves the same effect with a document called an 'abstract of particulars', which is included with the tender documentation.

The project-specific data contains such items as the starting date (or date for possession), completion date (or duration of the contract), defects liability period (rectification period or defects correction period) and the amount of liquidated damages. Sometimes default provisions specify what is to happen if the relevant entry is left blank. Some provisions are not applicable at all unless the particular entry is completed, such as the employer's right to delay possession of the site by up to six weeks (as in clause 2.5 JCT SBC 05).

The contents of the project-specific data are of fundamental importance in assessing the amount and duration of responsibility accepted by the parties. Therefore, close attention should be paid to these entries.

Drawings

The production of drawings is generally, although not always, the responsibility of the design team. The technical complexity of most modern buildings makes it essential that this process (including all associated information, as well as the drawings themselves) is properly planned and organized. Detailed guidance on this is available from numerous sources (for example, Thompson 1999, Wakita and Linde 2002). As with all types of communication, clarity is essential. The aim is to transmit the information in a way that can be understood. Good communication will inspire confidence; bad communication will cast doubt on the ability and integrity of the designer.

Drawings have many functions and a different drawing usually fulfils each function. First, they form a model of the designers' ideas and help to articulate and predict problems with fabrication and with appearances. Isometrics, scheme design drawings and 'artistic impression' drawings help to fulfil this function. Although there is little information on them, which would enable a builder to erect the proposed building, they amplify and explain the basic nature of what will be required in the finished building.

Second, drawings are the vehicle by which the designers' intentions are conveyed to the contractor. The detail design drawings contain information that shows how the separate parts of the building interact with each other. The detailed information from specialist sub-contractors and from other designers is co-ordinated and presented through such drawings.

Third, drawings form a record of what has been done. These 'as-built' drawings are essential to the building owner as a basis for future maintenance of the facility, and may not be the same drawings that were used for the purposes of fabrication. They are also a useful control document to compare what has been done with what was planned.

The multiple uses which different drawings are called upon to perform, and the interaction between drawings and other types of documentation, have often been a source of problems. Thus, detailed guidance has been developed for ensuring that a common approach to describing different types of work is used, resulting in guidelines for the production of drawings, specifications and bills of quantities. These currently represent best practice in the industry and should be essential reading for everyone concerned with the documentation of construction projects (Allott 1998).

The designer usually retains copyright in the design and, on completion of the work, can insist that all drawings are returned. When this is the case, neither the contractor nor the client is entitled to use the information again, for example to construct an identical building, without first obtaining the permission of the designer.

Schedules

There are at least two relevant definitions of schedules. First, schedules can be tables of information that summarize the quantities and dimensions of certain generic items, for example windows or ironmongery. Although they will rarely be cited as contract documents, they will often be bound up and included with the sets of drawings, as they are invaluable in summarizing detailed information. Clearly, a drawing will show all the windows in a building, but it is very useful to extract all the details of sizes and numbers on to one table. Although dimensions can be summarized in tabular form, some information is best conveyed on a dimensioned sketch, such as the arrangement of opening lights in a window. Such schedules provide information to quantity surveyors, builders, buyers and others in a form which has advantages over drawings (Hackett *et al.* 2007: 173): mistakes are easier to check; getting estimates and placing orders for similar items is simplified; prolonged searching through a specification is avoided. The aim of the schedule is merely to simplify the retrieval of information. It is not intended to supplant the bills of quantities, which will contain the full and final description of the work.

Second, schedules are also a specific section in many contracts, including the newer JCT forms. In this context, schedules cover very specific contract data, such as JCT SBC 05, which uses schedules for defining the contractor's design portion (schedule 1), insurance options (schedule 2) and forms of bonds (schedule 6). According to the guide to JCT SBC 05, these schedules form part of the conditions (Joint Contracts Tribunal 2005b), a fact that is not explicit in the standard form itself. However, this explains why the schedules are not specified as a distinct part of the 'contract documents' in the definition of this term (clause 1.1 of JCT SBC 05).

Specifications

Drawings provide information about the shape, appearance and location of the various components that have to be assembled but they convey little about the methods to be used, the quality of finishes and the workmanship to be employed. Assuming that such things are to be specifically controlled by the contract, rather than relying on the terms that would be implied at common law, this is a matter for the specification. This will include information about the materials and components to be used, the standard of workmanship that will be required, any specific performance requirements as well as the conditions under which the work is to be executed.

Where bills of quantities are used for a project, information of this kind will be usually contained in the bills, and thus there will be no 'specification' in the sense of a separate document. However, not all projects need full quantification with bills of quantities; many projects are let by drawings and specification alone. Where this is done, the specification may be called upon to perform an even wider role. In order to prepare a useful and accurate specification, it is essential to be systematic and methodical.

Bills of Quantities

The purpose of bills of quantities, and their status, may vary under different standard form contracts. However, when used, they are almost invariably specified as a contract document.

In JCT contracts, when bills are used, they are called 'contract bills' and they have a wide-ranging role because they define the whole of the contractor's obligations for quality *and* quantity of work (for example JCT SBC 05: 4.1, JCT IC 05: 4.1.1). In the ICE 7 form (measurement version), by contrast, the bills have a narrower role. Quality is required by clause 36 to be of the kinds described in the contract, meaning the whole of the combined documentation. In any event this is subject in all cases to such tests (at the contractor's expense) as the engineer may require. The quantity of work described in the bills is only an *estimate* of what is required of the contractor (clause 55). The contractor is obliged to undertake as much work as is necessary, and this obligation is not affected by anything in the bills.

The contracts also govern the preparation of the bills, typically specifying that bills have been prepared in accordance with the relevant standard method of measurement. Because the bills have such contractual significance, especially under the JCT series of contracts, it is often necessary to be fully cognisant of what the relevant Standard Method of Measurement (SMM) contains. Any item of work that is not measured in line with the principles in the SMM must be expressed categorically in the bills, in terms of both the nature of the departure from the SMM and the items affected.

Bills of quantities originated for tendering purposes; indeed, engineering contracts still retain this emphasis. In many JCT contracts such as JCT SBC 05 (i.e. the version 'with quantities'), the bills of quantities have additional roles to play in the valuation of variations and interim certificates, and in the control of the works. In addition, various other categories of information are often gleaned from the bills, such as the location of work and basic cost planning data. Not surprisingly, bills of quantities are increasingly being seen as hopelessly inadequate for all of these conflicting needs. There are many ways of choosing a contractor, and many ways of fixing a price for the work to be undertaken. To use the bills as the sole vehicle for both these purposes is an over-simplification that can lead to many problems (Skinner 1981).

The bills typically consist of preliminaries, preambles and measured works. These terms have specific meanings and it is useful to clarify their meaning:

- **Preliminaries** contain the definition of the scope of the works. In the Common Arrangement of Work Sections (Allott 1998), section A is given over to defining the preliminaries and general conditions. It includes, among other things, particulars of the project, lists of drawings, description of the site, scope of the work, details of documentation and management arrangements. The preliminaries often contain items for pricing which will be for general use for the whole project such as huts, fences and security.
- **Preambles** were once a separate description of the materials and workmanship to be employed in assembling the building and this was known as the 'specification' section of the bill. In the common arrangement, this is Section A33 of the work sections and, as a result, it is likely that 'preambles' will disappear from the vocabulary of building contracts.
- **Measured works** are the detailed quantification of the works. This section should be presented according to the rules dictated in the contract.

Two further terms worthy of mention are prime cost and provisional sums. These terms are antiquated and their origins are now obscure but their use is widespread and seems likely to continue:

- **Prime cost (PC) sums** are used for works under the building contract that are to be carried out under the direction of the main contractor by certain other persons, namely, nominated sub-contractors, nominated suppliers and statutory authorities. The nominated sub-contractors will usually have been selected prior to the selection of the main contractor. Since the contractor has no control over the pricing of this work, an amount of money is simply included in the bills, to which the contractor can add a price for attendance (i.e. supervision, accommodation and plant) and profit. Prime cost is taken to mean that the contractor will be fully reimbursed for any valid expenditure.
- **Provisional sums** are used for work that has not been finalized or for costs unknown at the time the bills are prepared. They may simply be contingency sums, and their presence does not necessarily imply any obligation on the part of the employer to spend them. In effect, they are simply a method for the employer to express part of the budget for the project. The contract administrator must issue instructions to spend these sums.

Certificates

There are several types of certificate in building contracts. They fall into two main categories: those which certify the quantity of work done to date, on which payment is calculated; and those which certify that an event has taken place. The former are called **interim certificates**, and their use is one of the most distinctive features of building contracts.

Interim certificates are used in the majority of building contracts because it is not intended that the contractor should have to finance the whole of the building work. The contractor will be paid for the amount of work done at regular periods, usually one month. When each payment is made, a small percentage of the money due to the contractor (called **retention**) is kept back, and this is released when the work is completed. It is also possible for other amounts of money to be withheld, under a right known as **set-off**, for delay or for defective work. The issues of retention and set-off, which are of great practical importance, are dealt with in Chapter 15.

The second category of certificates consists of those that record an event (or non-event). They vary from one form of contract to another, but basically they are:

- Certificate of practical completion;
- Certificate of final completion;
- Certificate of non-completion.

The full effect of these certificates is covered in Chapter 18.

Other documents

Additional contract information is contained in **instructions**, which contain information more detailed than that found in the tender documents. There are also other documents, such as descriptive schedules, further drawings and setting-out information. These may be necessary for the contractor to execute the work but such additional information must not impose extra work on the contractor. There will usually be an express term to this effect. If the contract administrator wishes to impose changes when issuing these instructions then a variation order must be issued to cover the changes.

There will typically be an obligation on the contractor to provide a **programme of work**. Although certain dates in the contract are legally binding, the programme is not. All the contracts make it clear that, even though the contractor must provide a programme soon after being awarded the contract, there can be no extra obligations imposed by it. JCT 98 made it clear in a footnote that the employer can strike out the requirement for a programme, but this practice is not recommended (JCT SBC 05 does not contain this suggestion). The programme is essential as a management control mechanism. It is also extremely useful to all participants in resolving claims for extra time. In order for it to be effective in these roles it must be revised as the project progresses and this is also usually a requirement.

Interaction between documents

The relationships between the various forms of documentation are worthy of comment. These documents may have to stand on their own for certain purposes but for the purpose of procuring the building they must interact and this interaction must be consistent and dependable. One result of the JCT approach is that the specification can no longer be a separate document in its own right; the bills are the only vehicle for defining quality or quantity of the contract work. Because the specification is embodied within the bills, any specification notes on the drawings will have no contractual significance, unless they are incorporated into the bills either directly or by reference.

10.1.3 Priority of documents

On a construction project of any substantial size, the sheer weight of detail found in the contract documentation gives ample scope for discrepancies and inconsistencies. These may arise within one particular document, as where the contract bills contain two inconsistent provisions. They may also arise between two documents, as where something in the bills conflicts with something in the conditions.

The general approach of most standard form contracts to this problem is that in the event of finding a discrepancy, the contractor should refer it to the contract administrator (JCT SBC 05: 2.15, JCT IC 05: 2.13, ICE 7: 5). The latter must then issue instructions so as to resolve the discrepancy and, if these cause delay or disruption, the contractor will be entitled to an extension of time and to compensation for loss and expense (JCT SBC 05: 2.29.2.1 and 4.24.2.3; ICE 7 clause 13(3)). Furthermore, if the instructions constitute a variation, the contractor will be paid at the appropriate rate for the work involved.

When called upon to resolve a discrepancy of this kind, the contract administrator should normally act in accordance with the general principles of law that govern the interpretation of contracts. However, this is subject to the terms of the contract itself, and it is noticeable that these general principles are expressly modified by a number of standard form building contracts. For example, one of the common law rules is that, where there is a conflict, written words prevail over typed words, and typed words in turn prevail over printed words. That rule is based on the sensible ideas that documents which are prepared for a specific job, rather than being taken 'off the shelf', are more likely to reflect the parties' true intentions. If it were to be applied to a building contract, it would mean for example that provisions in bills of quantities would override the printed conditions of contract.

This is emphatically *not* the case, at least where the main standard forms are concerned. While ICE 7 is fairly neutral, stating in clause 5 that all documents are to be taken as mutually explanatory, GC/Works/1 provides categorically in clause 2(1) that the conditions shall prevail over all other documents. Similarly, JCT SBC 05 clause 1.3 states that nothing in the contract bills shall override or modify what is contained in the agreement or conditions. This wording does not prevent the bills from imposing *extra* obligations on the contractor; it means that, where a particular matter is dealt with in the conditions, any special provisions on that subject in the bills are to be ignored.

In cases dealing with similar wording in earlier JCT forms of contract, the courts have ruled that the clause can operate to defeat what is the clear intention of the parties as expressed in the bills. The most striking example of this is the case of *MJ Gleeson (Contractors) Ltd v Hillingdon London Borough*,[2] where a contract for the provision of a large number of houses gave a single completion date of 24 months after the date for possession. In reality, as the preliminaries bill showed, the parties' intention was that blocks of houses were to be handed over at 3-month intervals from 12 months onwards. When the first blocks of houses were not completed after 12 months, the employer deducted liquidated damages at the

---

[2] (1970) 215 EG 165.

contract rate, but this was held to be invalid. The contract made no provision for sectional completion, and it could not be varied by the contract bills.

The *Gleeson* principle has in other cases forced the courts to block the clear intentions of the parties by ignoring provisions in the contract bills which purported to deal with the contractor's duty to insure,[3] the nomination of sub-contractors[4] and the early taking of possession by the employer.[5] On the credit side, it also enabled a court to hold that, under a contract in which a contractor undertook merely to construct in accordance with the employer's design, an obligation to carry out design work could not be imposed on the contractor by the 'back door' method of inserting a performance specification in the contract bills.[6]

10.1.4 Interpretation of contracts

Where a contract is made in writing, a court will seek to give effect to the intention of the parties *as expressed in the written documents*. This means that, in general, oral evidence is not admissible to contradict, vary, add to or subtract from the written terms. However, oral evidence may be brought to explain the customary or technical meaning of a particular word in the contract, or to establish the background circumstances in which the contract was made.[7]

A particular question of interpretation, to which the courts have given conflicting answers, concerns the common practice of making alterations to, or deleting clauses from, standard form contracts. The question is to what extent, in interpreting the words that are present, a court should be influenced by what has been deleted. In one case,[8] the Court of Appeal held that deletions should be completely ignored, and the remaining words interpreted as if the deleted clause had never existed. However, the opposite view (which seems rather more convincing) is that the parties must have had a reason for amending or deleting the standard form, and a court should seek to identify and give effect to that reason.[9] Indeed, in another Court of Appeal case[10] this was taken one stage further; a majority of the court interpreted what appeared to be a 'one-off' contract by comparing it with the JCT form on which it had clearly been based, and interpreting it so as to give full effect to the differences.

10.2   EXEMPTION CLAUSES

The law imposes certain controls on contractual terms that seek to exclude or restrict a party's liability for breach of contract. Those controls are in principle fully applicable to construction contracts, but one or two specific points are worth making.

---

[3] *Gold v Patman & Fotheringham Ltd* [1958] 2 All ER 497.
[4] *North West Metropolitan Regional Hospital Board v TA Bickerton & Son Ltd* [1970] 1 All ER 1039.
[5] *English Industrial Estates Corporation v George Wimpey & Co Ltd* (1972) 7 BLR 122.
[6] *John Mowlem & Co Ltd v British Insulated Callenders Pension Trust Ltd* (1977) 3 Con LR 64.
[7] *Pigott Foundations Ltd v Shepherd Construction Ltd* (1993) 67 BLR 48.
[8] *Wates Construction (London) Ltd v Franthom Property Ltd* (1991) 53 BLR 23.
[9] *Mottram Consultants Ltd v Bernard Sunley and Sons Ltd* (1974) 2 BLR 28.
[10] *Team Services plc v Kier Management & Design Ltd* (1993) 63 BLR 76.

First, the *contra proferentem* principle lays down that, where one party seeks to rely on a contract term that it has drafted, any ambiguity in the words used is interpreted in favour of the other party. This principle will clearly apply where, for example, a supplier of materials has its own standard terms of business. However, although the courts have never had to reach a definitive decision on the issue, it seems that the *contra proferentem* principle would not apply to a standard form contract produced by a negotiating body such as the JCT, whose membership reflects different interest groups within the construction industry.[11]

The second point concerns the application of the Unfair Contract Terms Act 1977. Where it applies, this Act requires a party who seeks to rely on an exemption or limitation clause to prove that the clause meets a standard of 'reasonableness', which the Act defines. If the reasonableness test is not satisfied, the clause in question has no effect on that party's liability.

The 1977 Act applies in two situations. The first is where the party that seeks to rely on the clause in question has entered into a contract on its own 'written standard terms of business'. Whatever is meant by this phrase, it would seem that, as with the *contra proferentem* rule, it will not include a contract made on an industry-negotiated standard form. However, the Act has been held to apply to a management contractor's own form of management contract,[12] a supplier's 'general conditions of sale' for the supply and installation of an overhead conveyor system in a factory[13] and the supply on standard terms of pipe work for the construction of a tunnel.[14]

The fact that an exemption clause is subject to the Unfair Contract Terms Act does not automatically mean that it will be struck down as being unreasonable. The courts start from the sensible position that, if a contract is entered into between two commercial organizations of roughly equal bargaining power, the terms of that contract should normally be regarded as 'reasonable'. This approach has led the courts to uphold a clause providing that a management contractor should not be held responsible for a trade contractor's default,[15] and another restricting a pipe work supplier's liability for defects to the cost of replacing the defective materials.[16] On the other hand, the Court of Appeal has struck down as unreasonable a supplier's clause which sought to exclude the client's right to set off a claim for defects against the price.[17]

The second situation in which the 1977 Act applies the requirement of 'reasonableness' is where one of the parties 'deals as consumer' (which means that it does not make the relevant contract in the course of business). As one might expect, the courts have appeared generally more protective towards private clients than towards commercial organizations, who can be expected to look after their own interests. However, even where consumers are concerned, the crucial test is that of reasonableness, and a clause restricting an architect's liability to £250,000

---

[11] See *Tersons Ltd v Stevenage Development Corporation* (1963) 5 BLR 54 and Section 8.1.
[12] *Chester Grosvenor Hotel Co Ltd v Alfred McAlpine Management Ltd* (1991) 56 BLR 115.
[13] *Stewart Gill Ltd v Horatio Myer & Co Ltd* [1992] 2 All ER 257.
[14] *Barnard Pipeline Technology Ltd v Marton Construction Co Ltd* [1992] CILL 743.
[15] *Chester Grosvenor Hotel Co Ltd v Alfred McAlpine Management Ltd* (1991) 56 BLR 115.
[16] *Barnard Pipeline Technology Ltd v Marston Construction Co Ltd* [1992] CILL 743.
[17] *Stewart Gill Ltd v Horatio Myer & Co Ltd* [1992] 2 All ER 257.

(roughly the expected cost of the works the architect was to supervise) was held to satisfy that test against a private client.[18]

The position of consumers has been further strengthened by the Unfair Terms in Consumer Contracts Regulations 1994 and 1999, which were enacted to give effect to EU Directives. The Regulations, which apply to all contracts for the supply or sale of goods and services by businesses to consumers, enable a court to strike out any contract term which has not been individually negotiated and which, contrary to the requirement of good faith, causes 'a significant imbalance in the parties' rights and obligations arising under the contract'.

10.3   INCORPORATION BY REFERENCE

Many construction contracts, and certainly most of those which concern sizeable projects, are drawn up and executed by the parties in a formal manner. The terms of those contracts are therefore easily identifiable and, if the parties intend to incorporate other documents by reference, this will be made clear in the main contract document. This is commonly done, for example, in relation to such documents as the contract drawings, bills of quantities or specification.

This principle of incorporation by reference is also of great importance where (perhaps through meanness, laziness or simply an over-casual approach to business procedures) no formal contract is executed at all, despite the fact that the parties intend their relationship to be governed by one of the standard forms with which they are familiar. The question that then arises is whether the terms found in the relevant standard form contract are applicable.

It may be stated at the outset that there is in general nothing to prevent the incorporation of an entire standard form contract by simply referring to it, in either a written or an oral agreement. Indeed, this is precisely what happened in *Killby & Gayford Ltd v Selincourt Ltd*.[19] In that case an architect wrote asking a contractor to price certain work and stated: 'assuming that we can agree a satisfactory contract price between us, the general conditions and terms will be subject to the normal standard-form RIBA contract.' When the contractors submitted an estimate that the architect accepted, and the work was done, it was held that the current RIBA form of contract was incorporated, although no formal contract was ever signed.

One danger in the practice of incorporation is that not all references to well-known forms of contract are as clear and unequivocal as in the *Killby* case and, if the parties fail to express properly what they mean, it cannot be assumed that a court will always rescue them. In particular, it has consistently been held that a sub-contractor's undertaking to carry out work 'in accordance with the main contract' does not necessarily incorporate *all* the terms of that main contract, so that in *Goodwins, Jardine & Co v Brand*[20] for example, an arbitration clause was not included.

---

[18] *Moores v Yakely Associates Ltd* (1998) 62 Con LR 76.
[19] (1973) 3 BLR 104.
[20] (1905) 7 F (Ct of Sess) 995.

An even more graphic illustration of the dangers is *Chandler Bros Ltd v Boswell*.[21] The main contract in that case provided that, if a sub-contractor was guilty of delay, the employer could instruct the main contractor to dismiss that sub-contractor. The sub-contract expressly covered many of the matters in the main contract, but did not specifically give the main contractor a power of dismissal for delay. When the main contractor (acting on instructions from the employer) purported to dismiss a defaulting sub-contractor, it was held by the Court of Appeal that the main contractor was guilty of a breach of the sub-contract. The sub-contractor's undertaking to carry out the work in accordance with the terms of the main contract was not enough to incorporate the power of dismissal for delay.

Notwithstanding these two decisions and others to similar effect, there have been many cases in which the courts tried very hard to make sense of what the parties said or wrote, and to give effect to what they really intended. For instance, where sub-contractors were appointed 'in full accordance with the appropriate form for nominated sub-contractors RIBA 1965 edition', it transpired that, not only was there no RIBA 1965 edition of any contract (the current form of main contract was the 1963 version), there were no RIBA forms of sub-contract at all! Despite this appalling drafting, the Court of Appeal was convinced on the evidence that what the parties had in mind was the FASS 'green form' of sub-contract; it was accordingly held that that form should be incorporated.[22]

The decision in *Brightside Kilpatrick Engineering Services v Mitchell Construction Ltd*[23] is perhaps even more generous in rescuing parties from the consequences of sloppy drafting. A sub-contract was there placed on an order form that identified the main contract and concluded: 'The conditions applicable to the sub-contract shall be those embodied in RIBA as above agreement'. Printed references on the order form to the 'green form' of sub-contract had been deleted, although the standard-form tender, to which reference *was* made, stipulated that a 'green form' sub-contract should be used. The main contractors argued that the *whole* of the RIBA 1963 form of contract was to be incorporated into the sub-contract but the Court of Appeal, after some hesitation, agreed with the sub-contractors that what had to be incorporated was merely those terms of the main contract which dealt with nominated sub-contractors. And, since these contemplated a 'green form' or equivalent sub-contract, this is what would be implied.

10.4   IMPLIED TERMS

Construction contracts are, like other contracts for work and materials, subject to the implied terms contained in the Supply of Goods and Services Act 1982.[24] In addition, it may be possible for terms to be implied *by the courts* into such contracts, although this is less likely to happen where the contract document itself is detailed and apparently exhaustive.

---

[21] [1936] 3 All ER 179.
[22] *Modern Building (Wales) Ltd v Limmer & Trinidad Co Ltd* [1975] 1 WLR 1281.
[23] (1973) 1 BLR 62.
[24] As amended by the Sale and Supply of Goods Act 1994. See Section 11.1.

In considering the question of implied terms, it is important to appreciate that two different kinds of implication may be involved. First, certain terms will be automatically implied *as a matter of law* into certain *categories* of contract, provided only that they are not inconsistent with any express terms. Secondly, and far less common, a term may be implied *as a matter of fact* into an *individual* contract where that contract would be commercially unworkable without it. Again, of course, there can be no implication of a term that would conflict with an express term. These two kinds of implication may be separately considered.

10.4.1 Implication in law

The law regards certain terms as 'usual' in building contracts, so much so that, if the contract is silent on these matters (as may well be the case with small informal agreements), these terms will be implied. However, it must be stressed once again that the court will not re-write a contract freely entered into, and so there will be no implication of terms that would be inconsistent with the express agreement. This will have a direct practical effect: Where, as under the main standard-form building contracts, all the 'usual' matters are covered in some detail, implied terms are rendered largely (though not entirely) irrelevant.

The terms which are regarded as 'usual' in this respect, and which will therefore be implied into construction contracts, will be dealt with in more detail in Chapters 11–15. Nonetheless, a brief summary of the most important terms may be useful at this stage.

Employer's obligations

The implied obligations of the employer, though capable of appearing as a list, can effectively be reduced to two: a general duty not to hinder the contractor's efforts to complete the work and, more positively, a duty actively to co-operate with the contractor. The duty of non-hindrance has been breached, for example, by an employer causing delay via servants or agents (who include, for this purpose, the contract administrator, but not independent third parties or even nominated sub-contractors); interfering with the supply of necessary materials to the contractor; and interfering with a contract administrator's function as independent certifier under the contract.

As to the duty of positive co-operation, this involves such matters as giving possession of the site; appointing an architect and nominating sub-contractors and suppliers; supplying the necessary instructions, information, plans and drawings; and, if the contract administrator persists in applying the contract in a wrongful manner, dismissing him or her and appointing a replacement. All these things must be done and, what is more, they must all be done without undue delay.

Contractor's obligations

Where a building contract does not specify a date for completion or, more commonly, where a contractual completion date has passed, the contractor's implied duty is to complete within such a time as is reasonable in all the circumstances. However, whether or not the contract specifies a completion date, it seems that there will not normally be an implied obligation on the contractor to carry out the work with due diligence so as to maintain any particular rate of progress.

As to the *standard* of the work, it is implied that the contractor will use proper workmanship and, in most cases, that the materials used will be of good quality and fit for their purpose. However, the 'fitness for the purpose' warranty will be excluded where it is clear that the employer has placed no reliance on the contractor's 'skill and judgment' in selecting the materials, such as where the employer specifies the material and nominates the supplier. Further, there may be circumstances in which even the warranty as to quality will not be implied, for example where the employer knows that the contractor will have no right of recovery against the actual supplier of the materials.

Apart from the fitness for the purpose of the *materials* used, there may in certain cases be an implied warranty that the completed *works* will be fit for their purpose. Such a term depends upon the employer's *reliance* on the expertise of the contractor. Accordingly it will not be implied where the contractor is merely to build in accordance with detailed plans and specifications, nor where the contractor is under the supervision of an architect. A warranty of fitness will be fairly readily implied, for example, into a contract to buy a house that is in the course of being built. However, the sale of an already completed building is subject to an ancient legal principle known as 'caveat emptor' ('let the buyer beware') and consequently attracts no such implication.

10.4.2 Implication in fact

We saw earlier that a court will strain to make sense of the parties' agreement, for example in identifying the form of contract to which they have inaccurately referred. However, the courts have repeatedly stated that they will not make or improve contracts for the parties. The principle of **freedom of contract** means that, if parties have entered into a contract that is unreasonable, inconvenient or commercially unwise, it is not for the courts to change their arrangement. They must simply be left to bear the consequences. A term will not be implied into a particular contract just to make it more convenient, reasonable or sensible; it will only be implied if its absence is so glaringly obvious that both parties *must* have intended to include it.

A good example of the courts' extreme reluctance to imply a term just to 'improve' a contract is the decision of the House of Lords in *Trollope & Colls Ltd v North West Metropolitan Regional Hospital Board.*[25] A contract for the construction of a hospital extension in phases provided that Phase III should

---

[25] [1973] 2 All ER 260.

commence six months after the issue of the certificate of practical completion of Phase I, but that Phase III should itself be completed by a fixed date. There were express provisions for extensions of time to be granted in respect of Phase I, but no express provision for this to have a 'knock on' effect on Phase III. Delays in completing Phase I (for virtually all of which extensions of time were granted) effectively reduced the period for Phase III from 30 months to 16 months, whereupon the contractors argued that a term should be implied permitting the Phase III time to be extended in accordance with any extensions to Phase I. However, the House of Lords refused to make any such implication, ruling that, since the contract was clear and unambiguous in fixing a time for completion for Phase III, the parties must live with the agreement that they had made.

Another example of this judicial reluctance to intervene is *Bruno Zornow (Builders) Ltd v Beechcroft Developments Ltd*,[26] a case in which sloppy drafting made it very difficult even to identify the *express* terms of the contract. The judge was ultimately persuaded to imply a completion date (on the basis that no commercial party would contemplate a contract without one); however, he refused to imply a term for sectional completion, even though it seemed fairly clear that this was what the parties had intended.

10.5   LIABILITY IN TORT FOR NEGLIGENCE

The possibility of bringing an action in the tort of negligence is important in the construction context in two distinct situations. The first is where there is no contract between the party who has suffered loss and the party who caused that loss. We consider this problem in relation to claims by a contractor against the contract administrator (Section 18.2.3), claims by an employer against a sub-contractor (Section 19.5.1) and claims by a person who acquires a defective building against the designer or constructor responsible for the defects (Section 22.1).

The second situation in which negligence becomes relevant is where the parties *are* linked by a contract, but the claimant sees an advantage in framing the claim in tort. In practical terms, the most important difference between contract and tort that might prompt such a choice is undoubtedly the different time periods within which each type of claim must be brought if it is not to be statute-barred.

Under the Limitation Act 1980, an action for breach of contract must normally be commenced within six years of the relevant breach (or twelve years if the contract is made by deed). An action in tort, on the other hand, must be commenced either within six years from the date on which the claimant suffers damage or within three years from the date on which any latent damage is or ought to be discovered. Tort claims within these rules are also subject to a final 'longstop' time bar 15 years after the defendant's breach of duty. It is clear from these different rules that there will be many situations in which an action in contract would be barred by the Limitation Act, but an action in tort would not.

---

[26] (1989) 51 BLR 16.

After many years of uncertainty, it was settled by the House of Lords in *Henderson v Merrett Syndicates Ltd*[27] that, where the facts of a case satisfy the requirements of claims in both contract and tort, the claimant is in principle free to choose which cause of action to pursue.

In seeking to establish concurrent liability, it is obviously important to know whether the relationship between a claimant and a defendant is one of sufficient 'proximity' to create a duty of care in the tort of negligence. In this connection, the law currently appears to require that the claimant has *relied* on the defendant for *advice*, in the sense in which those concepts were used in the case of *Hedley Byrne & Co Ltd v Heller & Partners Ltd*.[28] And here the courts appear to be drawing a distinction (on dubious grounds) between consultants and contractors. Hence, while it is clear that the relationship between an employer and an architect or other design consultant will found a duty of care in tort, there is considerable doubt as to whether the relationship between employer and contractor will do so. In *Barclays Bank plc v Fairclough Building Ltd*,[29] the Court of Appeal held that a sub-contractor owed a duty of care in tort to the main contractor to avoid causing financial loss to the latter (in the form of damages which the main contractor had to pay to the employer under the main contract). However, in *Nitrigin Eireann Teoranta v Inco Alloys Ltd*,[30] it was held that a specialist manufacturer of pipe work could be liable only in contract to the employer who purchased pipes; there was no parallel duty of care in tort.

One final point that is worth making about concurrent liability concerns the defence of contributory negligence. This defence, which operates in cases where the claimant's loss is attributable partly to the claimant's own fault, results in a proportionate reduction of the damages payable by the defendant.[31] The defence is certainly available where an action is brought in the tort of negligence. It is also available where an action for breach of contract is based on facts that *would* have supported a tort claim (i.e. where there is concurrent liability).[32] However, it has been confirmed by the Court of Appeal that it does *not* apply to a claim for breach of a *strict* contractual obligation.[33] In such a case, the court is forced (somewhat unrealistically) to rule that the employer's loss has been caused *entirely* by either the contractor's breach of contract or the employer's fault.

---

[27] [1994] 3 All ER 506.
[28] [1964] AC 465.
[29] (1995) 76 BLR 1.
[30] [1992] 1 All ER 854.
[31] Law Reform (Contributory Negligence) Act 1945, Section 1.
[32] *Forsikringsaktieselskapet Vesta v Butcher* [1988] 2 All ER 43.
[33] *Barclays Bank plc v Fairclough Building Ltd* (1994) 68 BLR 1.

11   Contractor's obligations

The contractor's obligations begin with the obligation to construct the works in accordance with the documents within the required time. This Chapter deals with the fundamental obligations and associated obligations as to workmanship, quality of materials, co-ordination and management of the site. Obligations as to time are dealt with in Chapter 14 and obligations as to insurance and bonds are dealt with in Chapter 17.

The exact nature of what the contractor undertakes depends upon the way the contract is written. If the contract is for a complete building, such as a dwelling, then it is implied that the contractor will provide everything that is indispensably necessary to achieve the result, regardless of what is contained in the specifications.[1] Where the contract makes clear that the bills of quantities contain an exhaustive definition of the works (as under JCT SBC 05, clause 4.1), the contractor's obligation is limited to providing what is described in the bills. If this is not sufficient for completion of the building, the employer must pay for extra work.[2] ICE 7 also limits the contractor's obligation to what can be inferred from the contract documents (clause 8).

11.1   STANDARD OF WORK

The contractor's basic obligation, so far as the standard of work is concerned, is to comply with the terms of the contract. This includes both express terms (such as the requirement of JCT SBC 05, clause 2.3 that work shall be of the standards described in the bills) and implied terms (such as the principle that all materials shall be of 'satisfactory quality').

In addition, there are often provisions in the bills for work to be to the satisfaction of the contract administrator. This might seem to suggest that the contract administrator has an absolute right of rejection of such work, but JCT SBC 05 places a limit upon this power. Clause 2.3.3 makes clear that, where approval of workmanship or materials is a matter for the opinion of the contract administrator, 'such quality and standards shall be to his reasonable satisfaction'. So, where work is subject to the subjective test of the contract administrator's satisfaction, the wording of the JCT contract applies *reasonableness* to the test of satisfaction. This can be contrasted sharply with ICE 7, clause 13 in which the engineer has complete authority to approve or reject work, regardless of what the bills might say about quality or workmanship, with no apparent test of reasonableness.

---

[1] *Williams v Fitzmaurice* (1858) 3 H & N 844.
[2] *Patman & Fotheringham Ltd v Pilditch* (1904) HBC 4th ed, ii, 368.

Where a contract requires work to be to the satisfaction of the contract administrator, an important issue is whether this becomes the *only* contractual standard, or whether the contractor must also satisfy whatever *objective* standards are contained, expressly or impliedly, in the contract documents. In other words, is it open to an employer to argue that, notwithstanding its approval by the contract administrator, certain work does not comply with a standard set out in the contract bills, or falls below the standard of 'satisfactory quality' laid down in the Supply of Goods and Services Act 1982? In short, is the contractor under one obligation or two?

The answer to this question, as is so often the case, is that everything depends upon the wording of the particular contract. The weight of authority has traditionally been in favour of making the contractor satisfy *both* standards although, under an early version of JCT 98, the position appeared to be different. In *Crown Estate Commissioners v John Mowlem & Co Ltd*,[3] it was held by the Court of Appeal that the clause making the *final certificate* conclusive as to the contract administrator's 'satisfaction' applied to *all* items of work under the contract, not just to those for which that satisfaction was *expressly* required. This ruling is discussed in Section 18.2.1, under the heading 'Conclusiveness of certificates'.

The phrase 'standard of work' in fact contains three separate, though linked, ideas, which are discussed below.

11.1.1 Workmanship

The standard of workmanship may be defined in considerable detail by the contract, for example by requiring it to comply with an appropriate code of practice. Under JCT SBC 05, such a requirement would appear in the bills, and would have contractual force by virtue of clause 2.3.2. This provides that 'workmanship for the works ... shall be of the standards described in the contract bills'. Clause 2.3.3 adds that, where the bills do not set a standard, the workmanship 'shall ... be of a standard appropriate to the works'. A similar, if not identical, obligation would in any case be implied by law since, in the absence of any express terms covering this issue, the courts will imply a term in the contract that the work will be carried out with proper skill and care, i.e. in a 'workmanlike' manner. If part of the work is to be designed by the contractor, then the standard of workmanship will be set out in the Employer's Requirements for the Contractor's Design Portion (CDP) or otherwise in the Contractor's Proposals.

It is at least arguable that 'workmanship' refers only to the standard of the finished item, and not to the method used to achieve that standard, certainly in building contracts. If this is so, then the contract administrator has no power to control the manner of working at the time it is being carried out. However, JCT SBC 05 clause 2.1 gives the contract administrator such power by providing that 'the contractor shall carry out ... the works in a proper and workmanlike manner'. If the contractor breaches this, then clause 3.19 empowers the contract administrator to issue whatever instructions are necessary. Even if these constitute

---

[3] (1994) 70 BLR 1.

a variation, the contractor will not be entitled to any payment nor to any extension of time. By contrast, the custom in civil engineering is for a more open and involved relationship between engineer and contractor as regards methods of work and temporary works. For example, clause 14 in ICE 7 make specific reference to the flow of information between contractor and engineer, particularly with regard to programme, methods of construction and design criteria for temporary or permanent works that are designed by the contractor.

It is important to note that, in any contract, the contractor is responsible, not only for personally performing unsatisfactory workmanship, but also for that of any sub-contractor, either domestic or nominated.

11.1.2 Standard of materials

Clause 2.3.1 of JCT SBC 05 states that *all materials and goods shall ..., so far as procurable, be of the respective kinds and standards described in the contract bills*. This wording allows for the eventuality that some things available at the time of tender may become unavailable. Again, ICE 7, JCT IC 05 and GC/Works/1 all place a similar obligation upon the contractor, but none of them makes allowance for materials that may have become unprocurable. The GC/Works/1 form covers this point at clause 7(1)(h), where the contractor is assumed to have assessed everything that might have an influence on the execution or pricing of the work. This would include the relevant market for labour and materials.

It sometimes happens that the contract bills (inadvertently or deliberately) fail to specify the quality of the materials and goods. JCT 98 made no express provision for this situation, leaving the issue to be decided on the basis of a term, implied by statute, which requires all goods and materials supplied to be of 'satisfactory quality'. This phrase, which replaced the time-honoured expression 'merchantable quality',[4] means that they are to be as free from any defects as it is reasonable to expect, given such factors as their price and the way they are described.

The situation has changed, at least in theory, under JCT SBC 05. Clause 2.3.3 provides that where the contract bills are silent as to standards, materials and goods supplied must be 'of a standard appropriate to the works.' However, it is not easy to see what, if anything, this express term adds to the normal implied term.

It has been held by the House of Lords that a contractor will be liable if materials are unsatisfactory, even where it is the employer who has selected those materials (for example by nominating a particular supplier).[5] However, the contractor will *not* be liable for defective materials where forced by the employer to obtain those materials from a supplier who, to the employer's knowledge, excludes or limits liability for defects.[6]

---

[4] Sale and Supply of Goods Act 1994.
[5] *Young & Marten Ltd v McManus Childs Ltd* [1969] 1 AC 454.
[6] *Gloucestershire CC v Richardson* [1969] 1 AC 480.

11.1.3 Suitability of materials

Even where materials are of satisfactory quality, in the sense of being free from defects, they may still not be fit for the purpose for which they are used. Where this is the case, it is again possible for the contractor to be liable, but the term to be implied here is of a more limited kind. Section 4 of the Supply of Goods and Services Act 1982 implies a term that goods shall be reasonably fit for the purpose for which they are supplied, but *only* where it is clear that the recipient is relying on the skill and judgement of the supplier. In practical terms, such reliance on the contractor (and thus liability for unfit materials) can only be found where the contractor has the choice of materials. There will be no such implied term in respect of materials specified by the employer or the architect, for here there is no reliance on the contractor. However, the mere fact that the contractor is given a limited choice does not necessarily mean that the employer is relying on the contractor's skill and judgment; all the circumstances of the case must be taken into account. Thus, where bills of quantities specified the type of material to be used as hardcore, but left the contractor some discretion as to what was supplied, it was held that the employer had not intended to rely on the contractor's skill and judgment, but had believed that it was unnecessary to lay down a more detailed definition. There was accordingly no implied guarantee by the contractor as to the suitability of the hardcore.[7] The implied term as to the fitness of the materials used has also been made express by clause 2.3.3 of JCT SBC 05 through the wording 'a standard appropriate to the works'.

11.1.4 Suitability of the building

Quite apart from the obligation to provide suitable materials, a contractor may in some circumstances be subject to an implied term that the *building* itself, when completed, will be fit for its intended purpose. Such a term, which will normally be implied only into 'design and build' contracts, is dealt with in Chapter 13.

11.2   STATUTORY OBLIGATIONS

Statutory obligations are the obligations imposed by acts of parliament and other legislation such as government regulations. All parties must of course obey the law and are thus bound by these obligations. However, failure by the contractor to comply with a statutory obligation would not in itself amount to a breach of contract for which the employer could claim damages. That will only be the case where the contract specifically requires the contractor to comply with the relevant statutory rules.

The reason for making compliance a contractual requirement is that, where the contract works are carried on in such a way as to breach a statute, the effect may well be that the employer becomes liable to criminal prosecution or to civil actions brought by third parties who are adversely affected. By placing a contractual

---

[7] *Rotherham Metropolitan Borough Council v Frank Haslam Milan & Co Ltd* (1996) 78 BLR 1.

obligation of compliance on the contractor, the contract seeks to ensure that, in such circumstances, the employer will be entitled to claim an indemnity from the contractor.

11.2.1 Contractor's duties

The most important provisions controlling building work are likely to be the Building Regulations 1985, the Health and Safety at Work Act 1974 and the Construction (Design and Management) Regulations 2007, but the contractor's obligation extends much wider than these. JCT SBC 05 clauses 1.1 and 2.1 make it clear that there must be compliance with local authority byelaws, and also with regulations made by statutory undertakers such as electricity, gas and water boards, to whose systems the works are to be connected.

In relation to any fees, charges, rates or taxes payable in respect of the works, JCT SBC 05 clause 2.21 again places the primary obligation of paying these upon the contractor. However, the ultimate cost will be borne by the employer, for the amounts are to be added to the contract sum.

The position outlined above is in general terms followed by ICE 7 in clauses 26 and 27. However, no doubt because the nature of civil engineering projects makes it more likely that work will be done in streets or other highways, the clauses cover these eventualities in considerable detail.

11.2.2 Divergence between statutory requirements and contract

The obligations of the contractor described above are subject to an important qualification under JCT SBC 05 clause 2.17.3 (JCT IC 05 clause 2.15.3). If the reason why there is a breach of statute is that the work, as designed, does not conform to statutory requirements, then responsibility for this falls upon the employer and not upon the contractor. The contractor's only obligation is to notify the contract administrator of any apparent divergence between the contract documents and the statutory requirements. Provided the contractor does this (and it is to be noted that, somewhat strangely, the contract does *not* require a positive search for such divergences), the contract administrator must then, within seven days, issue instructions so as to bring the works into line with the statutory requirements. Such instructions will be treated as a variation, which means that the contractor will be paid for extra work. Further, where compliance with the instructions causes delay or disruption, the contractor will be entitled to claim an extension of time and for loss and/or expense.

The treatment of divergences described here applies to most JCT contracts and also to ICE 7. However, the JCT form DB 05 is very different. Except where the Employer's Requirements are specifically stated to be in accordance with statutory requirements, it is the contractor's responsibility to ensure that the work as designed and built does not contravene the law. As a result, any divergence must be put right at the contractor's own expense.

11.2.3 Emergency work

JCT SBC 05 clause 2.18 makes provision for the circumstance where emergency work is necessary in order to comply with statutory requirements. This is likely to occur where work is necessary to ensure the continuing safety of people and property; for example, where a neighbouring property has become unsafe and may collapse on to the site. The contractor must immediately (i.e. without waiting for instructions from the contract administrator) undertake such limited work and supply such limited materials as are necessary to ensure compliance with statutory obligations. Provided that the emergency is then shown to have arisen because of a divergence between the contract documents and the statutory requirements, the contractor's work will be valued as a variation.

The ICE 7 form does not mention emergency work at all, which suggests that the cost of such work is to be borne entirely by the contractor. Where the emergency arises because of a divergence between the contract documents and statutory requirements, the engineer must issue a variation instruction and the contractor will be paid for this work. However, even here it is arguable that the contractor will remain uncompensated both for the earlier abortive work and for the cost of removing it.

11.2.4 Health and safety

The contractor's obligations for health and safety will depend upon whether the contractor under the construction contract is the *principal contractor* for the purposes of the Construction (Design and Management) Regulations 2007. These obligations are dealt with in Sections 19.3.4 and 19.3.5.

11.3   CO-ORDINATION AND MANAGEMENT

Apart from the main contractor's role as builder (which may be very limited where extensive use is made of sub-contractors, or even non-existent under a management contract), the main contractor has an important part to play in managing the site and the persons who work on it. This covers several issues, dealt with below.

11.3.1 Control of persons on the site

The main contractor is responsible for programming the overall project and co-ordinating the contributions made by various other persons and organizations, notably sub-contractors. As to the extent to which the contractor bears responsibility for the defaults of such sub-contractors, even when they are nominated, this is dealt with in Chapters 19 and 20.

Apart from sub-contractors, an important group of participants in construction projects consists of statutory undertakers. There are two types of these: those who exercise some sort of power over the works, and those who have to undertake work on the site. Examples of the first type are the public bodies exercising supervision over the design, quality or safety of the works by the enforcement of planning legislation or building regulations. Examples of the second type are those public bodies or private firms providing utility services such as gas, water, electricity and drainage. With the latter, the contractor must co-ordinate and connect with their work, but will not have the same powers of control as over sub-contractors.

In the UK it is often the case that statutory undertakings will undertake work that is not part of their statutory powers. Similarly, private firms may undertake work subject to statutory regulation and also work that is not. For example, an electricity board may be employed to carry out sub-contract works. In this case, they are not treated as a statutory undertaking, but are in the same position as any other sub-contractor (whether nominated or domestic). It is obviously of great importance to know in what capacity a statutory undertaking is operating in any particular case, but unfortunately this is not always clear, especially where they exceed their powers.[8]

Although overall responsibility for the way that work is carried out lies with the contractor, the contract administrator has a supervisory role to play. To enable this to take place, JCT SBC 05 clause 3.1 provides that the contractor must allow the contract administrator to have access to the works and to the contractor's workshops at all reasonable times. The contractor is also required to ensure that such access is also possible to the workshops of sub-contractors. Furthermore, if the employer exercises the right to appoint a clerk of works (see Section 18.1.1) to act as inspector under the directions of the contract administrator, JCT SBC clause 3.4 requires the contractor to afford every reasonable facility for the performance of that duty.

It may finally be noted, as a practical feature of the contractor's responsibility for management and co-ordination, that JCT SBC clause 3.2 specifically requires the contractor to keep constantly upon the works 'a competent person-in-charge'.

11.3.2 Exclusion of persons from the works

All the main standard-form contracts give power to the contract administrator to order the exclusion from the works of specified persons. Under clause 3.21 of JCT SBC 05, such an instruction, which must not be issued 'unreasonably or vexatiously', extends beyond the contractor's own employees to include the employees of sub-contractors. In earlier editions of the form, the clause appeared immediately after the contract provisions in respect of bad workmanship which suggests that its purpose is to exclude persistent bad performers, although this was not made explicit. The strict wording would appear to entitle the contract administrator to exclude militant shop stewards and the like, but this would probably be a misuse of the provision. However, the question of unreasonableness, or vexation, is one which can only be decided by an arbitrator/judge.

Under ICE 7, clause 16 requires the contractor to employ only those people who are properly skilled and undertake their duties with an appropriate level of care. The engineer has the power to order the removal from site of any of the contractor's employees, but only for lack of diligence or workmanship.

---

[8] *Henry Boot Construction Ltd v Central Lancashire New Town Development Corporation* (1980) 15 BLR 1.

The equivalent clause in GC/Works/1, clause 26, gives the project manager extremely far-reaching powers to exclude any individual from the site. No reason need be given, and the contractor cannot question the exclusion. This is clearly and unambiguously disconnected from workmanship and diligence, and appears under the general heading of security.

Under the Construction (Design and Management) Regulations 2007, the contractor must exclude all unauthorized persons from the site. This additional requirement applies to all construction work as defined by the regulations and it places a very strict obligation on a contractor to maintain proper security and careful governance of admissions to site. This particular regulation (regulation 10) carries civil as well as criminal liability.

11.3.3 Antiquities

Contractual provisions dealing with the discovery of interesting objects, new or old, during the course of the work are necessary for various reasons. One important reason is that the removal or disturbance of such objects may actually destroy their worth. One example of this was the discovery of Shakespeare's Globe Theatre in London, where an archaeological find of great importance delayed a very large and expensive development project for a considerable period of time. More generally, fossils may be valuable, and some types may completely degrade on exposure to atmosphere for even a short length of time. Further, one of the most significant features of a fossil is the exact location in which it is found. It must be left in place, for a random fossil is of little use to a palaeontologist without data about its location. A third problem relates to the ownership of treasure trove.

JCT clauses 3.22 to 3.24 contain extensive provisions to cover these issues, beginning with a general statement that *all fossils, antiquities and other objects of interest or value which may be found on the site or in excavating it during the progress of the Works shall become the property of the employer*. This provokes three comments:

- The phrase *all fossils* is a rather unfortunate one because, if the site is on chalk or limestone, it will include every single shovelful of material!
- The phrase *other objects of interest or value* means that this clause does not only apply to old objects; it can equally apply to new things, provided that they are of interest or value.
- It is important to appreciate that this clause only settles the issue of ownership as between the employer and the contractor. It can have no bearing on the relative rights of, say, landlord and tenant, where the site is held under a lease. In such circumstances it is necessary to refer to the general law governing the right to any objects found.

The remainder of clauses 3.22 to 3.23 (also followed in general terms by ICE 7 clause 32) lays down detailed provisions as to what shall happen where such an object is discovered. The contractor is not to disturb the object, but must take all necessary steps to preserve it in the exact position and condition in which it was found. The contract administrator is to be informed forthwith of the discovery, so that instructions can be given as to what to do. If this delays progress of the works then the contractor is entitled to an extension of time under clauses 2.28, 2.29, 3.23. Any loss/expense incurred by the contractor is also reimbursable.

Apart from the contractual provisions governing antiquities, attention must be paid to the Ancient Monuments and Archaeological Areas Act 1979. This empowers the Secretary of State and local authorities to designate areas of archaeological importance, which will be registered under the Local Land Charges Act 1979. Under the Act, if any operations are to be undertaken on a designated area, which would disturb the ground, an 'Operations Notice' must first be served on the local authority. The Secretary of State then has a power to delay the building work, and no compensation is payable unless the archaeologists damage the site. If the provisions of this Act are contravened, criminal penalties may be imposed, and there is also power to issue injunctions preventing any further contravention.

These statutory provisions are not covered directly by clauses 3.22 to 3.24, which only apply to objects found during the progress of the works. But there are provisions in JCT SBC 05 that seem to address this situation. When the UK government, by exercising its statutory powers, directly affects the execution of the works, the contractor is entitled to an extension of time (clause 2.29.12) and, if the operations are suspended for the period of suspension stated in the contract particulars, both parties are entitled to terminate the contract (clause 8.11.1.5).

11.3.4 Testing and approvals

The question of whether or not a contractor's work is up to the contractual standard may arise in arbitration or litigation, in the course of a claim for damages for breach of contract. More frequently, however, it will be something for the contract administrator to decide through the certification procedure, for which see Chapter 18. Whatever the precise nature of the contract administrator's role in approving work, many contracts oblige the contractor to open up or uncover work for inspection or testing (JCT SBC 05 clause 3.17, ICE 7 clause 36, GC/Works/1 clause 31). Such clauses usually make provision for the contractor to be paid compensation in the event that no defects are found.

11.4   TRANSFER OF MATERIALS

The common law rules that govern the transfer of ownership in materials from contractor to employer are fairly straightforward. They will operate as implied terms in any contract where they are not overridden by express provisions. Such express provisions, known as **retention of title clauses**, are sometimes found in sub-contracts or suppliers' standard terms.

11.4.1 General position

At common law, as soon as any materials or goods are incorporated into a building, they cease to belong to the contractor and become the property of the employer.[9] However, until the materials are built into the works, even though they have been delivered to site, they remain the property of the contractor. Even the fact that the employer has paid the contractor for the materials will not make a difference, unless (as is usually the case) the contract makes express provision for this.

The question of whether ownership lies with the employer or the contractor becomes most important where one of the parties becomes insolvent. As a general principle, all the creditors of an insolvent person or company are entitled to share equally in the remaining assets, and these will include all property owned by the insolvent person at the time. Thus, if the contractor becomes insolvent, materials on site that have not yet passed to the employer may be seized by the contractor's receivers. What is more, unless the contract provides otherwise, this will hold good even where the value of those materials has been included in interim certificates.[10]

Somewhat strangely, the courts have not always applied the same principle to cases where it is the employer who becomes insolvent. It has for instance been held that, although the ownership of building materials may have passed to the employer under a 'vesting clause', the contractor's continuing rights to use those materials in constructing the building are strong enough to override the claims of the employer's creditors.[11]

11.4.2 Contract provisions

JCT SBC 05 does not alter the principle that the ownership of materials is transferred when they are incorporated into the building, but clauses 2.24 and 2.25 provide a method by which the ownership may pass to the employer at an earlier stage. The value of unfixed materials and goods intended for the works, whether they are on- or off-site, may be included in an interim certificate. If this is done, then the ownership of those materials and goods will pass to the employer as soon as the amount is duly paid.

Some contracts, notably those used on civil engineering projects, depart further from the common law position. ICE 5 and 6, for example, contained a clause stating that, not only goods and materials, but also the contractor's equipment and temporary works, should be 'deemed to be the property of the employer'. However, despite the sweeping words used, it has been held that this does not literally transfer legal ownership of the contractor's plant to the employer, but merely creates a charge over the plant (so that, if the contractor becomes insolvent, the employer can use the plant in order to complete the works).[12] In any event, clause 54 of ICE 7 has dropped the 'deemed' transfer of ownership, providing instead that such materials and equipment may not be removed from the site

---

[9] *Sims v London Necropolis Co* [1885] 1 TLR 584.
[10] *Hanson (W) (Harrow) Ltd v Rapid Civil Engineering Ltd and Usborne Developments Ltd* (1987) 38 BLR 106.
[11] *Beeston v Marriott* (1864) 8 LT 690.
[12] *Cosslett (Contractors) Ltd v Mid-Glamorgan CC* (1997) 85 BLR 1.

without the written consent of the engineer (which is not to be unreasonably withheld).

11.4.3 Retention of title

The contract may deal satisfactorily with questions of ownership as between the employer and the contractor; however, problems may arise in cases where the contractor brings on to the site materials that are still in the ownership of a supplier. In particular, many suppliers operate under standard conditions of sale, which provide that they shall retain the ownership of goods until full payment is made. There is an obvious possibility of conflict between the terms of the main contract and those of the contract of supply.[13] The problems that can then arise are dealt with in Chapter 19.

---

[13] See *Dawber Williamson Roofing Ltd v Humberside CC* (1979) 14 BLR 70; *Archivent Sales and Developments Ltd v Strathclyde Regional Council* (1984) 27 BLR 98.

12 Employer's obligations

The most important of the employer's obligations under a construction contract are monetary: to pay the contractor what is due for work done, and in certain circumstances to compensate the contractor for loss and expense. These obligations will be discussed in detail in Chapters 15 and 16.

This Chapter considers a number of other obligations imposed on the employer. We look first at those obligations implied by law wherever they are not overridden by some express term of the contract. Second, a brief account is given of the obligations expressly imposed upon the employer under JCT SBC 05. These obligations are few in number, but very important to the effective discharge of the contract.

The reason why the employer appears to have relatively few express contractual obligations, and to play a merely passive role, is that the contract allocates numerous duties to the contract administrator. In truth, many of these duties are the employer's responsibility in the sense that, if the contract administrator fails to perform, the contractor may claim against the employer for breach of contract. The extent of the employer's responsibility is also considered in this chapter, although the actual content of the contract administrator's duties is dealt with in more detail in Chapter 18. The fact that so many of the employer's obligations are carried out by the contract administrator leads to an important practical point; an employer who tries to circumvent this relationship, and to communicate directly with the contractor, runs a serious risk of causing confusion and resulting problems. A wise contractor will decline to accept direct instructions from the employer, requesting instead that they be channelled through the contract administrator. In any event, the contractor should always ask the contract administrator for written confirmation of the employer's direct instructions.

12.1 IMPLIED OBLIGATIONS

It is obvious to everyone that the contractor has a *duty* to carry out and complete the contract works. What is sometimes overlooked is that the contractor also has a *right* to do this. Unless the contract provides otherwise, the contractor is entitled to carry out the whole of the contract works within the contract period, and the employer must co-operate to enable this to be achieved.

12.1.1 Non-hindrance and co-operation

It was acknowledged in *Merton LBC v Stanley Hugh Leach Ltd*[1] that two general obligations on the part of the employer are to be implied into all building contracts. They are expressed as follows:

1. The employer will not hinder or prevent the contractor from carrying out all its obligations in accordance with the terms of the contract, and from executing the works in a regular and orderly manner.
2. The employer will take all steps reasonably necessary to enable the contractor to discharge all its obligations and to execute the works in a regular and orderly manner.

Although they are expressed as two separate obligations, these are in reality the positive and negative aspects of the same thing. Together they make it the employer's duty to *co-operate* with the contractor in all aspects of the contract work. In considering what this means in practice, we shall accordingly not attempt to divide our examples into those involving 'non-hindrance' and those requiring 'co-operation'.

12.1.2 Specific examples of non-hindrance and co-operation

One aspect of the contractor's right to do the work tendered for is that the employer cannot order the omission of work from the contract, with the intention of giving this work to another contractor.[2] This principle is so fundamental that such conduct could well constitute a repudiatory breach by the employer (see Chapter 23), which would entitle the contractor to terminate the contract.[3] It is thought that the employer is likewise unable to take away work for which the main contractor has priced and instead nominate a sub-contractor to carry it out.

The contractor's right to carry out the work means that, as a general rule, the employer cannot unilaterally decide to abandon the project altogether. This is certainly the position under JCT SBC 05, where the employer's right to terminate the contractor's employment (considered in Chapter 23) arises only in certain very limited circumstances. However, some other standard form contracts (such as JCT MC 98 and GC/Works/1) contain an express clause allowing the employer to terminate the project at any time without reason. Naturally, under such a contract, an employer who abandons the project will not be liable for damages to the contractor. It is important for a contractor to ensure that, where the main contract contains a unilateral termination clause, every sub-contract contains a similar provision. If this is not done, the contractor will be liable to sub-contractors when the project is abandoned and will not be able to pass responsibility for this on to the employer.[4]

The principles which govern abandonment also apply to the temporary suspension of the work, although it is far more common to find an express term of

---

[1] (1985) 32 BLR 51.
[2] *Commissioner for Main Roads v Reed & Stuart Pty Ltd* (1974) 12 BLR 55.
[3] *Carr v JA Berriman Pty Ltd* (1953) 89 CLR 327.
[4] *Smith and Montgomery v Johnson Brothers Co Ltd* [1954] 1 DLR 392.

the contract giving the employer powers in this respect. The question of suspension is considered in more detail in Chapter 18.

As will be seen in Chapter 14, the employer's duty of co-operation includes giving the contractor possession of the site at whatever time the contract states. If the contract is silent on this matter, then the employer's implied obligation requires the contractor to be given possession at such a time as will enable the work to be finished by the specified completion date.[5]

Whether or not the employer's obligation extends to the obtaining of any official permits for the proposed work, such as planning permission, depends essentially on the terms of the contract. Under JCT SBC 05, for example, the overall effect of clause 2.17.3 is that the employer is responsible for ensuring that the works as designed comply with all statutory requirements; it is thereafter the contractor's responsibility to serve whatever notices and obtain whatever permissions are necessary (clause 2.21). By contrast, many design and build contracts, such as JCT DB 05, place the entire responsibility on the contractor, except in so far as the Employer's Requirements are specifically guaranteed to comply with the statutory requirements.

An important part of the employer's duty of co-operation concerns the appointment of a contract administrator and, where appropriate, the nomination of sub-contractors. These matters are usually covered by express terms of the contract but, if not, the implied obligation certainly extends to them. Indeed, where the contract envisages the use of an architect or an engineer to supervise the works and act as the employer's agent, it is a condition precedent to the performance of the contractor's obligations that the appointment is made.[6] This means that a contractor can refuse to carry out any work until the employer makes such an appointment. If, however, the contractor proceeds without the appointment of a contract administrator, this would constitute a waiver of the right to insist on an appointment.[7]

Similarly, where the contract makes provision for sub-contractors or suppliers to be nominated, the employer must make these nominations within a reasonable time. Should it become necessary to renominate (or to reappoint a contract administrator), this too must be done within a reasonable time.

As far as the general running of the project is concerned, the employer may be put in breach of the duty of co-operation by what is done by a number of other parties. For example, the employer will be liable to the contractor for breach of contract if the contract administrator delays unreasonably in giving necessary instructions; if materials which the employer has undertaken to supply are delivered late; or if the contractor's work is impeded by other contractors working directly for the employer. However, the employer is *not* responsible in this way for the defaults of nominated sub-contractors or suppliers; any responsibility the employer has for such defaults must be based on an express term of the contract (such as JCT 98 clause 25.4.7).

Before leaving the subject of the employer's implied obligations, it is important to note that these do not require co-operation to an extent that would enable the contractor to do *more* than the contract specifies. Thus, while it is the contractor's

---

[5] *Freeman v Hensler* (1900) 64 JP 260.
[6] *Coombe v Green* (1843) 11 M&W 480.
[7] *Hunt v Bishop* (1853) 8 Exch 675.

right to complete *on or before* the contractual completion date, the employer's duty of co-operation extends only to ensuring that the completion date is achieved. A contractor who seeks to finish early cannot force the employer to assist.[8] Similarly, where sub-contractors undertake to carry out work 'at such time or times as the contractor shall direct or require', there is no implied term requiring the main contractor to make sufficient work available to the sub-contractors to enable them to work in an efficient and economic manner.[9]

12.2   EMPLOYER'S OBLIGATIONS UNDER JCT SBC 05

In addition to the implied obligations described above, the express terms of any construction contract will undoubtedly impose duties on the employer. Some of these (such as the duty to give possession of the site) will go no further than the term that would in any event be implied. Others, such as duties to insure, break completely fresh ground. Naturally, the range and content of the employer' express obligations will vary from one form of contract to another, and it would not be practicable here to deal with all the possibilities. We shall therefore look at those obligations contained in JCT SBC 05, in order to give some idea of what is commonly found.

12.2.1 Payment

Undoubtedly the most important of all the employer' express obligations is to pay the contractor the sum of money which forms the consideration for the contract, known as the *contract sum*. The extent of this obligation, and the way in which the contract sum is assessed and can be altered, are dealt with in Chapter 15.

12.2.2 Necessary nominations

Article 3 of JCT SBC 05 names the architect or contract administrator who is to be responsible for performing all the functions which the contract conditions allocate to such a person. Similarly, if a quantity surveyor is to be appointed, this is done by Article 4. The relevant article provides in each case that, if the person named ceases to hold this post, it is the employer's duty to nominate a replacement. According to clause 3.5 of the conditions, this renomination must take place within a reasonable time, which must in any event not exceed 21 days. It is also made clear that, as a general rule, the contractor has a right of objection on reasonable grounds to the renomination. However, there is no right of objection where the employer is a local authority and the architect, contract administrator or quantity surveyor is an employee of that authority.

---

[8] *Glenlion Construction Ltd v Guinness Trust* (1987) 39 BLR 89.
[9] *Martin Grant & Co Ltd v Sir Lindsay Parkinson & Co Ltd* (1984) 29 BLR 31.

12.2.3 Site obligations

As we will see in Chapter 14, the need to give the contractor possession of the site at the right time is fundamental to the contract. Failure to do so will cause the employer to forfeit any claim for liquidated damages on late completion.[10] It will also render the employer liable to pay damages to the contractor.[11] Clause 2.4 alludes to this important obligation with the text, *on the Date of Possession possession of the site, or in the case of a Section, possession of the relevant part of the site, shall be given to the Contractor.* Although the employer is not specifically mentioned, a failure to secure possession for the contractor will constitute a breach by the employer, but liability for prevention by an independent third party depends on the specific wording of the contract.[12]

Unless otherwise agreed, the employer's obligation is to give possession of the whole of the site from the outset, and not just those parts where work is to begin immediately. It is further accepted that the contractor must be given, not only the actual area to be built on, but also sufficient surrounding space to enable the work to be properly undertaken.[13] This would include, for example, room to erect temporary buildings and compounds, to store equipment and so on. However, the employer will only be obliged to provide access to the site across adjoining land where this is specifically stated in the bills.

12.2.4 Insurance of the works

As we shall see in Chapter 17, clauses 6.7 to 6.10 of JCT SBC 05 offer a choice of insurance arrangements, depending among other things on whether the work consists entirely of new buildings or whether existing buildings are involved. All that need be said here is that, if the option chosen for inclusion in the contract is either option B or option C, then it is the employer's duty to procure the necessary insurance.

12.2.5 Confidentiality

One final obligation imposed upon the employer by JCT SBC 05 is found in clause 2.8.4. This forbids the employer to divulge or use any of the rates or prices in the contract bills, except for the purposes of the contract.

12.3   RESPONSIBILITY FOR THE CONTRACT ADMINISTRATOR

It has been emphasized above that many construction contracts allocate to the contract administrator the performance of what are really the employer's obligations. These will be considered in more detail in Chapter 18; our interest for

---

[10] *Holme v Guppy* (1838) 3 M&W 387.
[11] *Rapid Building Group Ltd v Ealing Family Housing Association Ltd* (1984) 29 BLR 5, CA.
[12] *Rapid Building Group Ltd v Ealing Family Housing Association Ltd* (1984) 29 BLR 5, CA.
[13] *R v Walter Cabott Construction Ltd* (1975) 21 BLR 42.

the moment is in the extent to which the employer can be held responsible to the contractor for the contract administrator's actions.

It is crucial in this context to distinguish between two separate aspects of the contract administrator's role. First, there are the numerous things done in the capacity of agent of the employer (such as supplying information, giving instructions as and when necessary, and so on). In *Merton LBC v Stanley Hugh Leach Ltd*[14] it was held that the employer impliedly guarantees to the contractor that these functions will all be performed *with reasonable diligence and with reasonable skill and care*. This means that any negligence on the part of the contract administrator will render the employer liable for breach of contract to the contractor. The employer will then be able to claim in turn against the contract administrator, for breach of the terms of engagement.

The second aspect of the contract administrator's role concerns certain 'independent' or 'discretionary' functions, such as certification. As to this, it was said in *Merton v Leach* that the employer *does not undertake that the architect will exercise his discretionary powers reasonably; he undertakes that, although the architect may be engaged or employed by him, he will leave him free to exercise his discretion fairly and without improper interference by him*. As a result, an employer who obstructs or interferes with the issue of a certificate, or who puts improper pressure on the contract administrator, will be liable to the contractor.[15]

Notwithstanding this statement, there are circumstances in which the employer's duties in respect of the contract administrator's 'discretionary powers' go further than mere non-interference. In *Perini Corporation v Commonwealth of Australia*[16] an employer became aware that the architect was acting improperly, by refusing to deal at all with applications for extension of time (as opposed to merely making incorrect decisions on applications). The Australian court held that it was the employer's obligation in such circumstances to order the architect to carry out his duties under the contract and, if that failed, to dismiss and replace him. In *Perini*, the architect was an employee of the employer, a Government department, but it is thought that the same principle would apply in cases where the architect is 'independent'.

12.4   RESPONSIBILITY FOR SITE CONDITIONS

A question which has arisen many times over the years concerns the extent to which an employer who initiates a project bears legal responsibility in respect of its feasibility. In particular, if site conditions such as the subsoil turn out to be unexpectedly adverse, or the proposed method of working proves impracticable, is there any way in which the contractor can claim redress? The answer to these questions naturally lies within the terms of the contract, but a fairly clear overall picture emerges from the leading cases.

---

[14] (1985) 32 BLR 51.
[15] *John Mowlem & Co plc v Eagle Star Insurance Co Ltd* (1992) 62 BLR 126.
[16] (1969) 12 BLR 82.

12.4.1 Contractor's risk

As a basic principle, it is clear that the risk of adverse site conditions rests with the contractor. This was firmly established in *Bottoms v York Corporation*,[17] which concerned the carrying out of sewerage works. Because the soil was softer than anticipated, the contractor had to carry out considerable extra work. When the engineer refused to authorize extra payment for this work as a variation, the contractor left the site and claimed a reasonable sum for the work done. No boreholes had been sunk in advance by either party, but the employer had received reports before signing the contract that the contractor was almost certain to lose money in the type of ground to be expected. It was nevertheless held that the employer owed no duty to disclose these reports to the contractor, whose claim accordingly failed.

The principle that adverse site conditions are the contractor's risk is not altered merely because the employer provides plans or specifications at the time of tender. The mere fact that these are provided does not imply any warranty by the employer as to their accuracy. In *Sharpe v San Paulo Brazilian Railway Co*,[18] for example, a contractor undertook to build a railway in Brazil for a lump sum. The engineer's plans proved to be hopelessly inadequate and, as a result, the contractor was forced to excavate about twice as much as had been anticipated. It was held that, since the accuracy of the plans was in no way warranted, the contractor was not entitled to any extra payment in respect of this work.

Similar principles apply to working methods. In *Thorn v London Corporation*,[19] a contract was let for the demolition and replacement of Blackfriars Bridge. The plans and specifications prepared by the engineer (whose directions the contractor was required to obey) featured the use of caissons to enable work to be done whatever the state of the tide. Unfortunately, these caissons proved to be useless, with the result that the contractor suffered considerable delay and extra expense. Once again, however, it was held that there was no implied guarantee by the client that the bridge could be built in the manner specified, and so the risk lay on the contractor.

12.4.2 Employer's responsibility

Notwithstanding the general principle outlined above, there may be situations in which an employer incurs liability when the project proves to be unexpectedly difficult or expensive to carry out. The major ways in which this occurs are as follows:

- *Implied warranty*. Although the mere fact that an employer provides tendering information does not automatically mean that its accuracy is guaranteed, it may occasionally be possible for a warranty of accuracy to be implied. In *Bacal Construction (Midlands) Ltd v Northampton Development Corporation*,[20] for example, a contractor tendered to design and build six

---

[17] (1892) HBC 4th ed, ii, 208.
[18] (1873) LR 8 Ch App 597.
[19] (1876) 1 App Cas 120.
[20] (1975) 8 BLR 88.

blocks of dwelling houses, under instructions to design the foundations on certain hypotheses as to ground conditions. These hypotheses, which were based on borehole data, subsequently proved inaccurate. It was held that, in these circumstances, the employer must be taken to have warranted that the ground conditions would be as they were hypothesized to be.

- *Misrepresentation*. Even where there is no *warranty* as to site conditions, a contractor may be able to show some *misrepresentation* by or on behalf of the employer. This will certainly be the case if there is any deliberate fraud in covering up the true conditions.[21] In the absence of fraud, it is no doubt possible to base a claim on a *negligent* misrepresentation.[22] However, this will not be easy to establish, especially where (as is common) the contract provides that it is the contractor's responsibility to check such matters.[23]

- *Standard method of measurement*. Building contracts frequently state that the bills of quantities on which they are based have been prepared in accordance with a specified Standard Method of Measurement. For example, JCT SBC 05 clause 2.13.1 refers to SMM 7, and ICE 7 clause 57 refers to CESMM3. Where this is so, a contractor who unexpectedly encounters rock may claim that this should have been a separate item in the bills and that the employer must therefore pay the extra cost. Such claims have been described as without foundation and without merit (Wallace 1986: 448), but there is some legal authority to suggest that they are valid.[24]

12.5   HEALTH AND SAFETY

The employer's obligations under the Construction (Design and Management) Regulations 2007 are extensive. Most importantly, the employer of almost any contractor has criminal liability to ensure that a CDM co-ordinator is appointed early in the process and to ensure that no construction work commences until a Health and Safety Plan is in place. The obligations on the employer (the regulations refer to the 'client') include ensuring that any person appointed to the role specified by the regulations is competent to carry out that role. Any person who will carry out design or construction work must be either competent or under the supervision of a competent person. Furthermore, the client has an obligation to ensure that the work can be carried out without risk to the health or safety of any person. The obligation extends to ensuring that all relevant pre-construction information is provided promptly to designers and contractors. Civil proceedings are specifically ruled out for most of the obligations under these regulations, which means that there is little point including reference to them in contracts, since breaching the regulations would incur criminal liability only.

---

[21] *S Pearson & Son Ltd v Dublin Corporation* [1907] AC 351.
[22] *Morrison-Knudsen International Co Inc v Commonwealth of Australia* (1972) 13 BLR 114.
[23] *Dillingham Ltd v Downs* [1972] 2 NSWLR 49.
[24] *Bryant & Sons Ltd v Birmingham Hospital Saturday Fund* [1938] 1 All ER 503.

13   Responsibility for design

Traditional building procurement systems draw a strict dividing line between the functions of design and construction. Design, which includes not only the broad concept of the building but also matters of considerable detail, is the responsibility of the employer's design team. This normally consists of an architect (or civil engineer), backed up where necessary by other specialist consultants such as structural engineers. Construction, on the other hand, is the responsibility of the contractor, whose obligation is simply to construct in strict accordance with the contract documents provided. Under such a procurement system, it is naturally important to decide whether any defect in the finished building is the result of a design fault or whether it arises out of bad construction, since this will determine who is to be legally responsible for the defect.

Of course, not all procurement methods are of this traditional kind. Design and build contracts (dealt with in Chapter 4) are well-established arrangements under which legal responsibility for both design and construction lies with the same party. In this situation, the *nature* of a defect in the finished building may be less important. Even here, however, it cannot be completely ignored, for reasons that will appear below.

In dealing with liability arising out of defective design, we must consider first, the nature of design and its complexity; second, how the law actually defines design obligations; and third, which participants in the construction process may be liable in respect of design faults.

13.1   DESIGN MANAGEMENT

Design management is best seen as an information processing system driven by innovative and/or creative solutions to problems of the client organization. One of the most difficult parts of construction design, particularly in architecture (which is typically more complex than civil engineering), is in defining the problem to which the construction project is the solution. This process is called briefing. It is not simply a question of the client specifying what is wanted so that the design team can get on with instructing the construction team. The development of a good brief is an active process on behalf of the designer and the client. Such a document sets out a design philosophy that should be sufficiently comprehensive to guide all design decisions on a project but this ideal is rarely achieved in practice. The complexity of the information processing exercise is brought about because of the involvement of many specialist designers, each contributing a small part to the overall picture. Moreover, each of these parts interacts with the other. The co-ordination and integration of such a diverse range of inputs is a daunting task. It is no surprise that many projects are tendered on incomplete information!

After the initial briefing stage, construction projects typically move through a series of stages: feasibility, outline design (sketch scheme), detail design, contract preparation, construction and commissioning of the complete facility (Hughes and Murdoch 2001). Some projects also involve a stage prior to all of these, called inception. In this stage the client's organization goes through internal processes connected with defining needs and assembling finance or funding. The point about these various stages is that each involves different participants. Moreover, the nature of the task changes at each stage. Thus, as the mass of project documentation grows, different specialists come and go, decisions are made and recorded, the relative importance of different criteria changes and various interest groups partake in decisions that affect the nature and scope of the project. The concept of a design team containing a few consultants for the duration of the project is a gross oversimplification of an extremely difficult issue (Gray and Hughes 2000).

One of the most important difficulties is the overlap between design and construction. There is a distinction drawn between design and workmanship, which is more to do with who is taking decisions than with the nature of decisions. For example, a decision about fixings for a skirting board would be a design decision if it were specified in the bill of quantities, and a workmanship decision if the bill were silent on the issue. The decision itself is the same in both cases. Because of this, any arguments about design being complete at the time of tendering are specious. It must be recognized that in all but the simplest projects, the complexity of the process demands a continuing interaction between design and construction teams until the project is completed.

13.2   DESIGN DUTIES IN LAW

Under this heading we look at some of the factors a court will take into account in deciding whether or not a designer is in breach of legal obligations. We are not concerned at this stage with detailed questions of how those obligations arise or to whom they are owed.

13.2.1 Standard of liability

From a legal point of view, one of the most interesting questions concerning a designer's duty is whether it is limited to an obligation to use reasonable care and skill, or whether it goes beyond this to a *guarantee* that the design will be fit for its purpose. If the former is correct, it means in effect that a designer will only be liable if 'professional negligence' can be proved. The latter interpretation, on the other hand, would impose on the designer a type of liability equivalent to that of a seller or other supplier of goods.

Apart from its legal interest, this question also has practical implications, which are of critical importance. As we shall see, the courts have made it clear that a guarantee of 'fitness for intended purpose' will fairly readily be implied into a design and build or package deal contract. Under such a contract, therefore, the main contractor will be strictly liable to the client for any defect resulting from an error of design. Now if, as is frequently the case, the actual design has not been carried out 'in-house' by the main contractor, but has been sub-contracted to either an architect or a specialist sub-contractor, the main contractor will seek to pass liability down the line to the actual designer. This, of course, can only be done if the designer's liability under the sub-contract is at the same level as the contractor's liability under the main contract.

An excellent illustration of the problem, and the strongest authority for subjecting a designer to strict liability, is the case of *Greaves & Co (Contractors) Ltd v Baynham Meikle and Partners*.[1] The claimants there were employed under a package deal contract to construct a warehouse, the first floor of which was to be used for the storage of oil drums, stacked and moved by forklift trucks. The claimants sub-contracted the design of this warehouse to the defendants, a firm of consultant structural engineers, who knew precisely the purpose of the finished building. The defendants' design did not make sufficient allowance for vibrations from the forklift trucks; as a result, the floor cracked and became dangerous, and the claimants became liable to the clients for the cost of replacement. The claimants accordingly sued the defendants to recover this cost, claiming that it was an implied term of their sub-contract that the design would be fit for its intended purpose.

It was held by the Court of Appeal that, on these facts, the defendants had been guilty of negligence and were accordingly liable. However, the importance of the case for present purposes lies in the court's ruling that, even if there had not been negligence, the defendants would still have been liable for the failure of their design. This was because, while it could not be assumed that *every* designer would be taken to warrant the fitness of his or her design (that is, as a term implied by law into every design contract), such a term could be implied into *this* contract as a matter of fact. The defendants had always regarded their brief as being the design of a warehouse for a particular purpose, and could be taken as guaranteeing that it would be fit for that purpose when completed.

Notwithstanding this decision, it must be recognized that the traditional legal position favours liability on the part of a 'pure' designer only where there is negligence, a position reaffirmed by the Court of Appeal in *Hawkins v Chrysler (UK) Ltd and Burne Associates*.[2] The first defendants there, who had provided showering facilities for employees at their factory, accepted liability to one of their employees who was injured when he slipped on wet tiles. The first defendants then claimed against the architects who had designed, specified and supervised the installation of those showers. The Court of Appeal held that there was no reason on these facts to imply any warranty by the architects other than that they would use reasonable care and skill; there was no warranty that the materials selected would be fit for their intended purpose. While recognizing that this might create an anomalous distinction between a pure designer and a designer/builder, the Court of Appeal did not feel justified in raising 'professional liability' to a new level.

As mentioned above, the legal standard of design liability becomes important partly because of the law's readiness to impose strict liability in design and build

---

[1] [1975] 3 All ER 99.
[2] (1986) 38 BLR 36.

contracts. The basic legal position was described in *Francis v Cockrell*[3] where it was said that:

> ... *when one man engages with another to supply him with a particular article or thing, to be applied to a certain use and purpose, in consideration of a pecuniary payment, he enters into an implied contract that the article or thing shall be reasonably fit for the purpose for which it is to be used and to which it is to be applied.*

This principle was applied to a design and build contract in *Viking Grain Storage v TH White Installations Ltd*,[4] which concerned a defective grain storage and drying installation designed and erected by the defendants. It was also accepted, though less conclusively, by the House of Lords in *Independent Broadcasting Authority v EMI Electronics Ltd and BICC Construction Ltd*.[5] The contract in that case was for the design and erection of a 381 metre television mast at Emley Moor, Yorkshire. This was let to EMI as main contractors, and they sub-contracted the work to BICC on virtually identical terms. The mast, which was of a novel cylindrical design, collapsed due to vortex shedding (induced by wind) and asymmetric ice loading. It was held by the House of Lords that, at the very least, EMI must be taken to have warranted that BICC's design would not be negligent. Since it *was* negligent, EMI were liable to the clients and BICC were in turn liable to EMI. As a result of this finding of negligence, it was unnecessary to decide whether there would have been any strict liability for the defective design. However, the court stated that there probably *would* have been such liability.

The implication of a strict 'fitness for the purpose' obligation into design and build contracts will not, it appears, be negated by the mere fact that the actual design is carried out by a sub-contractor nominated by the employer. However, all the circumstances of the case must be examined. Where it is clear that the employer has placed no reliance whatsoever on the main contractor in respect of design, there will be no implied term. This occurred in the Irish case of *Norta Wallpapers (Ireland) v Sisk & Sons (Dublin) Ltd*,[6] in which the roof of a factory was supplied and erected by a specialist sub-contractor nominated by the employer. Since the main contractor in this case had no option but to accept this sub-contractor *and to adopt the sub-contractor's design*, it was held that no warranty of fitness for the purpose could be implied into the main contract. (In *IBA v EMI*, by contrast, the main contractors were not bound to accept any particular design produced by the nominated sub-contractors.) The question of reliance on the discretion and skill of a contractor was also raised in *Rotherham Metropolitan Borough Council v Frank Haslam Milan & Co*[7] in which a contract specified that hardcore included slag. When the type of slag used in the hardcore expanded and caused the building to fail, the judge at first instance found the contractor liable for breach of the implied term that all materials will be fit for their purpose. However, the Court of Appeal overturned this on the ground that, at the time, no contractor knew that the type of slag used was unsuitable, and the contractor's freedom to choose was

---

[3] (1870) LR 5 QB 501.
[4] (1985) 33 BLR 103.
[5] (1980) 14 BLR 1.
[6] [1978] IR 114.
[7] (1996) 78 BLR 1.

based upon the assumption that all types were suitable. Therefore, the contractor's discretion was not being relied upon.

Before leaving the 'fitness for purpose or negligence' issue, three other points are worth making. First, while it may indeed be easy for a court to imply the higher standard of obligation into a package deal contract, this cannot override any contrary *express* terms. In this connection, it should be noted that JCT DB 05 makes clear that the main contractor's liability goes no further than that which would be incurred by *an architect or ... other appropriate professional designer holding himself out as competent to take on work for such design* (clause 2.17.1). Clause 2.19.1 of JCT SBC 05 deals in the same way with CDP works, i.e. the contractor's designed portion. Similar terms are found in ICE 7 and GC/Works/1. Thus, where these forms of contract are chosen, the contractor's design duty is merely to use reasonable care and skill, and does not extend to a guarantee that the completed works will be fit for their intended purpose. The ACA Form of Contract clause 3.1, by contrast, is one under which 'fitness for purpose' *would* be guaranteed.

The second point arises out of the increasingly common practice whereby nominated sub-contractors design their own work. If the main contract is of the design and build type, then the legal situation is as outlined above. However, where the main contract is of a conventional 'build only' type, it is highly unlikely that a court would hold the main contractor responsible for any defect resulting from the nominated sub-contractor's design. In view of this possibility, it is of course very important for the employer, or the employer's professional advisors, to ensure that the sub-contractor undertakes direct design responsibility, for example by entering into a collateral agreement.

The third and last point concerns the limits of professional indemnity insurance. Many of the policies held by architects and other specialist designers do not cover any form of liability stricter than negligence, so that liability under an implied warranty could well be left unprotected. Further, many contractors' liability insurance policies do not cover the 'design' element at all. It is of course vital that such matters be checked thoroughly at the start of a project.

13.2.2 Duration of liability

A designer, such as an architect or civil engineer, is often involved in superintending the process of construction. Where this is so, it is clear that the designer's responsibility in respect of that design does not end when the contractor receives the necessary documentation and begins to build what has been designed. The architect remains under a continuing obligation to see that the design will work.[8] As to the precise duration of this obligation, it undoubtedly lasts until the date of practical completion, almost certainly throughout any defects liability period, and probably until the issue of the final certificate.[9] Indeed, the judge in one case suggested that consulting engineers might remain under a duty *after completion* to check and recheck their design in the light of new knowledge, and to

---

[8] *Brickfield Properties Ltd v Newton* [1971] 3 All ER 328.
[9] *Merton LBC v Lowe & Another* (1981) 18 BLR 130.

inform their clients if new sources of danger come to light. Fortunately for engineers, this suggestion was unanimously rejected by the Court of Appeal.[10]

13.2.3 Techniques and materials

One of the most important aspects of 'design' is the selection and specification of materials. In this, as in all other aspects, the designer is required to exercise reasonable care and skill. It should be appreciated that, where an architect specifies materials, there will be no implied warranty from the contractor that the materials used will be fit for their purpose (although the contractor's implied warranty as to the 'quality' of the materials will normally be unaffected). It is therefore important for the architect to take such steps as are possible, such as testing or examination of other sites, to ensure the suitability of any new product to be specified. It would also be sensible to obtain, wherever possible, some collateral warranty from the supplier as to the product's performance.

As with new materials, so with new techniques. The standard of care demanded of a designer is judged in the light of professional knowledge at the relevant time, what is called the 'state of the art'. However, this does not mean that a designer may simply leap into the unknown without any legal responsibility. Indeed, the very fact that there *is* no general experience and expertise to draw on in relation to a novel form of design may require the taking of *extra* precautions. As has been said in the House of Lords:[11]

> *The project may be alluring. But the risks of injury to those engaged in it, or to others, or to both, may be so manifest and substantial and their elimination may be so difficult to ensure with reasonable certainty that the only proper course is to abandon the project altogether ... Circumstances have at times arisen in which it is plain commonsense and any other decision foolhardy. The law requires even pioneers to be prudent.*

13.2.4 Compliance with statutory requirements

The proper carrying out of a design function clearly includes the task of seeing that the designed works can be carried out lawfully. This means that the works will not contravene Building Regulations, planning law or other relevant legal requirements. In this respect, as with other aspects of design, the law draws a distinction between someone who merely designs and someone who operates under a design and build contract. A mere designer, it appears, impliedly undertakes only to use professional skill and reasonable care, and does not warrant that the design will not contravene any relevant legal principle. Thus, where an architect-designed building failed to qualify for the office development permit that the client wanted, the architect nonetheless avoided responsibility. The court stated that the legal

---

[10] *Eckersley v Binnie & Partners* (1990) 18 Con LR 1.
[11] *Independent Broadcasting Authority v EMI Electronics Ltd and BICC Construction Ltd* (1980) 14 BLR 1.

rules governing the matter were complicated and that the architect's advice, though wrong, was of a kind that a reasonably competent architect might give.[12]

By contrast, a contractor under a design and build contract will be *strictly* liable to the employer for any breach of the Building Regulations. This point was established in *Newham LBC v Taylor Woodrow (Anglian) Ltd*,[13] a case which arose out of the notorious collapse of a block of flats at Ronan Point.

Responsibility for a building which contravenes the law will usually fall, in the final analysis, upon the designer rather than the builder, at least where a traditional form of contract is used. Under JCT SBC 05 clause 2.1, for example, the contractor is made *generally* responsible for ensuring that all statutory rules are complied with. However, a contractor who has merely worked in accordance with the contract drawings or bills is protected from liability by clause 2.17.3. This protection is subject to the proviso that, upon discovering any discrepancy between these documents and the legal requirements, the contractor immediately notifies the contract administrator.

Not surprisingly, JCT DB 05 gives no such protection. Clause 2.1 makes it clear that the contractor is responsible for ensuring that the works comply with the law. This is so even where the Employer's Requirements are at fault, unless the contract specifically states that these are in accordance with the law.

Quite apart from the designer's potential liability to the client for breach of contract, there is the possibility that defective design may lead to *personal* liability for breach of Building Regulations.

The Construction (Design and Management) Regulations 2007 impose severe and wide ranging obligations on all designers, the breach of which carries criminal penalties. The most important of these obligations is that designers must do everything reasonably practicable to avoid danger to the health and safety of anyone working on the site, or affected by the work. This obligation extends to anyone cleaning the finished project, or affected by cleaning, and further to anyone demolishing the facility when it comes to the end of its life. There are additional obligations to co-operate with other designers and with the planning supervisor and to ensure that the client is aware of any health and safety obligations.

13.3   LEGAL RESPONSIBILITY FOR DESIGN

Whether a person can be liable for breach of contract in respect of a design fault is, in theory at least, a simple one to answer. It depends upon the terms, express and implied, of the contract under which the design function has been carried out. However, as we shall see, the application of this straightforward principle is not always obvious.

13.3.1 Architect

The overall responsibility for the design of the project will usually be borne by the architect, where one is involved. Indeed, the basic rule is that the architect cannot

---

[12] *BL Holdings Ltd v Robert J Wood & Partners* (1979) 12 BLR 1.
[13] (1981) 19 BLR 99.

delegate *any* part of the design work. Such delegation, without authority, will render the architect personally responsible for any defects in the design that arise out of negligence.

It was on this basis that liability was imposed upon an architect in *Moresk Cleaners Ltd v Hicks*.[14] The claimants in that case, who wanted an extension to their laundry, employed the defendant architect for the project, which involved designing a reinforced concrete structure on a sloping site. Feeling that this was beyond his competence, the architect gave the job to a contractor on a design and build basis. When the design proved defective, the architect argued that either he had implied authority to delegate specialist design tasks, or he had acted merely as the employer's agent in employing the contractor. However, it was held that the architect was liable in respect of the design, which he had been wrong to delegate. The judge stated that an architect who lacks the ability or expertise to carry out part of a design job has three choices:

- to refuse the commission altogether;
- to persuade the employer to employ a specialist for that part of the work; or
- to employ and pay for a specialist personally, knowing that any liability for defective design can then be passed along the chain of contracts.

Notwithstanding the basic principle applied in *Moresk*, the complexity of modern construction technology has led to an increased dependence on specialists at both design and construction stages of a project. As a result, authority for an architect to delegate specified parts of the design will frequently be given by the employer, and it may even be implied from the circumstances of the case. In *Merton LBC v Lowe*,[15] for example, an architect was held entitled to appoint a specialist sub-contractor who specialized in the use of a certain proprietary ceiling.

In recommending the appointment of a particular specialist (or, for that matter, a particular contractor), architects owe their clients a duty to use reasonable care and professional skill. They will not be *automatically* responsible for the defaults of the people they recommend. But if their recommendations are negligent, they may become liable for their clients' losses. In *Pratt v George J Hill Associates*,[16] for instance, a contractor who was strongly recommended by the defendant architects proved to be highly unsuitable. He became insolvent leaving a trail of defective work. It was held that the defendants were liable to the client for the money she was unable to recoup from the contractor.

Even assuming that architects act reasonably in recommending a specialist, this is not the end of their duty to the client. The Court of Appeal has summed up the legal position as follows:

> In relation to the work allotted to the expert, the architect's legal responsibility will normally be confined to directing and co-ordinating the expert's work in the whole. However ... if any danger or problem arises in connection with work allotted to the expert, of which an architect of ordinary competence reasonably ought to be aware and reasonably could be expected to warn the client, despite the employment of the expert, and

---

[14] (1966) 4 BLR 50.
[15] (1981) 18 BLR 130.
[16] (1987) 38 BLR 25.

despite what the expert says or does about it, it is ... the duty of the architect to warn the client. In such a contingency he is not entitled to rely blindly on the expert, with no mind of his own, on matters which must or should have been apparent to him.[17]

13.3.2 Contractor

As already mentioned, traditional methods of procurement regard design as the exclusive province of the architect, plus such specialists as are necessary. Design and build or package deals may be different but, under general contracting, the contractor's responsibility is merely to build in strict accordance with the designer's specification. This frequently includes a considerable measure of detail as to the quality and standards of materials and workmanship. Thus one might expect the contractor to be free from any form of responsibility for design.

This indeed does represent the basic legal position, as appears from the case of *Mowlem v British Insulated Callenders Pension Trust*.[18] This concerned a contract let under JCT 63, where a performance specification in the bills of quantities purported to impose a measure of design responsibility on the contractor. It was held that this was ineffective because of clause 12(1) (now clause 1.3 of JCT SBC 05), which prevents anything contained in the bills from overriding the contract conditions.

In theory, then, a contractor is not involved in design in any way that may create legal liability. In practice, however, even under a traditional form of contract things are not quite so clear-cut. Contractors and sub-contractors tend to take on a measure of design responsibility in the following ways:

- Where the contract documents do not give sufficiently fine detail, a contractor who exercises discretion is effectively taking on a design function. This may shade into questions of 'workmanship', or it may be very small-scale, such as a decision on how far apart to place fixing screws. In any event, it appears that a contractor who uses initiative in such circumstances, instead of seeking an architect's instruction, will incur responsibility for any defects that ensue.
- Contractors and sub-contractors are often asked, as a project progresses, for their opinion as to the best means of overcoming a particular problem that has arisen. There can be no doubt that, if such advice is given (at least where the person giving it is a specialist), a duty of care will arise.
- Where contractors, sub-contractors or suppliers are required to produce drawings for the architect's approval, any matters of design that are included may be a source of liability. This is despite the possibility that the architect may also be liable.
- Most importantly, it has been held by the Court of Appeal that a term will be implied in some, if not all, building contracts, requiring the contractor to warn the employer of any defects in the design.[19] The extent of such an

---

[17] *Investors in Industry Ltd v South Bedfordshire DC* [1986] 1 All ER 787, 808.
[18] (1977) 3 Con LR 64.
[19] *Plant Construction plc v Clive Adams Associates and JMH Construction Services Ltd* [2000] BLR 137.

obligation, which is based on the contractor's duty to exercise reasonable skill and care, has not yet been fully worked out by the courts; it has so far been applied only to defects of which the contractor is actually aware (as opposed to those of which he ought reasonably to be aware) and only where the defects in question are *dangerous*.[20]

Apart from these hidden forms of design liability, there are of course certain situations in which a contractor specifically undertakes responsibility for the design as well as the construction of a building. If the parties do indeed intend their relationship to be on a design and build or package deal basis, then it is important to adopt an appropriate form of contract. Such contracts would include JCT DB 05, ACA/2 or (where only *part* of the work is to be contractor-designed) JCT SBC 05 with its – newly incorporated – option of a contractor's designed portion (CDP Works). However, as we have already noted, an important drawback from the point of view of an employer is that the JCT forms limit the contractor's design liability to 'reasonable care and skill' level. There is no warranty that the design will be fit for its intended purpose.

13.3.3 Sub-contractor

Even more common, perhaps, than cases of 'contractor's design' are cases in which part or all of the design is allotted to a specialist as *sub-contractor*. Where the main contract is JCT SBC 05 and the specialist in question is a sub-contractor, the employer is well advised to enter into a collateral warranty with the sub-contractor. It should be borne in mind that the sub-contractor, under the appropriate form for this warranty agreement (JCT SCWa/E) does not warrant that the sub-contract works will be fit for their purpose, but merely that reasonable skill, care and diligence will be exercised regarding their design. In addition, the agreement provides that the sub-contractor owes the same obligation to the employer as to the main contractor, subject to certain exceptions which are stated in the agreement.

Where the main contract is JCT 98, clause 35.21 makes clear that the main contractor is under no responsibility to the employer for any design work carried out by a nominated sub-contractor.

---

[20] See *Equitable Debenture Assets Corporation v Moss* (1984) 2 Con LR 1; *Victoria University of Manchester v Wilson & Womersley* (1984) 2 Con LR 43; *University Court of Glasgow v Whitfield* (1988) 42 BLR 66; *Lindenberg v Canning* (1992) 62 BLR 147.

14   Time

Time is an extremely important issue in construction. Together with cost and quality, it is a primary objective of project management, and a major criterion by which the success of a project is judged (Charmer 1990). The scope of this subject may be seen from clause 2.4 of JCT SBC 05, which states:

> On the Date of Possession possession of the site or, in the case of a Section, possession of the relevant part of the site shall be given to the contractor who shall thereupon begin the construction of the Works or Section and regularly and diligently proceed with the same and shall complete the same on or before the Completion Date.

This identifies the three main time-related issues as **commencement**, **progress** and **completion**. There are two further issues: the contractor's continuing **obligations after completion**, and the **extensions of time** which may be available to the contractor when the work is delayed by certain specified causes.

14.1   COMMENCEMENT

The issues at the beginning of the contract involve giving possession of the site to the contractor, the timing of this possession and potential delays to the possession. Normally, possession should take place not more than two months after the successful contractor has been awarded the contract. Too speedy a start may cause extra work and delay, rather than hastening the construction period. This needs to be balanced against the needs of the client to avoid undue delay, which may cause extra costs (Joint Contracts Tribunal 2002).

14.1.1 Possession of the site

An employer who fails to give the contractor possession of and access to the site may be liable to pay damages for breach of contract.[1] This is so despite provisions in the contract for the contract administrator to postpone all or any part of the works, since it seems that such provisions may not be used to postpone the entire project.[2] However, the employer is not deemed to *guarantee* possession or access and will therefore not be liable if the contractor is prevented from gaining access by some third party, such as unlawful pickets, over whose activities the employer has no control.[3]

---

[1] *Rapid Building Group Ltd v Ealing Family Housing Association Ltd* (1984) 29 BLR 5, CA.
[2] *Whittall Builders Co Ltd v Chester-le-Street DC* (1987) 40 BLR 82.
[3] *LRE Engineering Services Ltd v Otto Simon Carves Ltd* (1981) 24 BLR 127.

Under JCT SBC 05 clause 2.4 and ICE 7 clause 42, the contractor is entitled to possession of the *whole* of the site, even though access to some parts may not be required until a later stage of the project. But there are exceptions to this default provisions. Under JCT SBC 05 clause 2.4, if the work is split into sections, the employer only has to give possession of the relevant sections to the contractor. Similarly, clause 42(1) of ICE 7 makes provision for access and possession to be prescribed explicitly, meaning that the parties can agree that only distinct portions of the site are to be handed over to the contractor at the beginning. In either case, what is given to the contractor must include not only the actual area to be built on, but also enough of the surrounding area to enable the work to be undertaken.

14.1.2 Date for possession

Most building contracts will name a date on which the contractor is to be given possession of the site, after which the contractor may commence the works. If possession is not then given on the date specified, the employer will lose the right to recover liquidated damages from the contractor in the event of late completion.[4] ICE 7 clause 41 gives the employer a little extra flexibility by providing that, if no date is specified from the outset, it is for the engineer to notify the contractor of the date for commencement of the works. This notice must be given in writing, and the date itself must be between 14 and 28 days of the award of the contract.

If the contract contains no specific commencement provision, then the contractor must be given possession at such a time as will enable the work to be completed by the completion date.[5] The contractor is not obliged to start work on the date for possession; however, a contractor who does not start reasonably quickly may be liable for not proceeding 'regularly and diligently' (JCT SBC 05, clause 2.4) or 'with due expedition and without delay' (ICE 7 clause 41(2)).

14.1.3 Deferred possession

Although the contract administrator's power to order the postponement of any work cannot be used by the employer so as to justify delay in giving the contractor possession of the site, there are specific provisions in JCT SBC 05 clause 2.5 and JCT IC 05 clause 2.5 under which the employer may defer the date for possession by up to six weeks, if the relevant entry is made in the Contract Particulars.

14.2   PROGRESS

Where a construction contract fixes a date for completion, but makes no provision as to the rate at which the works are to progress, it appears that the courts will not imply any such term. This is because, in the absence of any indication to the contrary, the contractor has absolute discretion as to how the work is planned and

---

[4] *Holme v Guppy* (1838) 3 M&W 387.
[5] *Freeman v Hensler* (1900) 64 JP 260.

performed, provided only that it is completed on time.[6] Furthermore, while many contracts require the contractor to submit a programme for the execution of the works, this in itself does not mean that there is a contractual obligation to keep to that programme.[7] Indeed, it should be appreciated that, if there *were* such an obligation, it would apply to both parties. Thus the employer would have to ensure that the contractor was provided with all necessary information at such a time as to enable compliance with the programme.

From an employer's point of view, it would be very inconvenient to have no control at all over the progress of the contract works. It is for this reason that most construction contracts require the contractor to maintain a satisfactory rate of progress throughout the project. For example, JCT SBC 05 clause 2.4 imposes an obligation on the contractor to proceed with the Works 'regularly and diligently', and JCT SBC 05 clause 8.4.1.2 makes failure to do so a ground on which the employer can terminate the contractor's employment. Similarly, under ICE 7 clause 41(2), the contractor is required to proceed with the works 'with due expedition and without delay'.

The meaning of the phrase 'regularly and diligently' was considered by the Court of Appeal in the case of *West Faulkner Associates v Newham LBC*.[8] The court suggested that the word 'regularly' means that the contractor must attend on a daily basis with sufficient labour and materials to progress the works substantially in accordance with the contract. 'Diligently', it was said, refers to the need for the contractor to apply that physical capacity industriously and efficiently. Taken together, the contractor's obligation is *to proceed continuously, industriously and efficiently with appropriate physical resources so as to progress the works steadily towards completion substantially in accordance with the contractual requirements as to time, sequence and quality of work.*

Progress is important in sub-contracts, because of the need to coordinate the sub-contractor's work with that of other sub-contractors and of the main contractor. In a case where a sub-contractor was expressly required to complete the work in reasonable accordance with the progress of the main contract, it was said that this did not mean compliance with every detail of the main contractor's programme, However, it meant more than merely not unreasonably interfering with progress under the main contract, in that the sub-contractor must make all reasonable efforts to keep up with the actual progress of the works.[9]

14.3 COMPLETION

In UK construction contracts, completion is a vague concept. The fact that building projects can be handed over in a less than perfect state is to the advantage of *both* parties. This is clear when the legal meaning of completion is considered.

---

[6] *GLC v Cleveland Bridge and Engineering Co Ltd* (1984) 34 BLR 50; *Pigott Foundations Ltd v Shepherd Construction Ltd* (1993) 67 BLR 48.
[7] *Kitsons Sheet Metal Ltd v Matthew Hall Mechanical and Electrical Engineers Ltd* (1989) 47 BLR 82.
[8] (1994) 71 BLR 1.
[9] *Ascon Contracting Ltd v Alfred McAlpine Construction Isle of Man Ltd* (2000) 66 Con LR 119.

14.3.1 Meaning of completion

A contractor cannot truly be said to have totally performed the contract if a single item of work is missing or defective. From a practical point of view, however, to delay the handover of something as complex as a large building for a trivial breach would cause enormous inconvenience. As a result, most building contracts require the contractor to bring the works to a state described by such expressions as **practical completion** (JCT contracts) or **substantial completion** (ICE 7).

Whether or not a building is 'complete' in this sense is normally a decision for the contract administrator, based on an inspection of the works and the exercise of judgement. As to precisely what is required, rulings handed down by the courts have ranged between two extremes. In *Westminster CC v Jarvis & Sons Ltd*,[10] Salmon LJ in the Court of Appeal adopted a very functional (and liberal) approach. He defined practical completion as 'completion for all practical purposes, that is to say for the purpose of allowing the employers to take possession of the works and use them as intended'. However, when *Westminster v Jarvis* was appealed to the House of Lords,[11] Lord Dilhorne took the much stricter line that 'what is meant is the completion of all the construction work that has to be done'.

Many in the construction industry would undoubtedly favour the first of these opinions as being the more practical. It is also worthy of note that the only standard form contract which seeks to define practical completion, the ICE Minor Works form, adopts this approach (see clause 4.5(1) of ICE Minor Works 2001). However, the current judicial view is much closer to that of Lord Dilhorne. It appears that a contract administrator may certify practical completion notwithstanding that there are trivial defects or omissions in the works, but should not do so where there are any patent defects which go beyond what is merely trivial.[12]

14.3.2 Date for completion

It is not essential for a building contract to specify a date for completion; if it does not, the contractor will have an implied obligation to complete the works within a reasonable time. To rely on such an obligation, however, is not very satisfactory, at least on a project of any appreciable size, for the client will normally need to have some degree of confidence about precisely when to expect completion. In addition, without a specific completion date there can be no provision for 'liquidated damages', that is a fixed sum to be paid by the contractor for every day or week of delay in achieving practical completion.

It is thus usual to name the date by which completion is required. In JCT SBC 05 this is done by means of an entry in the Contract Particulars. In ICE 7, it is done by inserting a time for completion in the Appendix to the Form of Tender. Moreover, even where no precise date has been included in the contract itself, a

[10] [1969] 3 All ER 1025.
[11] [1970] 1 All ER 942.
[12] *HW Nevill (Sunblest) Ltd v William Press & Sons Ltd* (1982) 20 BLR 78; *Emson Eastern Ltd v EME Developments Ltd* (1991) 55 BLR 114.

court may be persuaded to imply a term for completion by a certain date, on the ground that the parties must have intended this.[13]

JCT SBC 05 draws a distinction between the 'Date for Completion' and the 'Completion Date', both of which are defined in clause 1.1. The Date for Completion is defined as that date which is fixed and stated in the Contract Particulars. Thus, the Date for Completion can only be that date and no other. The Completion Date, on the other hand, is defined as the Date for Completion *or* any other date fixed under either clause 2.28 or by a Pre-agreed Adjustment (in the sense of clause 2.26.2). This means that, while the Completion Date can be changed, the Date for Completion cannot.

The practical result of this twin definition is that the contractor's obligation is to complete the work on or before the Completion Date. Thus, if an extension of time has been granted under clause 2.28, the obligation is to complete on or before the new date. However, even where the Completion Date is moved *forward* in time (for example, where work has been omitted), clause 2.28.6.3 makes it clear that the contractor can never be obliged to complete before the original Date for Completion.

The JCT approach to time may be contrasted with that of ICE 7, under which the contractor inserts the time required to carry out the work in the appendix to the tender. The contractor's obligation is then to complete the works within the stated time (as extended under clause 44, where this is appropriate), calculated from the 'Works Commencement Date'.

14.3.3 Delay in completion

The contractor's obligation to complete the works by the completion date is, like all such obligations, backed up by legal sanctions. If the contract is one in which time is 'of the essence', any lateness in performance will entitle the other party to terminate the contract. This will be the case where the contract expressly says that time is of the essence, and also where it is impliedly so, as for example in a contract for the sale of perishable goods.

Time will seldom be impliedly of the essence in a construction contract. However, once the contractor is in breach of contract through delay, it is possible for the employer to make time of the essence by giving notice to the contractor to this effect. Such a notice, which cannot be served until the contractor is in breach of contract,[14] must give a reasonable time for completion and state that, if the contractor does not comply with this, the employer reserves the right to terminate the contract.

Where late completion does not justify the employer in terminating the contract, the employer's legal remedy will be an award of damages for breach of contract. As to how such damages are to be measured, it is of course perfectly possible for the contract to say nothing, and to leave the assessment of the employer's loss (including any loss of profit) to an arbitrator or a court. However, it is standard practice in building and civil engineering contracts to state in advance what the damages shall be for delay, and this is usually done by specifying a fixed

---

[13] *Bruno Zornow (Builders) Ltd v Beechcroft Developments Ltd* (1989) 51 BLR 16.
[14] *Shawton Engineering Ltd v DGP International Ltd* [2006] BLR 1.

sum of money to be due for every day, week or month by which the contractor fails to meet the prescribed completion date. Such sums, which are called liquidated damages, liquidated and ascertained damages or LADs, are dealt with in Chapter 21.

14.3.4 Sectional completion and partial possession

If the intention of the parties is that the contract work should be completed and handed over in phases, it is essential that the contract documents make proper provision for this. Most standard forms enable different Sections to be identified in the respective project-specific data section, each with its own time for completion and its own liquidated damages. For example, ICE 7 contains a provision to this purpose in its Appendix, NEC 3 has the Option X5, and JCT SBC 05 provides for this in its Contract Particulars (in particular in the Sixth Recital). In terms of JCT's standard form, this is new because the JCT 98 Standard Form contained no such provision, although there was a Sectional Completion Supplement.

The inclusion in the main form of sectional completion dates has eliminated a longstanding source of problems. These had to do with the provision in older JCT conditions, for example, that nothing contained in the bills shall override or modify the terms of the contract (clause 2.2.1 of JCT 98, now clause 1.3 of JCT SBC 05). When the parties wanted to split the works into sections, they used a sectional completion supplement. But the use of this supplement would not be sufficient for projects in which the starting date of one section was dependent upon completion of a previous section. In such circumstances, if it was intended that an extension of time for the completion of phase 1 should delay the start of phase 2, then (under JCT 98) the contract would have needed to be specially amended to provide for this. Instead of drafting appropriate contractual terms, parties sometimes attempted to provide for sectional completion by listing separate phase dates in the bills of quantities. Under JCT 98 such attempts were doomed to failure due to clause 2.2.1.[15] Nor, it seems, could a similar effect have been achieved by expressing liquidated damages as '£x per week for each uncompleted house', since this was inconsistent with other provisions of the contract which already provided for liquidated damages to reduce as work is handed over.[16]

Generally, a contractor is entitled to keep the possession of the site or, in the case of a section, possession of the relevant part, until the contract administrator has issued the Practical Completion Certificate or the relevant Section Completion Certificate. However, JCT SBC 05 clause 2.33 enables this right to be given up by the contractor. If the employer requests it and the contractor agrees (and there must not be an unreasonable refusal), the employer can take possession of those parts of the works that are ready. Where this is done, the contract administrator must issue a written statement identifying the parts taken over and, for most purposes, practical completion of those parts is deemed to have occurred. It should also be noted that, under clause 2.6.1, the employer may be permitted by the contractor to use parts of the site for *storage or otherwise*, provided that this will not prejudice

---

[15] *MJ Gleeson (Contractors) Ltd v Hillingdon London Borough* (1970) 215 EG 165.
[16] *Bramall & Ogden Ltd v Sheffield CC* (1983) 29 BLR 73.

the insurance position. Once again the contractor's consent is not to be unreasonably withheld.

14.3.5 Effects of completion

When the contract administrator certifies that the works have been completed, a number of consequences will follow. Precisely what these are will depend upon the terms of the contract concerned, but the following are typical:

- The employer is entitled and obliged to take possession of the contract works.
- The contractor's responsibility (if any) for insuring the contract works comes to an end. This is often dealt with expressly, but it has also been held to apply by implication under JCT MW 80, notwithstanding an argument that responsibility for insurance should last until the end of the Defects Liability Period.[17]
- Any liability of the contractor to pay damages for late completion ceases. Moreover, this liability will not be revived if the work is later found to contain defects, for such a discovery will not retrospectively invalidate the certificate.[18]
- The contractor usually becomes entitled to the release of one-half of the accumulated retention money.
- The Rectification Period or Defects Correction Period (see below) begins.

Under the Construction (Design and Management) Regulations 2007, on completion of the project the Planning Supervisor must hand over to the client the Health and Safety Plan for the building, explaining its significance and any obligations that arise from it.

14.4 CONTRACTOR'S OBLIGATIONS AFTER COMPLETION

There are further obligations imposed on the contractor after completion, notably by JCT SBC clause 2.38, JCT IC 05 clause 2.30 and ICE 7 clause 49. Under the JCT contracts, the issue of the 'Practical Completion Certificate' marks the start of the 'Rectification Period', which is also known by its former name 'Defects Liability Period'. It lasts six months if no other period is specified in the Contract Particulars. Any defects, shrinkages or other faults arising during this period due to defective materials or workmanship must be put right by the contractor at its own expense.

The contractual procedure for dealing with defects arising during the Rectification Period is that the contract administrator should issue a schedule of such defects to the contractor not later than fourteen days after the end of the defects liability period, and the contractor then has a reasonable time to put them right. Once this has been done, the contract administrator will issue a 'Certificate of Making Good' (JCT SBC 05 clause 2.39, JCT IC 05 clause 2.31), following

---

[17] *TFW Printers Ltd v Interserve Project Services Ltd* [2006] EWCA Civ 875.
[18] *Westminster City Council v Jarvis & Sons Ltd* [1970] 1 All ER 942.

which the contractor becomes entitled to the remaining part of the retention money. It may be noted that, if no schedule of defects is issued, the employer retains the right to claim damages for breach of contract.[19] However, as will be seen below, the amount of those damages may be affected.

In the ICE form, the situation is similar, although the terminology is different. Here the issue of the 'Certificate of Substantial Completion' brings into operation the 'Defects Correction Period'. There is a similar requirement for the engineer to issue to the contractor a schedule of defects within 14 days of the end of the period. If the contractor does not make good the defects, after being served with a schedule of defects, then the employer can give notice that the work must be done. If the contractor does not comply with that notice immediately, then the employer can employ others to do the work, and recover the expense from the contractor.

Under the ICE conditions, the contractor's obligation goes beyond repairing and completing those items that are a direct consequence of its default. However, work done on other items (i.e. those which are not due to workmanship or materials being in breach of contract) entitles the contractor to extra payment.

It is sometimes said that, during a rectification period, the contractor has the *right* as well as the *obligation* to put right any defects that appear. What this means is that an employer who discovers defects should operate the contractual defects liability procedure, rather than appoint another contractor to carry out the repairs. In *William Tomkinson v The Parochial Church Council of St Michael and Others*,[20] an employer refused to allow the original contractor access to the site to remedy defects, but instead sued the contractor for the cost of having these rectified by another contractor. It was held that the employer's decision amounted to an unreasonable failure to mitigate the loss suffered, and the damages were reduced by the amount by which the employer's costs exceeded what it would have cost the original contractor to carry out the work. The Court of Appeal has since approved this decision.[21]

It is also worth noting that rectification clauses do not act as exclusion clauses. If a defect is not included on a schedule of defects, and is not noticed by the contractor or contract administrator before the end of the period, it is still a breach of contract for which the contractor is liable Since the period has expired, the contractor has no right to return to the site to repair the defect, but is liable to the employer for damages.

14.5   ADJUSTMENTS OF TIME

Most building contracts contain express provisions under which the period allowed for the contractor to undertake and complete the works can be adjusted. These provisions cater mostly for delays that are neither the fault nor the responsibility of the contractor. Such provisions obviously benefit the contractor, who will not be liable to pay damages for delay during the period for which time is validly extended. In addition, and less obviously, the power to extend time is also for the employer's benefit, for the following reason.

---

[19] *Pearce & High Ltd v Baxter* [1999] BLR 101.
[20] (1990) 6 Const LJ 319.
[21] *Pearce & High Ltd v Baxter* [1999] BLR 101.

At common law, the contractor's obligation to complete the works by the specified date is removed if the employer delays the contractor in the execution of the works. Thus if the contract administrator issues an instruction which increases the amount of work to be done, or is late in giving the contractor necessary instructions, the specified completion date no longer applies. In this situation, time is said to be 'at large', and the contractor's obligation is merely to complete the works within a reasonable time. In order to fix what is 'reasonable', all the circumstances of the particular project must be taken into account, but in many cases it will simply mean that the amount of delay for which the employer is responsible will be added to the old completion date.

The importance of losing the fixed date is that a contractor who has caused part of the delay is still liable to pay general damages for delay, but is not liable for *liquidated damages*. Even where the delay caused by the employer is a very small part of the overall delay, the employer cannot simply discount this and claim liquidated damages for the remainder.[22] The liquidated damages provision fails altogether, and the employer can claim only for those losses resulting from the delay which can actually be proved. Such proof is sometimes difficult, and in any event the amount recovered may be less than what was fixed as liquidated damages. An employer who has caused delay therefore has a very strong interest in being able to extend time for this, so as to retain the entitlement to liquidated damages from the revised completion date.

14.5.1 Grounds for adjustments of time

A fundamental point is that the time for completion can only be adjusted where the contract permits, and strictly in accordance with the contract provisions. If delay is caused by some event which the contract does not cover, then the contractor cannot claim more time, nor can the employer insist on giving any (in order to keep alive a claim for liquidated damages). For example, it was held that a power to extend time for delays caused by the ordering of extra work only applied where the extra work was *properly* ordered, and not where the architect gave the relevant instructions orally, instead of in writing, as the contract required.[23]

It is thus important that the likely causes of delay are covered by an extension of time clause, but the courts have made things more difficult by ruling that general words such as *other unavoidable circumstances* do not cover delay due to the fault of the employer.[24] Oddly, although this ruling is well known, many contracts continue to use general phrases of this kind. ICE 7 clause 44(1), for instance, speaks of *other special circumstances of any kind whatsoever*, while JCT MW 05 clause 2.7 uses *reasons beyond the control of the contractor*.

A further point that needs to be made is that there is often argument as to what is the actual cause of a particular delay. Suppose, for example, that the employer on a contract for refurbishment work is one week late in giving the contractor access to a particular part of the building, but that the contractor could not have started work anyway, as the necessary materials had not been ordered. Which of

---

[22] *Dodd v Churton* [1897] 1 QB 562.
[23] *Murdoch v Luckie* (1897) 15 NZLR 296.
[24] *Peak Construction (Liverpool) Ltd v McKinney Foundations Ltd* (1970) 1 BLR 111.

them has 'caused' the delay? And what if, during the week in question, the entire area was several feet deep in snow and work would have been impossible? The answers to such questions are not at all clear, since the only guidance from the courts is that it is *not* correct to identify one cause as 'dominant' and hold that cause entirely responsible.[25] It may be that since, in theory at least, it is always the contractor who claims an extension of time, the onus is on the contractor to prove that delay has been caused by an event for which an extension can be granted. Thus, if another sufficient 'cause' also exists, the contractor will fail.

14.5.2 'Relevant events' under JCT SBC 05

As a good example of the kind of grounds on which time for completion can be extended in building contracts, we can take the list of 'relevant events' which is found in clause 2.29 of JCT SBC 05. These are as follows:

Variations

The contractor is entitled to an extension of time in the event of a variation, whether the variation is required by an instruction of the contract administrator or in the event that any other matter has to be treated as a variation.

Instructions of the contract administrator

JCT SBC 05 gives the contract administrator wide powers to issue instructions to the contractor during the progress of the works. In relation to some, though not all, of these, clause 2.29.2 provides that any delay resulting from the contractor's compliance will carry an entitlement to an extension of time. Instructions on the following matters fall into this category:

- Discrepancy in or divergence between contract documents (clause 2.15).
- The expenditure of provisional sums (clause 3.16).
- The postponement of any work to be executed under the contract (clause 3.15).
- Any action to be taken concerning fossils, antiquities and other objects of interest or value (clause 3.23).
- Valuation of a variation in the event that the contractor does not provide a quotation (clause 5.3.2).
- The opening up and testing of work which has been covered up (3.17 or 3.18.4)

It is important to appreciate that, under JCT SBC 05 clause 3.10, the contractor must 'forthwith comply' with *all* instructions which the contract empowers the contract administrator to issue, subject to only a few exceptions stated in that clause. If the contractor fails to do so, the employer may employ others to do part of the works (clause 3.11). In these circumstances the contractor would not be entitled to claim an extension of time.

---

[25] *H Fairweather & Co Ltd v Wandsworth LBC* (1987) 39 BLR 106.

Deferment of Date of Possession under clause 2.5 (optional)

As we noted under Section 14.1.3, failure by the employer to give the contractor possession of the site on the agreed date is a serious breach of contract. However, if the optional clause 2.5 is used, then the employer may defer the giving of possession for whatever period (not exceeding six weeks) is specified in the appendix. If possession is deferred in this way, then naturally the contractor is entitled to an extension of time.

Execution of work under an Approximate Quantity

The contract JCT SBC 05 (i.e. the version 'with quantities') allows the parties to specify parts of the works to be let under approximate quantities. This has to be done in the bills (see footnote 3 to the second recital of JCT SBC 05). In the circumstance that the actual quantity of the works, for which approximate quantities are agreed, is not a reasonable forecast of the quantity of work required, the contractor will need longer than planned. Hence, the form contains an entitlement for an extension of time.

Suspension of work by the contractor

The contract period can be extended if a contractor, being not fully paid, suspends the work. This suspension has to take place in accordance with the requirements of clause 4.14 of JCT SBC 05 (in particular: seven days notice period). A suspension of the works which is validly based on this contractual right also entitles the contractor to claim for compensation under clause 4.23 of JCT SBC 05 (claim for loss and/or expense). It should be noted that the contractual right to suspend the works is also a statutory right of the contractor under Section 112 of the Housing Grants, Construction and Regeneration Act 1996.

Any impediment, prevention or default by employer or employer's agents

Clause 2.29.6 of JCT SBC 05 identifies the risk of anyone acting for the employer causing delay to the contractor. There is an exception: where the contractor or someone acting for the contractor contributes to such a delay. But this clause operates to the employer's advantage, in that although time would have to be extended for any delay caused solely by the employer's agents, any remaining delay is the contractor's responsibility, thus granting an extension of the contract period for this reason enables the completion date to be moved so that delays on the part of the contractor will attract liquidated damages. This clause gives full effect to this fundamental principle.

Delay in the supply of information would also, presumably, fall under this heading. It is expected that the contract administrator will keep the contractor supplied with the information needed to carry out the work. Failure to supply drawings or details (clause 2.12.1) or information on levels (clause 2.10) at the right time will be a breach of contract for which the contractor will be entitled to recover damages. Moreover, the contractor's claim for damages does not depend upon previous specific requests for the information; it is the contract administrator's responsibility to know when information is going to be required and to ensure that it is ready. Clause 2.29.6 means that, as well as leading to an action for damages, delay in supplying necessary information may in certain circumstances be a ground on which the contractor is entitled to an extension of time.

Clause 2.7 allows the employer to undertake parts of the works, either directly or through another contractor, where this is stated in the contract bills. Such work will not form part of the contract, but because of this clause a delay in that part of the work will give the contractor grounds for extension of time.

Another potential impediment that earlier versions separately identified is failure by the employer to give access over employer's land. This would only apply when land or buildings adjacent to the site are in the employer's possession and control, and the employer has undertaken to give the contractor access to the site through that land. It does not refer to the employer's obligation to give the contractor possession of the site itself; nor would it apply to any failure by the employer to obtain wayleaves over other people's property.

Statutory undertakings

This ground for extension of time refers only to the carrying out of work by a local authority or statutory undertaking *in pursuance of its statutory powers*. It does not cover the situation where a statutory undertaking is used as a sub-contractor. According to SMM 7, any work to be undertaken by a statutory undertaking should be included in the bills as a provisional sum.

Exceptionally adverse weather conditions

Exceptionally adverse weather is not restricted to bad weather. Since excessively hot and dry weather can also cause problems with progress of the works, this phrase was introduced in JCT 98 to include all exceptional weather conditions, rather than merely cold and wet weather.

Interestingly, while ICE 7 clause 44(1) uses the same phrase as JCT SBC 05, GC/Works/1 does not recognize any kind of weather conditions as grounds for extensions of time. This means that, under GC/Works/1, the entire 'weather risk' is borne by the contractor (who should price the tender accordingly). Under the other forms of contract, the risk is shared because while the contractor can have an extension of time, thus avoiding liquidated damages, there is no corresponding item that would also entitle the contractor to claim for loss and/or expense (see Chapter 15).

In a temperate and varied climate like that of the UK it can be difficult to establish exactly what constitutes 'exceptionally adverse weather'. An examination of local weather records should establish what is normal for the locality. This will provide a definition of what is 'usually adverse' weather and thus help to identify 'exceptionally adverse' weather. However, since the exact nature of the weather is crucial in deciding the validity of a claim, Meteorological Office records may be inadequate. It is wise to keep detailed site weather records because weather can vary greatly over short distances, especially where there are hills nearby.

The construction of the Humber Bridge provides an interesting example of this problem. It involved a complex in-situ cable spinning process to construct the main suspension cables between the towers of the bridge. Since work could only take place on these cables when the wind speed was low, weather records at the nearby Meteorological Office weather station were examined in detail before work commenced. Unfortunately, the site was a few miles from the weather station, and the effect of the hills channelling the wind up the estuary resulted in higher wind speeds than expected and led to considerable delays. However, these higher wind speeds were normal at the location of the bridge and therefore no extension of time was granted.

In considering weather problems, it must always be borne in mind that there can be no extension of time unless the whole project is actually or potentially delayed. If, for example, exceptionally adverse weather occurs at a time when most of the work is indoors, this is not a ground for an extension. It is also important to note that it is the *actual* effect of weather on the work which is relevant – thus what matters is the weather at the time when a particular part of the work was in fact carried out, not necessarily at the time when it was programmed to be carried out.

Loss or damage occasioned by the Specified Perils

The 'Specified Perils' are listed in clause 6.8 of JCT SBC 05 and include such matters as fire, lightning, explosion, storm, flood, etc. Clause 2.29 allows for an extension of the contract period if loss or damage caused by these things delays the contractor.

These perils are the subject of the insurance provisions in clause 6. Even if the loss or damage is brought about by omission or default of the contractor (or those for whom the contractor is responsible), it appears that clause 2.29 still applies.[26] Perhaps because of this result (enabling contractors to benefit from their own default), the courts tend to interpret the specified perils rather strictly. In one case, for example, a sub-contractor dropped a purlin on to a high pressure water pipe.[27] The pipe broke and the resulting high-pressure jet of water caused severe damage to the claimant's goods. Although the 'bursting of pipes' is a Specified Peril, so that the risk of damage is firmly with the employer, the judge in this case considered that what had happened did not fit into this category. A burst was held to be a disruption from within the pipe, and the cause of damage here was in fact the negligent dropping of a purlin. Thus the sub-contractors were found liable for the damage done.

---

[26] *Surrey Heath Borough Council v Lovell Construction Ltd* (1988) 42 BLR 25; *Scottish Special Housing Association v Wimpey Construction (UK) Ltd* [1986] 2 All ER 957.
[27] *Computer & Systems Engineering plc v John Lelliott (Ilford) Ltd* (1990) 54 BLR 1, CA.

Civil commotion, use or threat of terrorism

The risks associated with civil commotion or with terrorist acts are difficult to deal with. Given the potential for them to occur anywhere, with a low probability but high magnitude, they are very difficult for a contractor to assess. It may be possible to secure insurance cover for acts of terrorism, but, as pointed out in a footnote to clause 6.8, there can sometime be difficulties in getting terrorism covered in an all-risks policy. An employer may feel that inestimable risks of this nature are best transferred to the contractor. However, as with all inestimable risks, the contractor would not be able to assess how much contingency to include in a price. In this way, pricing would cease to be a commercial process and would become a game of chance instead. For extreme risks of this nature, it makes more economic sense to keep the risk with the employer, so that the contractor can focus on pricing the work, instead of trying to predict inestimable risks. Thus, the risk is included as a potential reason for extending the contract period.

Strike, lock-out etc.

This applies to delays caused by industrial action taken by any group of people who are employed on the works, are manufacturing items for inclusion in the works or are involved in transportation of goods to the works. It seems, however, that the wording of this clause would not cover a strike or industrial dispute affecting, for example, the supply of fuel to the transporters. Clearly, it makes no difference whether the strike is official or unofficial. Although a 'go-slow' or a 'work-to-rule' does not constitute a strike, it would presumably be covered by the phrase 'local combination of workmen'.

The exercise by the government of any power which directly affects the works

An example of the kind of exercise of government power which would be relevant here is the three-day week which resulted from the miner's strike in 1974. In an effort to break the miners' strike, the government decreed that people could only have electricity for three days per week, with a view to preserving coal stocks for as long as possible. In order for this clause to be used as a reason for claiming an extension of time, the contractor must have had no knowledge of the matter at the date of tender.

Force majeure

The phrase 'force majeure' derives from French law where it is used to describe situations where an unforeseeable event makes execution of the contract wholly impossible and is of such importance that it cannot be overcome. This is a narrower and stricter definition than the English doctrine of frustration, for which see Chapter 23. Its inclusion as a ground for extending the contract period is odd for two reasons. First, many of the events that might conceivably come within the definition are already specifically covered under JCT SBC 05 by other grounds. Second, any event which made execution of the contract wholly impossible might be a worthy candidate for terminating the contract, but it is difficult to see how extending the contract duration could be of any assistance unless the event merely delayed completion, in which case, execution of the contract is not *wholly* impossible but merely temporarily impossible.

Delay on the part of nominated sub-contractors/suppliers

There is no such thing as nomination of sub-contractors in JCT SBC 05. Nevertheless, it is likely that older versions of the JCT contract and, indeed, other forms of contract will continue to make reference to nomination. For this reason, it is worthwhile dealing with this difficult area of contract law. Before we look at the way in which the courts have interpreted this kind of clause, it is worthwhile considering its practical implications. If the main contractor is entitled to an extension of time when delayed by a nominated sub-contractor, there is no liability on the part of the employer for liquidated damages. This in turn means that, although the contractor may claim damages from the nominated sub-contractor for loss caused by disturbance of the progress of the works, these damages will not include the amount of the liquidated damages, because the contractor will not have paid these. As a result, unless the employer has a direct claim against the nominated sub-contractor for the delay (for example under NSC/W, the 1998 JCT standard-form employer/nominated sub-contractor warranty agreement), the employer will simply lose out and the nominated sub-contractor will evade responsibility!

The apparent strangeness and unfairness of this result may help to explain why the courts have interpreted JCT 98 clause 25.4.7 in what seems to be an artificially narrow way. It was held by the House of Lords in the case of *Westminster CC v Jarvis*[28] that *delay on the part of* is not the same as *delay caused by*. This phrase does not refer to any lack of diligence by the sub-contractor, but only to the sub-contractor's failure to achieve completion by the due date, which results in the main contractor being unable to achieve the overall completion date. Thus, in the *Jarvis* case, although the nominated sub-contractor's work was later found to be defective, so that the sub-contractor had to come back to the site to put it right, completion had apparently been achieved on the due date. It was accordingly held that there had been no *delay on the part of* the nominated sub-contractor. As a result, the main contractor was liable to the employer and the sub-contractor was in turn liable to the main contractor.

The contractor should closely monitor the progress of nominated sub-contractors to ensure that they remain on programme, and so that the contractor can warn the contract administrator in plenty of time of the likelihood of a nominated sub-contractor delaying the works. Similarly, supplies of goods and materials from nominated suppliers should be closely checked. Contractors who fail in this may lose their right to an extension of time, either because they have not taken *all practicable steps to avoid or reduce* the delay (JCT 98 clause 25.4.7) or, more

[28] *Westminster City Council v Jarvis & Sons Ltd* [1970] 1 All ER 942.

generally, because they have not constantly used their best endeavours to prevent delay (JCT 98 clause 25.3.4.1).

It is important to note that, if a nominated sub-contractor ceases work altogether (for example because of insolvency) this is not in itself grounds for the main contractor to claim an extension of time. The position is that the contract administrator has a duty to renominate another sub-contractor within a reasonable time, and that it is only delay in doing so that will entitle the main contractor to extra time.[29]

14.5.3 Procedures for claiming an extension of time

As a general principle, an extension of time can only be validly granted if the procedures which the contract lays down are strictly followed. Although JCT SBC 05 includes provisions to shorten the duration of the contract in certain circumstances, the procedure that begins with clause 2.27 is specifically about delay, not acceleration. However, an earlier completion date may be given as a result of omitting work (clause 2.28.4).

General

A point occasionally overlooked is that extensions of time can only be granted in respect of events likely to delay *completion*. Thus, if a contractor's schedule is geared to finishing before the completion date, or if the contractor is so far ahead of the contract programme when delay occurs that the 'slack' can be taken up, it appears that no extension should be granted at that stage.[30] However, where the contract permits the contract administrator to review the whole issue of extension of time at a later date, such delay can then be taken into account. This is the position under both JCT SBC 05 (clause 2.28.5) and ICE 7 (clause 44(4)).

An example of the importance of procedures is given by one case,[31] where the contract did not make clear that the architect could extend time retrospectively (i.e. after completion had been achieved). A court held that this was not possible, on the ground that the purpose of extending time was to give the contractor a date to work towards. However, most modern construction contracts (and certainly both JCT SBC 05 and ICE 7) avoid this problem, by making it clear that an extension of time can be granted retrospectively. In any event, it may be that the courts today are rather less strict in their interpretation of contractual procedures. For example, although JCT SBC 05 requires the contract administrator to grant an extension within 12 weeks of receiving the contractor's claim, it has been held by the Court of Appeal that an extension granted outside that period is still valid.[32]

An important question, which until recently remained unanswered by the courts, is whether it is possible to grant an extension of time at a point when the

---

[29] *North West Metropolitan Regional Hospital Board v TA Bickerton & Son Ltd* [1970] 1 All ER 1039; *Percy Bilton Ltd v GLC* [1982] 2 All ER 623, HL.
[30] *Hounslow LBC v Twickenham Garden Developments Ltd* [1971] Ch 253.
[31] *Miller v LCC* (1934) 50 TLR 479.
[32] *Temloc Ltd v Errill Properties Ltd* (1987) 39 BLR 30.

contractor is already in culpable delay and is thus liable to pay or allow liquidated damages. Furthermore, if this *is* possible, should the resulting extension be 'net' (so that the completion date is moved by the amount of the delay caused), or should it be 'gross' (so that the new completion date is set in the future and the contractor's accrued liability for liquidated damages is wiped out)? The former interpretation is clearly desirable, since it would seem unjust to excuse the contractor for past delays on the basis of something that happens much later. On the other hand, there is something illogical in the idea that a 'completion date' of say, April 10 can be moved to April 17 to take account of things which do not occur until the following September.

The problem is a difficult one but, for the moment at least, it has been solved by the case of *Balfour Beatty Building Ltd v Chestermount Properties Ltd.*[33] It was there decided that, under the form of wording used in JCT contracts, a contractor who was already in delay could be granted a 'net' extension of time for further delays for which the employer was responsible. However, the judge also ruled that no extension should be granted in such circumstances in respect of 'neutral' delays (such as those caused by exceptionally adverse weather conditions).

Procedures under JCT SBC 05

JCT SBC 05 clauses 2.26–2.28 lay down a specific procedure that must be followed in order to ask for the time for completion to be extended. This procedure can be utilized more than once during the contract; indeed it should be used on every occasion that delay occurs. Furthermore, clause 2.28.5 provides that the contract administrator shall within twelve weeks of practical completion make a final decision on extensions of time, whether or not notice has been given by the contractor, and may at that stage review any earlier decisions.

A contractor who wishes to claim an extension of time must at all times have used 'his best endeavours to prevent delay in the progress of the Works, how ever caused'. Subject to this, the procedure is as follows:

1. Whenever it becomes reasonably apparent that the progress of the works is being or is likely to be delayed for any reason (not just by an event listed in clause 2.29), the contractor is required to give written notice to the contract administrator. This notice should specify the cause or causes of the delay, and identify any 'relevant event'.
2. In respect of any 'relevant event' mentioned in the contractor's notice, the contractor must as soon as possible write to the contract administrator, giving particulars of the likely effects and an estimate of the expected delay. The contractor must from that time keep the contract administrator up to date with any developments.
3. On receipt of this information, the contract administrator is required to decide whether completion of the works is likely to be delayed beyond the completion date by one or more 'relevant events'. If this is found to be so, then a new completion date must be fixed, giving such extension as is fair and reasonable, attributing extensions to each relevant event, and the

---

[33] (1993) 62 BLR 1.

contractor must be notified in writing. If the contract administrator decides against the contractor, this too must be notified in writing. In either case, the decision is to be notified as soon as practicable, but in any case within twelve weeks, or before the completion date if this is less than twelve weeks away.

4. In fixing a new completion date on second or subsequent occasions, the contract administrator is entitled to bring the date forward to take account of any work ordered omitted. However, it is *never* possible to fix a completion date earlier than the original Date for Completion stated in the appendix, no matter how much work is omitted.

15 Payment

The provisions relating to payment concern the way the contractor is paid by the employer. The consideration given by the employer to the contractor is not always a fixed amount of money. However, there are only certain circumstances in which the contract sum can be altered. The most important of these is where there are variations, i.e. changes to the specification of work, but there are others. This Chapter addresses these issues, and also the way that mechanisms such as retention and set-off affect payment to the contractor.

15.1 EMPLOYER'S OBLIGATION TO PAY

The primary obligation upon the employer is to pay the contractor the sum of money which forms the consideration for the contract. Money must be paid promptly and fully unless there are specific reasons for withholding it.

15.1.1 Contract price

Different contracts deal with the contract price in different ways. Under JCT contracts with bills of quantities, the bid by the contractor is based upon the work described and quantified in the contract bills. If any quantities are altered because of changes (for example, to the client's requirements), then the contract sum will be altered accordingly. Otherwise, the contractor is paid the amount of the tender. This is by contrast with ICE 7 Measurement Version, which is often known as a 'remeasurement' contract. This means that, even though the bills priced by the contractor may contain quantities, these are provided purely for the purpose of the tendering process. The price to be paid for the works will not be fully established until the works are completed and the quantities actually carried out are ascertained. In this situation, the contractor takes the risks associated with the price per unit of work, whereas the employer takes the risks associated with the quantity of work differing from what was envisaged. A JCT contract of this type is JCT SBC 05 in its version 'with approximate quantities (also see Section 3.2.3).

These two types of contract are basic types, and naturally there are variations on these basic themes. The difference between them is of fundamental importance to any employer when considering which type of contract is appropriate for a particular project.

There are numerous provisions within the contractual conditions to ensure that the contract is for the whole of the work specified. The simple fact that the money is actually paid in stages does not alter this. This is a theme followed through in the contracts by the fact that only the final certificate is ever conclusive as to workmanship or quality of materials.

15.1.2 Time of payment

It has for many years been common practice in the construction industry for payment of the contract sum to be made by instalments, except on the smallest contracts and sub-contracts. One of the main purposes of this is to reduce the need for the contractor to fund the development of the project. This is because the total value of each contract forms a large proportion of a contractor's annual turnover. Payment by instalments should eliminate the need for the contractor to borrow money pending final payment.

Payment in construction contracts is now governed by Part II of the Housing Grants, Construction and Regeneration Act 1996, which requires every construction contract to which the Act applies (see Section 8.1) to contain certain provisions. To the extent that the terms of a contract fail to satisfy this requirement, the contract will instead incorporate implied terms which are set out in a statutory instrument called the Scheme for Construction Contracts.[1] It should be noted that, where a contract fails to make provision on a particular item, it is only the provisions of the Scheme relevant to that item that are implied into the contract, rather than the whole Scheme.[2]

Section 109 of the Act makes payment by instalments obligatory in every construction contract, except where either the contract specifies or the parties agree that the work is to be of less than 45 days duration. The parties are free to agree both the amount and the timing of interim payments. If they fail to do so, the Scheme for Construction Contracts provides that payment will be on the basis of work done and materials supplied during each successive period of 28 days.

Section 110 of the Act requires every construction contract to specify both the date on which each payment under the contract becomes due and a second final date by which each payment must be made. The contract must also make provision for the paying party to give written notice to the payee, specifying the amount of each payment and the basis on which it has been calculated. Section 111 lays down further rules to govern attempts by paying parties to withhold payment, for example on the ground of set-off or abatement. These rules are discussed in Section 21.4.4.

Unless the contract provides otherwise, the position (by virtue of the Scheme for Construction Contracts) will be as follows:

- payment of any instalment of the contract price (other than the final one) becomes due 7 days after the end of the relevant period or 7 days after a written demand from the payee, whichever is the later;
- payment of the final instalment of the contract price (or the whole of the contract price, in the case of a short-term contract without instalments)

---

[1] SI 1998 No. 649 Scheme for Construction Contracts (England and Wales) Regulations 1998 (SI 1998 No 649).
[2] *Hills Electrical & Mechanical plc v Dawn Construction Ltd* 2004 SLT 477.

becomes due 30 days after the end of the relevant period or 7 days after a written demand from the payee, whichever is the later;

- in each case, written notice specifying the amount of the payment and the basis on which it has been calculated must be given within 5 days of the payment becoming due; and
- the final date for making any payment is 17 days from the date on which the payment became due.

The amount of money due in each instalment is recorded by the contract administrator in an 'Interim Certificate'. The issue of such a certificate by the contract administrator imposes upon the employer a strict obligation to make payment (subject only to certain exceptions discussed in Section 22.4).

JCT SBC 05 (clause 13.1) obliges the employer to pay within 14 days the amount of money shown on an interim certificate. ICE 7, by condition 60(2), requires the contractor to submit a monthly statement to the engineer. Within 25 days of the submission of this statement, the engineer shall certify, and within 28 days the employer shall pay, the relevant amount of money. In the event of the main contractor defaulting in making payments to sub-contractors, there are frequently provisions in the main contract enabling the employer to pay them directly. This is dealt with more fully in Chapter 19.

Many construction contracts provide for the payment of interest upon any payments made late. Where they do not, there may be a right to claim interest at a rate fixed from time to time by the government under the Late Payment of Commercial Debts (Interest) Act 1998.

15.1.3 Effect of certificates

Interim Certificates are simply a mechanism for confirming that an instalment of the contract sum is due to the contractor. JCT SBC 05 and JCT IC 05 are based on a contract administrator's valuation of work done each month. ICE 7 is based on a contractor's claim for payment, which is evaluated by the engineer. GC/Works/1 is interesting in this respect as it is based on a graph of expected progress.

Whichever method is used to calculate the amount of money due, an interim certificate is not conclusive about anything. It says nothing about quality of materials or workmanship, nor does it indicate satisfaction with the work done to date. Anything included in such a certificate may yet be the subject of a later certificate. It is only the final certificate that is ever conclusive. As a result, the *only* obligation arising from an interim certificate is an obligation on the employer to make a payment within the stated time. Failure to do so is a serious breach of contract. (Certificates are considered in more detail in Chapter 18.)

15.2 THE CONTRACT SUM

The price for the work is typically referred to as the contract sum, contract price or tender total. The importance of the price is emphasized by the extent to which contracts have specific definitions of what is included and how it can be changed.

15.2.1 Definitions in the contracts

In JCT contracts, the **contract sum** is defined in the Articles of Agreement as the amount of money the employer will pay to the contractor. It is exclusive of VAT. Article 2 states that the contract sum is this amount of money or such other sum as shall becomes payable under the contract, at the times and in the manner specified in the contract. It is clear therefore that, under JCT terms, the sum is specified at the outset, although it may be altered as work proceeds.

In ICE 7, by contrast, the 'contract price' is defined in clause 1(1) as the sum to be ascertained and paid in accordance with the provisions. This concept of contract sum is very different from that in JCT contracts. Under the ICE form, the sum can only be ascertained when the contract is concluded. Although the employer covenants to pay the contract price, it is not a specified amount of money at the time the contract is signed.

Because of the intrinsic differences between JCT and ICE contracts in this respect, we shall concentrate in this chapter on JCT contracts, in order to explore the full detail of the payment provisions. The ICE conditions can then be understood by comparison.

In JCT SBC 05 and IC 05, the contract sum is fixed by clause 4.2, which states that it cannot be altered or adjusted in any way other than by the conditions. This also states that any errors in the computation of the contract sum, whether arithmetic or not, are deemed to be accepted by both parties. The effect of this clause is that the contract sum can only be changed when the conditions allow adjustments. Clause 4.4 of SBC 05 states that any changes to the contract sum, in the sense of amounts to be deducted or added, must take effect immediately. This means that, as soon as such a change has been computed or ascertained, it must be added to or deducted from the next interim certificate.

15.2.2 Permissible changes

The contract sum may be changed for a variety of reasons, which can be divided into three groups as follows:

1. Reimbursement of the contractor for certain expense caused by the contract administrator, employer, or certain events outside the control of the contractor. These matters are covered by clauses such as JCT SBC 05 clauses 4.23–4.26 under which the contractor can claim for 'loss and/or expense'.
2. Payment for extra work brought about by a contract administrator's instruction.
3. Reimbursement of extra expense brought about by market fluctuations affecting the contractor's inputs.

The loss/expense provisions will be covered in Chapter 16. This chapter concentrates on the second and third categories.

15.3   VARIATIONS

In order to change the specification of the work, a contract would, in principle, have to be re-negotiated. To avoid this, most contracts include a clause enabling the employer's design team to vary the specification. Such provisions are usually called **variations** clauses.

15.3.1 The need for variations

There are three ways in which a variation might occur:

1. Clients may change their minds about what they asked for before the work is complete.
2. Designers may not have finished all of the design and specification work before the contract was let.
3. Changes in legislation and other external factors may force changes upon the project team.

Although these three origins are very different, construction contracts tend to ignore these differences and deal with all variations in the same way. Clearly, buildings are so complex as to require changes to be made before they are completed. Some clients need to leave the specification of specialist equipment to the last possible minute before choosing it. Client-instructed variations are the prerogative of the client. The second group of variations arises because it is rare for design to be completely detailed at the time of tender; changes may have to be made simply to make the building work. However, there are cases where variations have to be instructed simply because designers have failed to complete their task and in their haste to get a project out to tender they leave portions of the work unspecified. The presence of a variation clause enables dilatory design teams to delay finalization of the design until late in the project. Such behaviour jeopardizes the success of general contracting because of the assumptions on which it is based (see Chapter 3). It is bad practice and should be avoided as far as possible. The third source of variations, external factors, is typically beyond the control of any of the parties. Since variations are so diverse, it seems odd that no distinction is made between them.

If there were no provision in the contract for varying the work to be done, then any attempt by the employer to vary it would require the agreement of the contractor. The lack of a provision for variations would effectively enable the contractor to negotiate a new price for the whole contract every time the employer tried to make a change. This apparent difficulty is not as rigorous as it sounds. The absence of a variations clause undoubtedly makes it difficult to vary the terms of the contract but it is at least possible that the courts would imply a term allowing *minor* variations to be made. In any event, it would of course be most unusual for a contractor to attempt to refuse to carry out small changes and even less likely that the contractor would go to court over an attempt to impose them.

By inserting a clause allowing for changes to be made to the works as they are being built, the employer, through the contract administrator, can alter the works as and when necessary. The purpose of the variations clauses is to allow such changes to be made, and also to permit any consequential changes to be made to the contract sum.

15.3.2 Definition of variations

The following discussion concentrates on the principles contained in JCT SBC 05 and IC 05, which have very similar provisions for variations. The only differences between JCT SBC 05 and JCT IC 05 relate to the way in which the pricing documents are referred to, the fact that IC 05 makes no reference to contractors' design portion work, and the exclusion from IC 05 of 'Schedule 2 quotations', a feature of SBC 05 that sets out a procedure for the contractor to provide a quotation for the time an price impact of proposed changes.

The definitions of 'variation' encompasses changes to the design, to the material specification, removal of work properly executed and changes relating to access to the site and working conditions. Thus, the definition is a wide one, applying to both the content of the work and the method of doing it. Many standard forms other than JCT make reference to 'any' work and may often state expressly that no variation instruction can vitiate (invalidate) the contract. However, bearing in mind the purpose of recitals in a contract, and the fact that unilateral changes that go to the root of a contract are not usually permissible in contract law, this is clearly not the true position (see Section 15.3.3). We may now look in more detail at the JCT SBC 05 provisions.

Alteration of the work

The basic definition includes any *alteration or modification of the design, quality or quantity of the Works*. The clause clarifies this further by stating that such alteration or modification includes additions, omissions or substitutions.

This part of the definition of a variation specifically includes the alteration of the kind or standard of any of the materials or goods to be used in the works. Such changes necessarily involve the contractor's head office in expenditure, such as the additional time needed to cancel orders, find alternative suppliers and re-order materials and goods. Detailed records may be needed to prove this type of expenditure, or alternatively it may be calculated on a pro-rata basis.

The third aspect of this part of the definition is the removal from the site of work or materials, which are there for the proper purposes of the contract. This is by contrast with clauses 3.18–3.20 (IC 05 clauses 3.15–3.16), which deals with the removal of things *not* in accordance with the conditions.

Alteration of working methods

In addition to allowing the works themselves to be altered, clause 5.1.2 allows for variations in the means of achieving those works. The means available to the contractor may be seriously affected by access to the site, working space, hours of work or the sequence in which work is to be carried out. Unlike the permitted variations to the Works; however, these definite and specific matters of working *method* are fairly limited in the way in which they can be varied. It must also be borne in mind that exercising further restrictions in these things could result in substantial money claims by the contractor under clause 4.23 (IC 05 clause 4.17).

One final aspect of the definition, that applies to earlier versions of the JCT contract, among others, is that a variation under JCT 98 may not include the nomination of a sub-contractor for measured work already priced by the contractor (JCT 98 clause 13.1.3). In accordance with the common law position,[3] the contract provides specifically that none of the contractor's work can be taken away and given to others while the contract is current.

Variations under ICE 7

The ICE definition of a variation appears in clause 51. It includes *additions, omissions, substitutions, alterations, changes in quality, form, character, kind, position, dimension, level or line, and changes in any specified sequence, method or timing of construction.* This list, which contains a number of synonyms, means in effect that the engineer may issue variations about almost anything. The clause also makes it clear that a variation order need not be issued merely for the purposes of altering quantities in the bills. This is because the bills are not intended as a full description of the work, but only as a mechanism for selecting the contractor.

15.3.3 Limits on the scope of variations

It is always possible for a contract to include a clause that fixes express limits on the amount of variations. In any event, it must be borne in mind that the existence of a variations clause does not entitle the employer to make large scale and significant changes to the nature of the works, as these are defined in the Recitals to the contract. In particular, variations that *go to the root of the contract* are not permissible. For example, if the Recitals state that 8 dwelling houses are to be built, then a variation altering this to 12 would possibly be construed as going to the root of the contract. However, if the Recitals state that the contract is for 1008 houses, then a variation changing this to 1012 would not go to the root of the contract, because it would be a minor change in quantity. If the quantity of work is not indicated in the Recital, then the question does not arise in the same way. What is probably more important is that if the contract is for the erection of a swimming pool, a variation which attempts to change it to a house would clearly be beyond the scope of the contract.

In *Blue Circle Industries plc v Holland Dredging Company (UK) Ltd*[4] the parties entered into a contract under which the defendants were to dredge a channel that served the claimants' docks in Lough Larne, Eire. The dredged material was to be deposited in areas of Lough Larne to be notified by the local authority. When the claimants instructed the defendants instead to use the dredged material so as to construct an artificial island, it was held that this could not be regarded as a

---

[3] *North West Metropolitan Regional Hospital Board v TA Bickerton & Son Ltd* [1970] 1 All ER 1039.
[4] (1987) 37 BLR 40.

variation. It was beyond the scope of the original contract altogether, and thus had to form a separate contract.

In *McAlpine Humberoak Ltd v McDermott International Inc*,[5] on the other hand, the claimants entered into a sub-contract for the construction of part of the weather deck of a North Sea drilling platform. The documents on which the claimants tendered included 22 engineers' drawings. However, when work began, a stream of design changes transformed the contract into one based on 161 drawings. The trial judge ruled that these changes were so significant as to amount to a new contract, but the Court of Appeal held that they could all be accommodated within the contractual variations clause.

As mentioned earlier, many contract forms contain words to the effect that *no variation ... shall vitiate this contract*. This is an unnecessary and somewhat misleading statement, which adds nothing to the general legal position. Permissible variations are not unlimited in scope, since the contracts tend to define them as changes to the Works (defined in the contract documentation). Extensive changes that alter the work beyond anything in the documents are excluded by the contractual definition, so an attempt to make such sweeping changes would be beyond the scope of the variations clause, despite the text asserting that it would not. Even without a variations clause, an attempt by the employer to vary the contractor's obligations in a minor way would not vitiate the contract; the contractor could simply refuse to carry it out.

15.3.4 Issuing variations

In the JCT forms of contract, the definition of variations is distinct from the power to issue them. A variation for this purpose is an instruction issued by the contract administrator. Hence, in order for such an instruction to be valid under the contract, it must be issued in accordance with JCT SBC 05 clause 3.10 (IFC 05 clause 3.8). This means, for example, that it must be in writing. It also means that the contractor has the right to object on reasonable grounds to a variation.

As well as issuing an instruction that may require a variation, the contract administrator may sanction a variation that has been made by the contractor. The contractor normally has no authority whatsoever to vary anything (except for the execution of emergency work by SBC 05 clause 2.18, or IC 05 clause 2.16). For example, the contractor cannot substitute higher quality work or materials than those specified. A contractor who *does* vary anything then can be instructed to remove it under SBC 05 clauses 3.18–3.20 (IC 05 clauses 3.15–3.16).[6] This power for the contract administrator to sanction an unauthorized variation by the contractor is thus an important one, since it allows the contract administrator easily to pick up and deal with minor items which are varied by the contractor, or which are directed by the Clerk of Works and not confirmed by the contract administrator. There are sometimes occasions when a contract administrator may be called upon to issue instructions to clarify what has to be done, even though this might involve extra work that turns out to be the responsibility of the contractor. In this situation, the fact that instruction have been issued need not constitute an

---

[5] (1992) 58 BLR 1.
[6] *Simplex Concrete Piles Ltd v St Pancras MBC* (1958) 14 BLR 80.

obligation on the employer to pay for a variation, provided that such instructions are clearly differentiated from those which follow the usual pattern.[7]

A variation instruction is not required simply to 'firm up' approximate quantities, even if approximate quantities are only used for part of the work. All approximate *or* varied quantities result in the contractor's payment being calculated on the actual quantity of work done whereas unvaried *and* firm quantities result in the contractor's payment being calculated on the quantities in the bills, regardless of the true quantity executed.

JCT SBC 05 and IC 05 both contain a clause under which the contract administrator issues instructions as to the expenditure of provisional sums (SBC 05 clause 3.15, IC 05 clause 3.13). It should be noted that this is a *duty* of the contract administrator, as well as a power. The definition and various uses of provisional and 'prime cost' sums are discussed in Section 9.1.2.

15.3.5 Valuation of variations

When a variation is validly issued, the extra work needs to be valued. There are three approaches to this. First, the contractor and employer might negotiate a satisfactory sum, on the basis that parties to a contract are free to renegotiate the terms at any time. JCT SBC 05, at Schedule 2, now provides a procedure to enable this to happen in a systematic way, but any contract can be renegotiated, bilaterally, at any point. Second, the contract administrator follows express terms laid down in the contract for deriving a value from the contract documentation. Third, the contractor follows express terms in the contract for deriving a value, subject to the approval of the contract administrator.

General principles

JCT contracts provide for certain work to be valued by the quantity surveyor in accordance with the 'Valuation Rules'. The work in question includes any work executed as a result of a variation, work sanctioned as if it were a variation and work done as a consequence of an instruction on the expenditure of a provisional sum.

In considering the valuation of variations under JCT contracts, three vital preliminary points must be made. First, the clauses make clear that, if a particular case does not fit into any of the specific rules provided, then the contractor is entitled to *a fair valuation*. This is very wide-ranging in scope, and is intended as a residual or 'catch-all' clause. Indeed, it may be noted that this provision is even wide enough to bring about a *reduction* in the contract price, if limitations on the contractor's working methods are removed under clauses 5.1.2.1 to 5.1.2.4.

The second preliminary point arises out of SBC 05 clause 5.9 (IC 05 clause 5.5). This states that, if compliance with an instruction results in a consequential change to the conditions of any other work, then all that other work shall also be valued as if it were a variation. In other words, if the conditions of work are altered

---

[7] *Howard de Walden Estates Ltd v Costain Management Design Ltd* (1991) 55 BLR 123.

because the contractor has to comply with an instruction, then the provisions of clause 5 will apply to all that work as well.

The third preliminary point concerns contractor's design portion work (JCT SBC 05 clause 5.8, no equivalent in IC 05). Valuations must include allowances for the design work. Rates and prices must be consistent with those of similar character set out in the pricing document, with due allowance for changed conditions. Consequential effects of changes to performance specified items must also be valued as if instructed as a variation.

Valuation rules

JCT effectively divides work to be valued into three categories: work that can be valued by measurement, omitted work, and work that cannot be measured. We shall look at these categories in turn. But first, attention should be given to three general points made in the valuation rules. These points relate to work that can be valued by measurement and to omitted work; they do *not* apply to work that cannot properly be valued by measurement. These points are:

- In relation to measurable work and omitted work, any measurement is to be carried out in accordance with the same principles as those used for the preparation of the contract bills. This cross-refers to the clause which refers to the Standard Method of Measurement (SMM), which in turn is specified in the definitions clause. The effect, therefore, is that all measurements taken to define the varied work must be taken in accordance with the relevant SMM.
- Allowance is to be made for any percentage or lump sum adjustments in contract bills. Such an allowance is necessary because it is common practice for contractors when tendering to allow for overheads and profit by a lump sum or percentage addition to the sum of the preliminaries, instead of allowing a proportion in each and every measured item. This provision allows the lump sum to be adjusted in line with the variation.
- An appropriate allowance is to be made in respect of certain preliminary items, the use of which would not be attributable in whole to the varied item, but which would be needed in a different way if items were varied. For example, changing the way in which brickwork is to be finished may involve keeping scaffolding on site for an extra period. The Standard Method of Measurement contains further examples of preliminary items that may be affected in this way.

We may now turn to the three categories of work to be valued which were identified above.

- *Measured work*: The basic rule applying to *additional or substituted work which can properly be valued by measurement* is that the rates and prices in the contract bills shall apply. However, this only applies to work of a similar character executed under similar conditions, which does not significantly change the quantity provided for in the bills. There is of course room for argument over exactly what is meant by such words as *similar* and *character*; nonetheless, most variations would be expected to fall within this principle.

If the varied work is of a similar character to work already in the contract bills, but is executed under different conditions and/or there is a significant change in the quantity, a different method of valuation must be used. Once again, the rates in the bills must be used as a basis, but here the valuation must include *a fair allowance* for the difference in conditions and/or quantity. The new rate, which is thus derived from a bill rate by making this fair allowance, is known as a **star rate**.

Examples of *different conditions* under this clause would include winter instead of summer working; night instead of day working; differences in levels (such as storey heights etc.) at which work is carried out; and the discovery of antiquities under SBC clauses 3.22–3.24. It is also worth making the point that changes in quantity alone must be *significant* in order to justify the adoption of star rates.

Work that does not fit into either of the previous two categories, because it is *not* of a similar character to work priced in the bills, is to be valued at *fair rates and prices*. It is not easy for a contractor to invoke this clause after the work has been done. The contractor should make it clear to the contract administrator before undertaking the work that it should be valued under this rule. If this is not done, the contract administrator can claim that the contractor's lack of protest shows that the work falls into one of the previous two categories, so that no further allowance need be made.

- *Omitted work*: It should be noted that this refers only to work that is being *intentionally* excluded under clause 5.1.1 from what was stated in the pricing document. Here the bill rates can and should be used for valuing the work omitted. Some contracts, such as the older IFC 98, do not expressly cover this point, but it would in any case be implied.[8]
- *Unmeasurable work*: In respect of *additional or substituted work which cannot properly be valued by measurement*, valuation may be on a *dayworks* basis. It is important to note that a decision as to whether or not this method should be adopted is for the quantity surveyor to make; it is not for the contractor to insist on valuation according to dayworks.

The clause lays down details as to how the dayworks items are to be calculated, both for the main contractor and for specialists. Certain *official* definitions are used to determine the prime cost, and to this is added the relevant profit percentages set out in the bills. There are strict procedures requiring the contractor to produce vouchers specifying the time spent each day on the work, the names of the workers employed on it, and details of the plant and materials used. All of these vouchers must be given to the contract administrator by the end of the week following that in which the work was executed. If the contractor does not follow these administrative procedures to the letter, the claim for valuation on a dayworks basis will probably fail.

---

[8] *Commissioner for Main Roads v Reed & Stuart Pty Ltd* (1974) 12 BLR 55.

15.4   FLUCTUATIONS

We may now look briefly at fluctuations, which are the third mechanism by which the contract sum may validly be adjusted. The purpose of a **fluctuations clause** is to provide a mechanism for reimbursing contractors for changes in input prices over which they have no control at all. Even if a contractor has caused the project to take longer than planned, the changes in market prices for supplies will be caught by a valid fluctuations clause, unless the clause expressly provides otherwise.[9] Both JCT SBC 05 and ICE 7 specifically provide that a contractor who is in delay shall *not* benefit from the fluctuations clause. The contract particulars must state which of the fluctuations clauses apply.

In JCT SBC 05, clause 4.21 simply brings into operation the fluctuations clauses. It does this by stating that fluctuations shall be dealt with according to whichever of three alternatives are identified in the contract particulars. These alternatives are clauses referred to as Options A, B and C. In the absence of any entry in the appendix, the basic minimum provision for fluctuations contained in Option A is stated to apply. The three fluctuation schemes are as follows:

- *Option A*: Contributions, levy and tax fluctuations. This clause applies to items which are affected by the government and are thus completely beyond both the control and the prediction of the contractor. The elements to which the clause applies are labour, materials and goods, electricity and fuels, but only to the extent that they are affected by tax etc. In some cases they are only covered as far as they have been listed in the contract documents. There are no methods for calculation set out in the clause, but obviously it would be necessary to take into account man-hours, quantities of materials and other directly ascertained costs. It is not intended that this clause should change the amount of contractor's profit.
- *Option B*: Labour and materials cost and tax fluctuations. This clause includes all the government-related items that are covered by Option A; it adds in labour and materials fluctuations. This covers the market costs of input, such as wage rates and prices of materials.
- *Option C*: Use of price adjustment formulae. This is a completely different type of calculation. It incorporates by reference a set of formula rules which define a technical financial calculation based on a wide variety of categories. The whole of the works is divided up into financial categories and a monthly published bulletin gives indices by which each sum should be multiplied. In order for this to work, it is necessary for the contract bills to reflect the categories used by the fluctuation categories. Since the index numbers reflect the market situation each month, they are deemed to include the matters covered by Options A and B, and so only one of the clauses has to apply. The purpose of this method is to reduce the amount of calculation that has to be undertaken by the project team.

These different types of fluctuations are referred to as **limited fluctuations** (Option A) and **full fluctuations** (Options B and C). In addition, it is of course possible to have no fluctuations provisions at all, by amending the contract. Such a

---

[9] *Peak Construction (Liverpool) Ltd v McKinney Foundations Ltd* (1970) 1 BLR 111.

procedure would however only be suitable for a small project under stable conditions.

ICE 7 approaches the question of fluctuations in a similar way to JCT. Again the fluctuations clauses are optional, but they are printed separately from the remainder of the contract and are based on published indices. However, to bring them into force requires a special condition added to the contract, rather than an entry in the appendix (equivalent to JCT's contract particulars).

15.5 RETENTION MONEY

It is common practice in the construction industry to withhold a small proportion of payments to a contractor until the work has been completed satisfactorily. These amounts are referred to as **retention** and are usually deducted from each interim certificate.

15.5.1 Nature and purpose of retention

The **retention fund** is intended to be available to the employer for the purposes of underpinning contractual performance, in particular rectifying, or inducing the contractor to rectify, any defects in the work appearing during the defects liability period (Hughes, Hillebrandt and Murdoch 1998). As we saw in Chapter 14, this period runs from the date of practical completion as certified by the contract administrator for whatever length of time is stated in the relevant project-specific data section (such as Appendix or Contract particulars). Six months is common.

Clauses 4.10.1 and 4.20 JCT SBC 05 illustrate the operation of a retention scheme. The first clause provides that, on the issue of every interim certificate, the employer is entitled to deduct the agreed retention percentage. This is deducted from the value of work that has not yet reached practical completion and from the value of any materials included in the certificate. Clause 4.20 specifies the rules as to its ascertainment.

It is stated that the retention percentage referred to will be whatever figure the parties have entered in the Contract Particulars. The default figure is 3% in SBC 05, but 5% in IC 05 and in JCT 98. Further, it is common to provide a *ceiling* beyond which no more will be retained, although JCT SBC 05 contains no such provision.

After the contract administrator certifies that practical completion has been achieved, but before the issue of a Certificate of Making Good, the deduction the employer is entitled to make is reduced by one-half. In effect, this means that the first interim certificate issued after practical completion will result in the release of one-half of the retention money currently held by the employer. The remainder will be released by the interim certificate required under clause 4.9.2, issued either at the end of the Rectification Period or upon the issue of the Certificate of Making Good, whichever is the later.

15.5.2 Status and treatment of the retention

Clause 4.18 of JCT SBC 05 contains rules regarding the treatment of the retention money. Contractors and sub-contractors are naturally anxious to ensure as far as possible that, if the employer becomes insolvent, their claim to retention money will take priority over the employer's general creditors. Clause 4.18.1 accordingly provides that *the employer's interest in the retention is fiduciary as trustee for the contractor (but without obligation to invest)*. The idea is to make the retention money a trust fund, though without imposing upon the employer all the investment and accounting duties of a trustee. If this is successfully achieved, a liquidator of the employer would be obliged to hand over the retention fund in full to the contractor.

It should, however, be noted that the words used in clause 4.18.1 are not in themselves sufficient to enable a priority claim to be made. This can only be done if the money in question has been set aside as a separate trust fund, usually in a separately identified bank account. Accordingly, JCT SBC 05 provides (in clause 4.18.3) that a contractor can require this to be done, albeit leaving the employer to keep whatever interest is earned. Interestingly, it appears that a contractor would be entitled to take such action even without a specific provision in the contract, on the basis that, if the retention money is a trust fund, it must be properly dealt with by the employer as trustee. This was laid down in *Rayack v Lampeter Meat Co*,[10] an unusual case in which the retention level was 50% and the Defects Liability Period lasted for five years.

The *Rayack* decision, upholding the contractor's rights, was followed in a later case where the predecessor to 4.18.3 (clause 30.5.4 of JCT 80) had been specifically deleted from the printed form of contract before it was executed by the parties.[11] It follows that an employer under JCT SBC 05 who does *not* want to deal in this way with retention money (usually because the employer intends to use that money as working capital), will have to delete clause 4.18.1 as well as clause 4.18.3.

The right of a contractor to secure retention money in the way described above is subject to two important qualifications. First, an injunction will not be granted once insolvency proceedings against the employer have already commenced.[12] This principle, which is based on the idea that the rights of all the creditors crystallize at that moment, was applied even where the bank which appointed receivers under a floating charge over the employer's assets had been given notice of the building contract terms at the time when the floating charge was granted.[13]

The second qualification is that a court will not grant an injunction against an employer who is making a *bona fide* claim against retention money (and the employer's right to set-off claims against the retention money is specifically preserved by JCT SBC 05 clause 4.13.2). Thus, for example, where the contract administrator has issued a certificate of delay and thus entitled the employer to liquidated damages in excess of the retention, the employer will not be compelled to pay such money into a separate account. This will be so, even if the contractor is

---

[10] *Rayack Construction Ltd v Lampeter Meat Co Ltd* (1979) 12 BLR 30.
[11] *Wates Construction (London) Ltd v Franthom Property Ltd* (1991) 53 BLR 23.
[12] *Re Jartay Developments Ltd* (1982) 22 BLR 134.
[13] *MacJordan Construction Ltd v Brookmount Erostin Ltd* (1991) 56 BLR 1.

seeking to challenge the certificate.[14] However, an employer can only adopt this course of action where the right of set-off is clear, for example where it is backed by a certificate. Where the employer's claims, although arguable, are speculative and unsubstantiated, an order to pay the money into a separate account *will* be made.[15]

[14] *Henry Boot Building Ltd v Croydon Hotel & Leisure Co Ltd* (1985) 36 BLR 41; *GPT Realizations Ltd v Panatown Ltd* (1992) 61 BLR 88.
[15] *Concorde Construction Co Ltd v Colgan Co Ltd* (1984) 29 BLR 120; *Finnegan (JF) Ltd v Community Housing Association Ltd* (1993) 65 BLR 103.

16 Contractors' claims for loss and expense

A contractor is of course entitled under the contract to be paid for work done, including where appropriate the ascertained value of any variations ordered. In addition, the contractor may be able to make other claims against the employer. In Chapter 14 we considered claims relating to an extension of the contractual *time* for completion. We now turn to claims that, if successful, will result in the employer having to pay *money* to the contractor.

16.1 CONTRACT CLAIMS AND DAMAGES

Most of the standard form construction contracts currently in use (with the exception of those designed for relatively small projects) contain detailed provisions under which the contractor can claim against the employer for any losses suffered if the work is disrupted due to certain specified causes. These provisions often bear some resemblance to those under which an extension of time may be claimed, but there are at least two important distinctions between the two issues. First, an extension of time will only be granted where the contract administrator believes that completion of the works is likely to be delayed, whereas financial compensation for disruption does not depend upon any such delay. Second, as we have already noted, clauses which deal with extensions of time for completion frequently apply to various 'neutral' events such as adverse weather, as well as to those causes of delay which are the employer's responsibility. By contrast, the vast majority of contractual provisions compelling an employer to pay financial compensation to the contractor relate only to disruption that is caused by the employer.

This feature of claims provisions means that, in many cases, an event that enables a claim to be made will also entitle the contractor to recover damages for breach of contract. In particular, it may amount to a breach of the employer's implied obligation of co-operation with the contractor, which was discussed in Chapter 12. If this is so, it is for the contractor to decide whether to sue for breach of contract at common law or to claim under the appropriate clause in the contract. The contractor's right to choose between these remedies can only be removed by clear words in the contract itself, and this would be most unusual; indeed, JCT contracts expressly state that the contractor's common law rights are preserved.

In deciding which is the better remedy to pursue, it is worth noting that the choice will not normally affect the *amount* of money the contractor is likely to receive. As we shall see, the courts have ruled that 'direct loss and/or expense' (the words used in many contracts to describe what can be claimed in such cases) is to be assessed on exactly the same basis as damages for breach of contract. However, there are certain respects in which the two remedies *are* different, and these differences must be borne in mind when a claim is under consideration.

The main distinctions between making a contractual claim and bringing a legal action for damages are as follows:

- Although there is a broad similarity between the grounds on which a contractual claim may be made, and the conduct of an employer that amounts to a breach of contract, the two things are not identical. For example, while the disruption caused by an architect's instruction to postpone any part of the work may give rise to a claim under JCT SBC 05, the giving of such an instruction would not normally amount to a breach of contract. Conversely, failure by the employer to provide continuous access to the site as promised is a clear breach of contract, but by no means all standard form contracts make this the subject of a claim.

- A claim under the contract will be settled by the contract administrator, sometimes with the assistance of the quantity surveyor, and will be paid through the normal contractual machinery. This means that the amount in question is added to the contract sum and will appear on the next interim certificate to be issued after the claim is settled. By contrast, *damages* can only be awarded by a court, arbitrator or adjudicator, whose judgment or award is then enforced in the normal way. The only exception to this is where the contract administrator is given express power to deal with claims for breach of contract; this is most unusual, although the ACA form of contract does give such power.

- The preceding point would tend to suggest that, where there is a choice, a contractor would be better off claiming under the contract. However, an important limitation on contractual claims is that any specific procedural provisions in the contract (such as the giving of written notice or the furnishing of sufficient supporting information) must be strictly complied with. If they are not, then the contractor is thrown back on common law rights and can only sue for damages. Some cynical observers have also pointed out that, since many of the events which justify a claim are based on some default by the contract administrator (such as late delivery of necessary information to the contractor), the latter may find it difficult to make a completely impartial decision on the contractor's claim.

16.2   GROUNDS FOR CONTRACTUAL CLAIMS

It cannot be too strongly emphasized that any *contractual* claim made must be based upon some specific provision of the contract in question. The mere fact that unexpected difficulties have been encountered, or that the work is proving far more expensive than was foreseen, does not entitle a contractor to be compensated by the employer. Naturally, there is considerable variation among construction contracts as to the permitted grounds of claim, although we can get a general idea from a brief consideration of some of the more important standard forms.

Before looking at specific grounds, however, an important general point must be made. Contractors' claims are often extremely complex affairs, in which it is often alleged that a large number of disrupting events have resulted in an equally large number of items of loss and/or expense, but where it is difficult or impossible to attribute every item of loss to an individual cause. In such circumstances it has been held that, so long as *all* the causes are qualifying events under the contract, and that the impossibility of separating them is not due to the contractor's own delay in bringing the claim, they may be presented as a 'global' or 'rolled up' claim.[1]

Some doubt was cast on the validity of 'global' claims by a decision of the Privy Council,[2] in which an action against an architect for negligence was struck out because the pleadings did not seek to demonstrate how the various acts of negligence had led to the various losses suffered. However, that case has now been explained merely as requiring a contractor to take all reasonable steps to itemize as much of the claim as is possible; provided that this is done, the unattributable remainder may be presented on a global basis.[3]

16.2.1 Loss and expense under JCT SBC 05 and IC 05

A claim for compensation under JCT SBC 05 clause 4.23 or IC 05 clause 4.17 requires the contractor to prove two things:

1. That the contractor has incurred, or is likely to incur, direct loss and/or expense which would not otherwise be reimbursed under the contract; and
2. that this loss arises from either:
   (a) deferred possession of the site, where this is permitted under clause 2.5; or
   (b) *...the regular progress of the Works or of any part of them has been or is likely to be materially affected by any of the Relevant Matters.*

The 'relevant matters' are as follows:

1. Variations, except when loss and/or expense relates to a confirmed acceptance of a Schedule 2 Quotation.
2. Instructions of the architect or contract administrator, issued in accordance with clause 3.15 (IC 05 clause 3.12), postponement of any work to be carried out under this contract. The generally accepted view is that this provision does not empower the contract administrator to postpone *all* the work, so as to enable the employer not to give possession of the site at the agreed time. It is concerned rather with instructions which alter the order in which the works are to be carried out, for instance to accommodate a sub-contractor. Such a procedure may of course cause the contractor loss through the need to reprogramme.

---

[1] *J Crosby & Sons Ltd v Portland UDC* (1967) 5 BLR 126; *Merton LBC v Stanley Hugh Leach Ltd* (1985) 32 BLR 51.
[2] *Wharf Properties Ltd v Eric Cumine Associates (No 2)* (1991) 52 BLR 1.
[3] *Mid-Glamorgan CC v J Devonald Williams & Partner* (1991) 8 Const LJ 61; *Imperial Chemical Industries plc v Bovis Construction Ltd* (1992) 32 Con LR 90; *Bernhard's Rugby Landscapes Ltd v Stockley Park Consortium Ltd* (1997) 82 BLR 39.

3. Instructions of the architect or contract administrator, issued in accordance with clause 3.16 (IC 05 clause 3.13), expenditure of provisional sums, excluding those for defined work.
4. The opening up for inspection of any work, or the testing of any work, materials or goods (including making good afterwards), under clause 3.17 (IC 05 clause 3.14). Naturally, there can be no claim where such inspection or testing reveals a breach of contract by the contractor.
5. Any discrepancy in or divergence between the contract drawings and the contract bills.
6. Suspension of the work by the contractor in accordance with clause 4.14 (IC 05 clause 4.11), which gives effect to the Scheme for Construction Contracts under the Housing Grants, Construction and Regeneration Act. If the employer fails to pay the contractor in full any payment due under this contract, and if the contractor has given seven days notice of the intention to suspend, then the contractor can suspend work until payment is made. Because this is a relevant matter, the contractor can also claim loss and/or expense resulting from this suspension.
7. Carrying out work where the quantity indicated in an approximate quantity was not an accurate forecast of the amount of work required.
8. There is a new clause in JCT SCB 05 and IC 05 that brings together a variety of events under an overall condition referring to *any impediment, prevention or default, whether by act or omission, by the employer* or various agents of the employer. Anything contributed to by the contractor's people is excluded. This clause will cover the late receipt of instructions, drawings or other details from the contract administrator. Unlike older versions of the contract, this is not dependent upon the contractor having specifically applied in writing for the information at a reasonable time. Contract administrators are expected to know when information will be required. Other contracts that mention this usually make this reason for loss and/or expense dependent on the contractor having asked for the information at a reasonable time. As to what is a 'reasonable' time, it was held in a civil engineering case[4] that this must have regard to the interests of the engineer, as well as to those of the contractor.
9. IC 05 contains a provision that does not appear in SBC 05, mention of delays related to instructions of the contract administrator in relation to named sub-contractors to the extent mentioned in clause 3.7 and Schedule 2.

16.2.2 Other contracts

In the case of GC/Works/1, clause 46(1) enables the contractor to claim if certain listed events result in *the regular progress of the Works or any part of them being materially disrupted or prolonged*, where this means that the contractor *properly and directly incurs any expense*. The relevant events do not cover a very wide area, although they *do* include such things as delay in providing drawings or other information and delays caused by other contractors. Moreover, while claims under

---

[4] *Neodox Ltd v Swinton & Pendlebury UDC* (1958) 5 BLR 34.

clause 46(1) can only be made where there is 'disruption or prolongation', any expense involved in complying with the project manager's instructions may qualify for payment under clauses 42 and 43. It should further be noted that this may in certain circumstances effectively entitle the contractor to be compensated in respect of 'unforeseeable ground conditions' (clause 7).

The approach adopted by the ACA form of contract is a more generally worded one. Clause 7 entitles the contractor to claim in respect of any 'damage, loss and/or expense' which is suffered when *any act, omission or default or negligence of the employer or of the architect disrupts the regular progress of the Works or of any Section or delays the execution of them in accordance with the dates stated in the Time Schedule.* However, this does not include delay or disruption resulting from compliance with the architect's instructions, something that is separately dealt with under clauses 8 and 17.

Finally, the approach adopted by ICE 7 may be noted. This makes no attempt to collect under a single clause the various grounds on which 'additional payment' may be claimed, although the *procedural* matters applying to all of them are found in clause 52(4). The grounds themselves are too varied to be summarized, but taken in total they appear to cover a wider area than the grounds under other forms, which means that more risks are borne by the employer. On the other hand, the ICE form makes frequent use of the formula ... *which could not reasonably have been foreseen by an experienced contractor,* so as to place limits on the circumstances in which successful claims may be made.

16.3   CLAIMS PROCEDURES

Every form of contract permitting the contractor to make money claims lays down certain procedural steps that must be followed if a claim is to succeed. These naturally vary between the forms, and space does not permit us to deal with all the variations. We shall therefore concentrate mainly on the provisions of JCT, merely drawing attention to some aspects of other contracts that are significantly different.

16.3.1 JCT SBC 05

The procedure in chronological order for making a claim for 'loss and/or expense' under JCT SBC 05 and IC 05 is laid down by clauses 4.23 and 4.17 respectively. It is as follows:

1. The contractor considers that:
   (a) regular progress of the work has been or is likely to be materially affected by a relevant matter;
   (b) because of this (or because late possession of the site has been given), direct loss and/or expense has been or is likely to be incurred; and
   (c) the contractor would not otherwise be reimbursed under the contract for this loss and/or expense.
2. The contractor makes written application to the contract administrator, stating the above. The contract does not prescribe any particular form of application, but it has been held in relation to JCT 63 that a valid

application must sufficiently identify the issue on which the architect's decision is required.[5] Naturally, in cases where the architect already knows the relevant facts a very brief and uninformative notice will suffice.

3. The contractor's application is expected to be made during the execution of the contract, when the disruption becomes apparent, rather than at the end of the contract. Any notice that does not comply with this time limit may simply be ignored by the contract administrator, in which case the contractor may be left to claim damages for any breach of contract that can be established.

4. The contract administrator must then decide whether or not the contractor's claim is well founded. In deciding this question, the contract administrator may request from the contractor any further information that is reasonably necessary, and the contractor must comply with any such request.

5. A contract administrator who agrees in principle with the contractor's claim must then ascertain or instruct the quantity surveyor to ascertain the amount of loss and/or expense incurred. Again, the contractor may be requested to furnish such details as are reasonably necessary.

6. Any amount duly ascertained under this clause is added to the contract sum and will therefore feature in the next interim certificate.

16.3.2 GC/Works/1

GC/Works/1 makes it abundantly clear that, unless a contractor complies strictly with the claims procedure laid down, there will be no entitlement to anything at all under the contract. Clause 46(3) makes it a condition precedent to an increase of the contract sum for disruption or prolongation expense that:

1. the contractor gives notice to the project manager *immediately* on becoming aware of likely disruption, specifying its cause and stating that an increase in the contract sum is being sought; and

2. the contractor furnishes details and documentary evidence of the expense *as soon as reasonably practicable* after it is incurred.

16.3.3 ACA form of contract

The ACA form of contract, too, imposes strict procedural burdens on a contractor who wishes to claim because the regular progress of the work has been disrupted. Notice must be given to the architect *immediately* it becomes apparent that a claimable event has occurred *or is likely to occur*. Furthermore, the contractor must, when next applying for an interim payment under the contract, submit a written estimate, supported by appropriate documentary evidence, of the damage, loss and/or expense so far incurred. If the architect accepts the estimate, or if it can be settled by negotiation or under the contractual adjudication procedure, then the contract sum is adjusted *and no further or other additions or payments shall be made* in respect of the claim.

---

[5] *Merton LBC v Stanley Hugh Leach Ltd* (1985) 32 BLR 51.

The contract specifically provides that, if the contractor fails to comply with the above procedure, there will be no entitlement to any adjustment of the contract sum until final certificate. Further, even when such an adjustment is made, it will not include any interest or financing charges for the intervening period.

16.3.4 ICE 7

The claims provisions of the current ICE contract (and of the FIDIC Conditions for Civil Engineering Contracts, 4th edition) appear to have overcome some at least of the problems caused by the tortuous wording of earlier editions. It is now clear that, if a contractor wishes to make a claim under any clause of the contract, other than those which deal with variations, the contractor must give written notice to the engineer *as soon as may be reasonable and in any event within 28 days after the happening of the events giving rise to the claim* (53(2)). The contractor must thereafter keep contemporary records to support any subsequent claim. The initial notice is to be followed, *as soon as is reasonable in all the circumstances*, by an account containing full and detailed particulars of the amount claimed.

The case law on early versions of the ICE conditions made it clear that any failure to comply with the provisions as to giving notice would be fatal to a contractor's claim. However, the 7th edition contains two provisions which mitigate the strictness of this principle. First, it is provided that, despite the contractor's failure, the engineer may still uphold a claim *to the extent that the engineer has not been prevented from or substantially prejudiced by such failure in investigating the said claim* (53(5)). Second, it is made clear that, where the contractor seeks interim payment on a claim but has submitted insufficient particulars, the engineer may nonetheless authorize payment *of such part of the claim as the particulars may substantiate to the satisfaction of the engineer* (53(6)).

16.4   QUANTIFICATION OF CLAIMS

If it can be established that a contractor is entitled to claim, a separate problem is the calculation of the amount of a claim. The issues of entitlement and magnitude should always be separated to make claims procedures easier to manage for everyone involved.

16.4.1 Nature of 'loss and/or expense'

When it comes to turning a contractor's claim into money, the courts have made it clear that 'direct loss and/or expense', and similar phrases found in other standard form contracts, require an assessment process equivalent to that for an award of damages for breach of contract.[6] This means that where appropriate the contractor may be compensated, not only for out of pocket losses, but also for loss of profit.

---

[6] *Wraight Ltd v P H & T (Holdings) Ltd* (1968) 13 BLR 26; *FG Minter Ltd v Welsh Health Technical Services Organization* (1980) 13 BLR 1.

However, such an interpretation depends entirely on the wording of the particular contract. For example, many of the clauses in ICE 7 which permit a claim to be made refer specifically to 'cost', which is defined in such a way as to rule out any profit element.

The courts have made it clear that, although claims clauses frequently use the word 'direct', the crucial question is whether a particular item of loss falls within the normal rules of remoteness of damage in breach of contract cases. Indeed, the courts appear to have adopted a fairly liberal approach to this question. In *Croudace Construction Ltd v Cawoods Concrete Products Ltd*,[7] for example, a contract to supply masonry blocks to the main contractors on a school project stated that the suppliers should not be liable for any 'consequential loss or damage' caused by late delivery or defects. When the blocks proved to be defective, the contractors claimed damages for loss of productivity, inflation costs resulting from delay and the cost of meeting a claim brought against them by sub-contractors. The Court of Appeal held that, despite the clause quoted above, the contractors were entitled to recover for all these items of loss.

It should at this point be stressed once again that a claim for 'loss and/or expense' is based, *not* on delay in completion of the works, but on the fact that the regular progress of those works has been disrupted. It is true that most cases do in fact concern delayed completion, and indeed some types of loss can *only* arise where this is so, but this must be seen as coincidental. Where, despite finishing a job on time, a contractor incurs additional expenses, such as the cost of management time to deal with difficulties caused by the contract administrator's instructions, or loss of productivity due to the unexpected operations of other contractors on the site, the contractor is fully entitled to claim for these losses.

16.4.2 'Immediate' costs

Where a job is prolonged or disrupted, certain types of loss are instantly recognizable as likely to occur. For example, where working conditions are rendered more difficult (for example because what was intended to be a summer job has now become a winter one), it may be necessary for the contractor to employ additional labour, use extra materials or hire extra plant, simply in order to achieve the same result. Similarly, the prolongation of a contract may mean that materials that would have lasted for the original period deteriorate during the overrun, requiring either replacement or expensive protective measures. Again, the natural effect of inflation may mean that the outlay on both labour and materials is increased because of a delay. Finally, it is fairly obvious that 'site overheads', that is those general expenses exclusively referable to the contract in question, will be greater if the contract period is lengthened.

All these items may in principle be the subject of a claim. However, three cautionary notes should be sounded:

1. It must be shown that the loss in question has actually been suffered. Thus, for example, any claim for increased labour or materials costs must give

---

[7] (1978) 8 BLR 20.

full credit for anything the contractor is entitled to receive under a fluctuations clause in the contract.

2. The loss must *as a matter of law* have resulted from the claim-provoking event. Thus an employer must pay compensation for turning a summer contract into a winter one, but will not be held responsible if the contractor is then caught by some totally independent and unforeseeable disaster, even though this would not have affected the project if the works had been completed on time.

3. Where the contractor's own plant stands idle in a period of delay, the courts will not normally uphold a claim based on current hiring rates, but will limit the contractor to a claim based on depreciation and maintenance costs.[8]

16.4.3 Head office overheads and profit

Any disruption to the regular progress of work under a contract may lead the contractor to incur administrative costs, such as the diversion of managerial time and effort, at head office. If so, these costs may justifiably be claimed, but it will not be simply assumed that such losses have been suffered. They must be specified and properly supported by the evidence, for example by records of the time spent by individuals in dealing with the particular problem.[9]

Where the contract period is *prolonged* by something for which the employer is contractually responsible, the contractor may also seek to claim in respect of *general* head office overheads. The contractor's argument here is either that the contract is making a smaller contribution to these business expenses than it should, or that the organization is being tied up so as to prevent it from earning the necessary contribution to head offices expenses elsewhere. Identical arguments are used to support a claim for lost or diminished *profits,* which, it is said, should have been earned from the contract in question.

In practice, because of the enormous difficulties of accurate quantification, contractors' claims under this heading tend to adopt a 'formula' approach. A notional daily or weekly contribution from the contract in question to general business overheads and profits is identified, and then multiplied by the number of days or weeks for which the contractor is entitled to claim. While the defects of such an approach are obvious, the courts regard it as generally acceptable, provided that any formula used is not applied blindly and without reference to the realities of the situation. Moreover, it is important to emphasize that the onus is on the contractor to prove that actual loss has been suffered. As a result, it was ruled in *Peak Construction (Liverpool) Ltd v McKinney Foundations Ltd*[10] that the contractor must be able to show that the organization could have worked elsewhere during the period of delay (that is, that there was work available in the construction industry generally).

---

[8] *B Sunley Ltd & Co v Cunard White Star Ltd* [1940] 1 KB 740.
[9] *Tate & Lyle Food and Distribution Co Ltd v GLC* [1981] 3 All ER 716; *Babcock Energy Ltd v Lodge Sturtevant Ltd* [1994] CILL 981.
[10] (1970) 1 BLR 111.

As mentioned, a 'formula' approach to this aspect of a claim is centred on the identification of a 'head office overheads and profit contribution' from the delayed contract. Where the three best-known formulae differ is in their method of identifying the appropriate percentage.

1. The *Hudson* formula (named after the leading textbook in which it first appeared) (Wallace 1995) reads as follows:

$$\frac{Overheads/profit\ percentage}{100} \times \frac{Contract\ sum}{Contract\ period} \times Period\ of\ delay$$

What this means is that the claim is based on the allowances actually made by the contractor in tendering for the contract, notwithstanding that these may of course have been unreasonably optimistic or pessimistic. This formula has been applied in a Canadian case,[11] and was also apparently approved by an English court in *JF Finnegan Ltd v Sheffield CC*.[12]

2. The *Emden* formula (named after another well-known textbook) (Bartlett *et al.* 2002) reads:

$$\frac{Total\ overheads/profit}{Total\ turnover} \times \frac{Contract\ sum}{Contract\ period} \times Period\ of\ delay$$

Here the figure used is the percentage that is relevant to the contractor's whole organization, found by dividing *total* overhead cost and profit by *total* turnover. This approach, which of course ignores the question whether the particular contract was more or less profitable than usual, has been accepted by an English court in *Whittall Builders Co Ltd v Chester-le-Street DC*.[13] Ironically, it was also actually applied in the *Finnegan* case,[14] although the judge there claimed to be using the *Hudson* formula.

3. The *Eichleay* formula (Wallace 1986), which is named after the US case in which it was first put forward, is more complex. It reads:

(a) $$\frac{Contract\ invoices}{Total\ invoices} \times Total\ overheads/Profit = Allocable\ overhead$$

(b) $$\frac{Allocable\ overhead}{Days\ of\ performance} = Daily\ contract\ overhead$$

(c) *Daily contract overhead* × *Days of delay* = *Amount recoverable*

This approach is widely used in the USA but does not yet seem to have appeared on this side of the Atlantic. Although more refined than the *Emden* formula, it is subject to the same criticism of ignoring any special features of the particular contract.

---

[11] *Ellis-Don Ltd v Parking Authority of Toronto* (1978) 28 BLR 98.
[12] (1988) 43 BLR 124.
[13] (1988, unreported).
[14] *JF Finnegan Ltd v Sheffield CC* (1988) 43 BLR 124.

16.4.4 Interest and financing charges

Common law has always refused to permit a creditor to claim either damages or interest for the mere fact of being kept out of money due by the debtor's late payment.[15] As a result, it was for a long time open for a debtor to avoid any further liability by paying the bare amount of a debt at any time before judgment was awarded.

The worst of this rule has now been removed by statute in respect of both litigation (Supreme Court Act 1981, section 35A) and arbitration (Arbitration Act 1950, section 19A). These provisions enable interest to be awarded wherever a debt was still unpaid when proceedings for its recovery were commenced. However, where payment is made late but before proceedings are launched, it has been reluctantly confirmed by the House of Lords that the common law rule still applies; the claimant in such circumstances is not entitled to either damages or interest.[16]

Rather questionably, perhaps, the lower courts have for some time been prepared to ignore the basic rule in cases where a claimant has suffered 'special damage' from late payment, for instance by having to pay interest on an overdraft which could have been cleared or reduced if the debtor had paid up. This escape route has now been seized upon in construction cases to justify claims by contractors, as part of 'direct loss and/or expense', for what are termed 'financing charges'. In *FG Minter Ltd v Welsh Health Technical Services Organization*,[17] a claim of this nature was upheld by the Court of Appeal. This was on the ground that, since *the loss of interest which he has to pay on the capital he is forced to borrow and on the capital which he is not free to invest would be recoverable for the employer's breach of contract*, it was equally recoverable under JCT 63.

The *Minter* decision was applied by the Court of Appeal in *Rees & Kirby Ltd v Swansea CC*,[18] where the court came to the realistic conclusion that such a claim should be assessed on the basis of compound rather than simple interest. It was also followed, after detailed analysis, in the Scottish case of *Ogilvie Builders Ltd v Glasgow City DC*.[19]

16.4.5 Costs of preparing a claim

In modern times, contractors' money claims have become of enormous practical importance, and something of a 'claims industry' has grown up to assist in the preparation and presentation of such claims. Where 'claims consultants' are employed in this way, their services may themselves constitute a considerable cost to the contractor, and a question which has arisen is whether these costs can then be recovered as part of the claim.

---

[15] *London, Chatham & Dover Railway Co v South Eastern Railway Co* [1893] AC 429.
[16] *President of India v La Pintada Cia Navegacion SA* [1985] AC 104.
[17] (1980) 13 BLR 1.
[18] (1985) 30 BLR 1.
[19] (1994) 68 BLR 122.

The answer in principle appears to be that, since a contractor is not required under most standard-form contracts to make a detailed claim for loss and/or expense, any costs incurred in so doing should not themselves be recoverable.[20] However, where compliance with the contract administrator's or quantity surveyor's request for further evidence involves an unusually heavy amount of managerial time, this can probably be recovered.[21]

---

[20] *James Longley & Co Ltd v South West Thames RHA* (1983) 127 SJ 597.
[21] *Tate & Lyle Food and Distribution Co Ltd v GLC* [1981] 3 All ER 716.

17 Insurance and bonds

One of the primary functions of a construction contract is to allocate certain *risks* to one or other of the parties. This may be done by providing that one party (X) shall be *liable* to the other (Y) if a particular kind of loss or damage occurs, thus placing the risk of that loss or damage on X. It occurs equally under a provision that X shall *not* be liable to Y for a particular kind of loss, since this effectively places the risk of that kind of loss on Y.

In view of the enormous size of some of these risks in financial terms, it is obvious that the party to whom a risk is allocated may want to cover it by means of insurance. What may not be so obvious, however, is that a party to whom a risk has *not* been allocated may still insist on insurance backing for that risk. This is simply because; while the contract may say for instance that the contractor shall be liable to the client for certain kinds of damage, the client's right to sue the contractor will be of little use if the contractor cannot afford to pay any damages awarded. In such circumstances, the client will only be protected if the contractor has either 'liability insurance' or some 'guarantee' of performance.

In this chapter we discuss briefly the legal principles on which insurance is based and the main types of construction insurance policy, illustrated by clauses from JCT SBC 05. We also consider professional indemnity insurance and the use of 'bonds' or 'guarantees' as a means of protection against contractual failure.

17.1 INSURANCE

It is essential to understand that all insurance policies in the construction industry (including the professional indemnity insurances of consultants) fall into one of two categories. JCT SBC 05 contains examples from each category.

17.1.1 Types of insurance

Under a **liability insurance policy**, an insurer undertakes that, if the insured person (the client) becomes legally liable to someone else (the 'victim'), the insurer will indemnify *the client* against damages and legal costs which become payable. A familiar illustration of this type of policy is 'third party' insurance for a car driver. By contrast, under a **loss insurance policy**, the insured person is entitled to be compensated by the insurers for loss or damage which *that person* has suffered, whether this is caused accidentally or by someone else's negligence. Such a policy may even provide cover in respect of loss or damage caused by the insured person's own negligence, although this is sometimes excluded or restricted by special terms in the policy. A 'comprehensive' motor insurance policy is an example of this type of policy.

Liability insurance

Examples of liability insurance commonly found in the context of construction include **public liability policies** for contractors and **professional indemnity policies** for architects and other consultants. What must be clearly understood is that, where an insurance policy is of this kind, the insurer's legal duty is owed to the client and *not* to the 'victim'. Indeed, the insurer is in theory not responsible at all until the insured person has actually been held liable to the victim by a court or arbitrator. In practice, however, the insurer will normally be brought into the picture as soon as a legal action is started against the client. Indeed, if the action is thought worth defending, the insurer may well take over the actual conduct of the case.

The fact that liability insurance is seen as something that affects only the insurer and the client places severe restrictions upon the protection it offers to victims. An important limitation was exposed by the case of *Normid Housing Association Ltd v Ralphs*,[1] where the claimants brought a professional negligence claim against the defendant architect for an amount of money far greater than the limits of the defendant's professional indemnity insurance policy. The wording of this policy was in fact not entirely clear as to the maximum the insurers could be made to pay, and the defendant agreed with the insurers to accept £250,000 in full settlement of all claims on the policy. (It should be pointed out that the defendant had no great interest in attempting to make the insurers pay more than this, since the excess, which he would be personally liable to pay to the claimants, was enough to render him bankrupt.) The claimants, who felt that the insurers could have been made to pay more, attempted to have this settlement agreement overturned, but their attempt failed. The Court of Appeal held that, since the defendant was under no legal duty to have liability insurance at all, any genuine settlement of the claim was a matter for the defendants and their insurers. The claimants had no rights whatever to intervene, unless they could establish that the settlement agreement between the defendant and the insurers had been reached in bad faith.

Had the defendant in this case actually been made bankrupt before settling with the insurers, the claimants could have utilized a procedure laid down by section 1(1) of the Third Parties (Rights Against Insurers) Act 1930. This provides that, where a person or company with liability insurance becomes insolvent, any claim the insured person could have made against the insurers is automatically transferred to the victim. As a result, the claimants here could have refused to settle with the insurers as the defendant had done, and could have instead sued the insurers for the higher amount they claimed was due under the policy.

Although the 1930 Act improves the position of victims, it is still subject to strict limits, as was demonstrated in the case of *Bradley v Eagle Star Insurance Co Ltd*.[2] The House of Lords ruled in that case that the 1930 Act only transfers to a

---

[1] (1988) 43 BLR 18.
[2] [1989] 1 All ER 961.

claimant those rights which the insured person already had against the insurers, and pointed out that these do not arise until the insured person has been held liable to the claimant. As a result, unless a successful claim is made against a company while it still exists (or at least while it can still be restored to the Companies' Register), the victim cannot take action against the company's liability insurers.

The *Bradley* decision carries an important implication, albeit on a point which did not arise in the case itself. It is that any defence an insurer would have against the client is equally available against a victim who takes over the client's rights by virtue of the 1930 Act. Thus, for example, where a person who has become bankrupt has failed to comply with some condition of his or her liability insurance policy (for example by failing to give notice of claims within a specified time), the insurers will be entitled to refuse payment to a victim, just as they could have refused to pay the client.

Loss insurance

As mentioned above, a loss insurance policy provides cover, not for a person's legal liability to others, but for losses which fall directly upon that person. In the construction field, such cover is required by a client on whom the contract places the risk of damage to the contract works or the existing structures.

There is an important general principle of the law governing loss insurance which is called **subrogation**. Under this principle, where a loss insurer pays the client in respect of a loss that has been caused by someone else's default (such as a breach of contract or a tort), the insurer is then entitled to take over any legal rights that the insured person could have exercised against the third party responsible. In such a way the insurer will seek, by suing in the client's name, to recover from that third party the amount that has been paid out to the client.

However, it must be stressed that there is *no* right of subrogation against any person who is also insured under the same 'loss insurance' policy. This is why many building contracts provide that any loss insurance policies which the contract requires are taken out in the joint names of the client and the contractor, and sometimes also of any sub-contractors. The legal effect of such a practice is shown by the case of *Petrofina (UK) Ltd v Magnaload Ltd*,[3] in which a **contractors' all risks policy** taken out in respect of an extension to an oil refinery defined the insured persons as including the employer, contractors and/or sub-contractors. The contract works were extensively damaged due to the negligence of sub-contractors engaged to lift heavy equipment, whereupon the insurers duly paid the owners. It was held that the insurers were not then entitled to recover their loss from the sub-contractors under the doctrine of subrogation.

17.1.2 Construction insurance under JCT SBC 05

The general scheme of the insurance provisions of JCT SBC 05 operates in two stages. First, it allocates as between client and contractor the risk of various kinds

---

[3] [1984] QB 127.

of loss or damage, which may arise in the course of building work. It then imposes responsibility for insuring against those risks.

Personal injury and damage to property

Where risks to third parties are concerned, the crucial provisions are clauses 6.1 and 6.2, under which the contractor undertakes to indemnify the employer against certain kinds of liability 'arising out of or in the course of or by reason of the carrying out of the works'. In particular, this indemnity applies under clause 6.1 to any liability the employer incurs (for example in the capacity of occupier of the premises) for *injury* to or *death* of any person. However, the contractor need not indemnify the employer to the extent that the latter's liability is due to any act or neglect of the employer or of any person for whom the employer is responsible.

As to the persons whose 'act or neglect' can be attributed to the employer for this purpose, these clearly include members of the employer's staff, any direct contractors on the site and (probably) the contract administrator. However, they do *not* include any sub-contractor, whether domestic or nominated. This means that where a person is injured or killed through a sub-contractor's negligence, any resulting legal liability for the employer may be passed on to the main contractor.

The contractor's duty to indemnify the employer also covers, under clause 6.2, *damage to property*, but in this instance only to the extent that the damage is due to *negligence, breach of statutory duty, omission or default* by the contractor or by someone for whom the contractor is responsible. In this context, the contractor bears responsibility for the conduct of anyone properly on the site in connection with the works, except for those persons who are the employer's responsibility (presumably the same as those falling within clause 6.1).

An earlier version of what is now clause 6.2 did not make clear whether the contractor's obligation was merely to indemnify the employer against any claims made by third parties whose property was damaged, or whether it also covered damage to the employer's own property (including the contract works). However, the current version expressly provides that the only parts of the *contract works* included in clause 6.2 are those in respect of which a certificate of practical completion has been issued or which have been taken over by the employer. The remaining parts fall instead under insurance option C, found in Schedule 3 of JCT SBC 05, which is discussed below.

As for other property of the employer, a distinction must be drawn between *existing structures* (in the case of a contract for extension or refurbishment work) and property which is completely separate from the contract. Damage to such independent property clearly falls within the contractor's obligation to indemnify the employer. However, in relation to the existing structures, it appears from Insurance option C that the contractor will not be liable for causing these to be damaged by a 'specified peril' (defined in clause 6.8 to include such things as fire, explosion, flood, burst pipes etc.). However, the contractor *will* be liable for negligently causing or permitting such property to be damaged in other ways, such as by impact, subsidence, theft or vandalism.

Duty to insure

Clauses 6.1 to 6.3 of JCT SBC 05 define the scope of the contractor's
responsibility to the employer for causing personal injury or damage to property.
Clause 6.4 then provides that the contractor shall *take out and maintain* certain
insurance policies, which will for the most part cover these forms of liability. In the
case of damage to property, cover is to be at a financial level imposed by the
employer by means of an entry in the Contract Particulars. In practice, this
obligation will normally be satisfied by a combination of the contractor's 'public
liability' and 'employer's liability' policies, although the 'third party' section of
the contractor's motor insurance policy may also be relevant in appropriate cases.

To protect the employer's position in respect of these insurances, clause 6.4.2
requires the contractor on demand to provide evidence that the policies are in force
and that they provide the necessary cover. If this is discovered *not* to be the case,
then the employer is empowered to take out personal insurance against any liability
or expense that may be incurred and to charge the premiums to the contractor.

A loophole in the protection effected by similar clauses in an earlier version of
the JCT form of contract was discovered in the case of *Gold v Patman &
Fotheringham Ltd.*[4] The employer was there held *strictly* liable in nuisance to the
owner of neighbouring property, for damage to that property due to subsidence
caused by the works. However, the contractor had not been guilty of any
negligence and was therefore not bound to indemnify the employer. To deal with
the result of this case, JCT SBC 05 clause 6.5 now permits the contract
administrator in certain circumstances to instruct the contractor to take out and
maintain an additional **joint names policy**. This policy will indemnify the
employer both against strict liability to third parties and against any damage to the
employer's own property (other than the contract works and materials on site)
resulting from collapse, subsidence, heave, vibration, withdrawal of support,
lowering of ground water and similar causes. It should be noted that this policy can
only be called for where the Contract Particulars so state; further, although it is the
contractor's responsibility to take out and maintain the policy, it is to be done at the
employer's expense.

Insurance of contract works in new buildings

If there were no specific contractual provisions on the matter, and assuming that
there had been no negligence by either employer or contractor, the risk of damage
to the contract works would lie on the contractor. This is because it is the
contractor's basic obligation to complete the job. On this basis it was held in
Charon (Finchley) Ltd v Singer Sewing Machine Ltd[5] that, where the works were
damaged by **vandalism**, the contractor's duty of substantial performance of the
contract included everything that was necessary to restore the works to the
contractual standard.

This basic legal position is substantially altered by JCT SBC 05 clauses 6.7–
6.10 and Schedule 3. These clauses are extremely complicated, but an important

[4] [1958] 2 All ER 497.
[5] (1968) 207 EG 140.

key to understanding them lies in the distinction that they maintain between two classes of insurable risk. On one hand are the 'Specified Perils', a list of catastrophes which includes damage from fire, lightning, explosion, storm, tempest, flood, burst water pipes, earthquake, aircraft and riot. On the other hand is 'All Risks Insurance', which covers 'any physical loss or damage to work executed and site materials', subject to some specific exceptions such as wear and tear, design defects, radioactivity or other nuclear-linked damage, sonic booms or terrorist action. This is not the place for a detailed comparison of the two lists, but 'All Risks' is clearly much wider than 'Specified Perils' and includes for example such causes of damage as impact, subsidence, theft and vandalism.

Where the contract in question is concerned with the erection of a new building, JCT SBC 05 offers a choice of two alternative clauses. These place the responsibility for taking out 'All Risks Insurance' in joint names on *either* the contractor (insurance option A) or the employer (insurance option B). Such insurance must be maintained until the issue of the Practical Completion Certificate. Whichever of these choices is adopted, it is provided by clause 6.9 that the party taking out the insurance must see to it that all sub-contractors are protected from any 'comeback' action by the insurers. However, it is important to note that this protection for sub-contractors extends only to damage which results from a 'Specified Peril'.

Under the Private editions of JCT SBC 05, any failure by either party to maintain the required insurance cover will entitle the other party to insure and charge the cost to the defaulter. In the Local Authorities editions, this remedy is available only to the employer. It is however specifically provided that the contractor may fulfil this obligation by maintaining a suitable annual policy to which the names of the employer and the sub-contractors can be added.

A significant difference between insurance options A and B lies in their treatment of any repair or restoration work which may be necessary and which it is the contractor's obligation to carry out. Where the *contractor* has insured, payment by the employer for this work will not exceed what is received from the insurer; by contrast, where the *employer* has insured, the contractor's work will be valued as if it were brought about by a variation under clause 15.1. The effect is that, if insufficient insurance cover has been taken out, the resulting loss will be borne by the party responsible for that under-insurance.

Insurance of existing buildings and of contract works in them

Where the contract works consist of alterations to or the extension of existing structures, neither of the above options should be used. The appropriate provision is option C, which effectively requires the employer to take out and maintain (until the issue of the Practical Completion Certificate) insurance to cover damage to both the contract works and the existing structures. In so far as it applies to the *contract works*, this provision broadly follows option B; it is for the employer to insure in joint names on an 'All Risks' basis, and to see that all sub-contractors are protected. However, in relation to the *existing structures*, option C differs from option B in two important respects: first, the employer need insure only against the 'Specified Perils', rather than against 'All Risks'; and, second, there is no obligation to protect sub-contractors against any 'comeback' claims by the insurers.

A further feature of option C is the provision that, where damage is caused to the contract works by an insured risk, either party may terminate the contractor's employment if it would be 'just and equitable' to do so. This provision, which is only intended to be used in cases of very serious damage or destruction, is discussed in Chapter 23.

Effect of insurance on liability of contractor

Where a contractor or sub-contractor negligently causes damage (either to the contract works or to existing structures) of a kind which falls within the insurance policies described above, the insurance provisions discussed above may have a significant effect on their liability. In *Co-operative Retail Services Ltd v Taylor Young Partnership*,[6] a disastrous fire in a partly constructed office block was assumed for the purposes of the case to have been caused by the combined negligence of the architects, consulting engineers, main contractor and electrical sub-contractor. When the employer sued the architects and consulting engineers for negligence, these defendants sought contribution (under the Civil Liability (Contribution) Act 1978) from the contractor and sub-contractor, on the ground that they would also, if sued by the employer, have been liable for the same damage. This claim was rejected by the House of Lords, who held that, since the contractor and sub-contractor were covered by a 'joint names' insurance policy taken out in accordance with the predecessor of option A, they could not have been made liable for the damage in question.

The effect of this principle on the liability of a sub-contractor is considered further in Section 20.5.1.

Insurance for loss of liquidated damages

Assuming that the insurance cover specified in options A, B or C is taken out and maintained at an adequate level, the costs of repair and restoration following damage to the contract works from the vast majority of causes should be recoverable. However, these costs are not the only ones that may result from damage to the contract works. In particular, where the cause of the relevant damage is one or more of the 'Specified Perils', the contractor may be entitled to an extension of time for completion, in which case the employer will lose the right to claim liquidated damages for the period of delay. This aspect of the employer's loss is clearly *not* covered by the insurance required by the clauses mentioned above.

In order to close this loophole, JCT 98 included an option for the employer's loss of liquidated damages to be separately insured. Clause 22D provided that, where the Appendix to the contract made it clear that such insurance might be required, immediately the contract is entered into, the contract administrator must

---

[6] [2002] 1 WLR 1419.

either instruct the contractor to arrange it or state that it was not needed. In the former case, the policy was to be kept in force until the date of practical completion, and the sum insured was to be based on the contract rate for liquidated damages.

Clause 22D was a rather odd provision. Since the insurance with which it dealt was designed entirely to protect the employer, and since it was the employer who paid for it through an addition to the contract sum, it was not at all clear why the contractor was called upon to arrange it. Not surprisingly, perhaps, this provision has been entirely omitted from JCT SBC 05.

17.1.3 Professional indemnity insurance

The various professional members of a design and construction team, such as the architect, specialist engineers and quantity surveyor, will each have their own insurance policy to indemnify them against liability for professional negligence. Such policies, which are often taken out as part of a scheme run by the relevant professional body, will of course vary greatly, but a few general points are nonetheless worth bearing in mind.

- A *contractor*, at least one who does not specialize in 'design and build' work, may well not have insurance cover of this kind; indeed, any limited company may find such insurance difficult or impossible to obtain. As a result, the client may have less insurance protection in respect of a contractor's design input than under a conventional set-up in which design is carried out by professional consultants.
- The cover provided by most professional indemnity insurance policies is subject to financial limits and, once those limits are exceeded, the professional's own money is at stake. And, of course, once *that* money is exhausted, the loss inevitably falls on the client.
- Professional consultants are normally required by their institutions to take out and maintain a specified level of indemnity insurance. However, while failure to do so would be a breach of the professional's code of conduct, it would not usually be a breach of contract with the client. This is because very few of the standard terms under which professionals are engaged in the construction industry (Section 2.2) contain any requirement to take out or maintain professional indemnity insurance (notable exceptions are the standard RICS terms). Indeed, even where they do, the client's only remedy for failure to insure would presumably be to claim damages, which, presumably, the uninsured professional would be unable to pay. In any event, as we noted at the beginning of this Chapter, a liability insurance policy of this kind is a contract between the insured and the insurer; the client has only very limited *direct* protection.
- Standard professional indemnity policies cover the insured against liability for *professional negligence*, but there are considerable doubts as to whether they extend to other forms of liability. Thus, if an architect is held to have *guaranteed* the suitability of a design, rather than merely undertaking that reasonable skill and care has been used, any resulting liability might well not be covered. Indeed, many policies specifically *exclude* liabilities that the

professional voluntarily assumes by way of such guarantees. As we shall see in Chapter 22, such considerations have caused major problems in the context of 'collateral warranties'.

- It is worth noting that most professional indemnity policies consist of a series of annual contracts written on a 'claims made' basis. This means that they cover claims actually made against the professional person during the period of insurance, regardless of when the negligent act took place. As a result, it is essential that insurance cover is maintained, even after the retirement or death of the professional concerned, and this can usually be arranged on a single-premium 'run-off' basis.

One might assume that the benefit of cover on a 'claims made' basis is that one is protected in respect of acts of negligence committed before the policy is taken out. However, while this is broadly speaking true, what should not be overlooked is that a person who seeks insurance has a legal duty to disclose all 'material facts' to the insurer. As a result, a professional who, on taking out a new policy or renewing an old one, fails to reveal to the insurers that a claim is pending, may find, when a quite separate claim arises at a later date, that the policy is voidable and that the insurers are under no obligation to pay.

17.2   BONDS AND GUARANTEES

A **bond** or **guarantee** is an arrangement under which the performance of a contractual duty owed by one person (A) to another (B) is backed up by a third party (C). What happens is that C promises to pay B a sum of money if A fails to fulfil the relevant duty. In this context A is commonly known as the **principal debtor** or simply **principal**; B is called the **beneficiary**; and C is called the **bondsman, surety** or **guarantor**.

In the construction context, such backup is likely to come from one of two sources. First, there are **parent company guarantees**, under which the contractual performance of one company within a corporate group is underwritten by other members of the group. The undoubted importance of such guarantees derives from the combination of two factors. First, many companies operating within the construction industry are seriously under-capitalized (despite forming part of a financially stable corporate group). Second, English company law does not treat a parent company as responsible for the debts of its subsidiaries, unless such responsibility is expressly taken on.

The second type of protection against contractual failure consists of **bonds**, which are normally provided (at a price) by a financial institution such as a bank or an insurance company. Some organizations, many of them from the USA or Canada, have specialized for many years in the provision of bonds for the construction industry. Unfortunately, and despite repeated criticism from the courts, these specialists continue to draft their bonds in a form that is archaic, obscure and full of traps for the uninitiated.

17.2.1 Nature of bonds

It is possible for almost any contractual obligation to be the subject of a bond, but in practice they are normally found in certain commercial fields where their use is well established. In the construction context, the obligations most commonly guaranteed in this way are the following:

- *Payment*: For example the employer's duty to pay the contractor or the contractor's duty to pay a sub-contractor. A contractor may also provide a bond in favour of the employer, in return for an early release of retention money or, indeed, instead of the normal retention provisions. If defects are then found in the building, the employer can call on the bond, rather than the retention money, to finance the necessary remedial work. There is an optional requirement in JCT SBC 05 for such a retention bond, and also for bonds covering advance payments to the contractor and payment for off-site materials or goods. These last two bonds, both of which are 'demand bonds' (see Section 16.2.2), provide the employer with some recourse in the event that the contractor fails to fulfil obligations that have been paid for.
- *A specific obligation*: For example a promise by a sub-contractor not to withdraw a tender. This may be of practical importance where, say, a main contractor tenders on the basis of bids received from domestic sub-contractors. If the main contractor, having been awarded the job, finds that a sub-contractor's bid is no longer open for acceptance, the main contractor may then have to pay a significantly higher price to another sub-contractor for that part of the work.
- *Performance of the contract in general*: This is the most common type of bond, in which every aspect of the contractor's performance is guaranteed. An optional requirement for the contractor to provide such a bond is found in both ICE 7 and in the FIDIC Conditions of Contract. JCT SBC 05 contains no such provision; however, the contract is frequently amended to require a bond, normally to a level of 10% of the contract sum. It should be noted that, where there *is* such a requirement, failure by the contractor to obtain a bond will be regarded as a sufficiently serious breach to justify the employer in terminating the contract.[7] Moreover, a contract administrator who fails to check that a required bond is in place may be liable to the client for professional negligence.[8]

It is perhaps not always fully appreciated that the true underlying purpose of a bond is to give financial protection where the principal debtor, whether it be the employer or the contractor, becomes insolvent. This is because a defaulting party who remains solvent can of course be sued directly. Moreover, even if the beneficiary chooses to sue the surety instead of the principal, the surety has an automatic right of recourse against the principal. Indeed, the surety's overall risk is further reduced by the possibility of seeking contribution from another defaulter. Thus, for example, if a defect in a building is partly the fault of the contractor and partly due to negligent design by the architect, a surety for the contractor can claim a contribution from the architect under the Civil Liability (Contribution) Act 1978.

---

[7] *Swartz & Son (Pty) Ltd v Wolmaranstadt Town Council* 1960 (2) SA 1.
[8] *Convent Hospital v Eberlin & Partners* (1990) 14 Con LR 1.

Given this intention to protect the beneficiary against another party's insolvency, the decision of the Court of Appeal in *Perar BV v General Surety & Guarantee Co Ltd*[9] came as something of a shock to the construction industry. The defendants in that case provided a general performance bond to guarantee the contractual obligations of a design and build contractor, in which they undertook to pay damages to the employer in the event of the contractor's default. The contractor went into administrative receivership, whereupon the contractor's employment was automatically determined. It was held by the Court of Appeal that, since 'default' meant simply 'breach of contract', the contractor's insolvency did not constitute default. Nor, once its employment had been determined, could the contractor be guilty of a breach of contract in failing to continue with the works. In effect, therefore, the bond provided no protection in precisely the circumstances where it was needed.

In an attempt to avoid the problem that arose in the *Perar* case, some bonds state expressly that the contractor's insolvency, as well as a breach of contract, shall trigger the bondsman's obligation. Unfortunately, however, it appears from the case of *Paddington Churches Housing Association*[10] that, where JCT SBC 05 is concerned, this wording may not necessarily have the desired effect. This is because the bondsman in that case promised to satisfy the employer's 'net established and ascertained damages', and it was held that those damages do not become 'established and ascertained' until the works have been completed by another contractor and a statement of account has been drawn up in accordance with what is now clause 8.7.

17.2.2 Types of bond

The type of bond normally found in the construction industry is the **conditional** bond, under which the surety agrees to pay if and when certain specified conditions are satisfied. In the case of a performance bond, the most likely such condition would be any default (i.e.breach of contract) by the contractor. In order to call for payment, the employer must provide evidence of both the contractor's default and the resulting losses suffered by the employer.[11]

A second type of bond is the **unconditional** or **demand bond**, something which has spread into the construction field from international trade. Such a bond entitles the beneficiary to call upon the surety for payment *whether or not there has been default under the principal contract*, provided only that the call is not fraudulent.[12] This means that, unless the surety has clear evidence of such fraud, payment must be made.[13] The use of these bonds in construction contracts is on the whole undesirable since, while the employer may not intend to call on such bonds irresponsibly, the contractor cannot rely on this and must therefore increase the tender price to cover the cost of something which is not really necessary. A

---

[9] (1994) 66 BLR 72.

[10] *Paddington Churches Housing Association v Technical and General Guarantee Co Ltd* [1999] BLR 244.

[11] *Nene Housing Society Ltd v National Westminster Bank Ltd* (1980) 16 BLR 22; confirmed by the House of Lords in *Trafalgar House Construction (Regions) Ltd v General Surety & Guarantee Co Ltd* (1995) 73 BLR 32.

[12] *Edward Owen Engineering Ltd v Barclays Bank International Ltd* [1978] QB 159.

[13] *Balfour Beatty Civil Engineering v Technical & General Guarantee Co Ltd* (2000) 68 Con LR 180.

preferable alternative, for an employer who requires extra security, would surely be to increase the size of the retention under the contract.

17.2.3 Creation of bonds

A contract of guarantee must either be made in writing or at least evidenced by writing, in order to satisfy Section 4 of the Statute of Frauds 1677. In practice it will frequently be made in the form of a deed, and it should certainly take this form wherever the bond is given *after* the main contract is entered into. This is because the beneficiary in such a case will not have given any consideration in return for the surety's promise and thus, unless made by deed, it will be unenforceable.

The *duration* of a guarantee depends upon the terms in which it is given. If no specific time limit is mentioned, then a surety for the contractor's performance is not released by completion or even by the final certificate but remains liable, as does the contractor, for any breach of contract which comes to light within the relevant limitation period. Perhaps because of the potential length of this period, some guarantors now make express provision for release when the final certificate is issued. Some instead provide for release on a fixed date, but this is undesirable from the employer's point of view since, if the contract period is extended, the bond may cease to operate before the works are completed.

The *financial limits* of liability are invariably expressed in the contract of guarantee. It should be made clear, in order to avoid disputes, whether the overall limit includes interest on money due and legal costs. It is also worth noting that some bonds provide for the entire sum guaranteed to become payable on any breach by the principal, regardless of how serious or trivial that may be. If this is the case, the provision is likely to be struck down as a 'penalty' and the beneficiary will be entitled only to so much of the sum as will compensate for the actual loss which has been suffered.

17.2.4 Release of surety

Apart from limits expressly contained in a bond or guarantee, there are a number of situations in which a surety is entitled to avoid liability.

General principles

Today, the giving of bonds and guarantees is a lucrative commercial enterprise for financial institutions. In Victorian times, however, bonds were more commonly given by benevolent uncles to guarantee the debts of extravagant nephews! Perhaps because of this, the law appears rather too ready to find a reason for releasing the surety from the guarantee.

It has on occasion been held, by analogy with the law governing insurance contracts, that the surety cannot be held responsible unless all material facts (such as unusually difficult construction conditions) were disclosed when the guarantee was entered into. However, it was confirmed by the House of Lords in the leading case of *Trade Indemnity v Workington Harbour & Docks Board*[14] that there is no general rule to this effect. In that case a firm of contractors submitted the lowest tender (£284,000) for a civil engineering contract which, as is common, provided that the contractors must satisfy themselves as to site conditions. The engineers warned the contractors that their excavation prices were far too low for ground with high water levels and gave them an opportunity to withdraw. When the contractors maintained their tender figure the employers, advised by the engineers, insisted on a bond of £50,000. The bondsman later sought to be released, claiming that these matters should all have been disclosed. However, it was held that, since the contract clearly showed that adverse ground conditions were solely at the contractor's risk, the employers owed the bondsman no duty of disclosure.

It should also be borne in mind that, while the general law does not require the beneficiary to sue the principal before claiming against the surety, nor even to give notice to the surety that the principal is in breach of contract, either of these requirements might be imposed by the contract of guarantee. If this is so, then non-compliance by the beneficiary may entitle the surety to be released from the guarantee. In *Clydebank District Water Trustees v Fidelity Deposit of Maryland*,[15] for example, an employer's failure to give written notice of a contractor's delay, which might lead to a claim for liquidated damages, proved fatal to a call on the bond. This was despite the fact that the employer's claim did not relate to liquidated damages at all, but to the cost of having the work completed when the original contractor became insolvent. This might seem a thoroughly unreasonable outcome, but it has been held that a term requiring the employer to give written notice of every breach of contract by the contractor should not be struck down as 'unreasonable' under the Unfair Contract Terms Act 1977.[16]

Alteration of contract terms

Any material variation in the terms of the main contract releases the surety's obligation, since it alters the nature of what is guaranteed. Oddly, perhaps, this applies even to alterations that appear to be for the benefit of the principal, such as where the beneficiary waives or compromises a claim against the principal. The reason why this releases the surety is said to be that the removal of pressure from the contractor may make it more likely that the contractor will breach the contract and so render the surety liable.

In accordance with this approach, guarantors of payment obligations have been held to be released where the creditor agreed to accept a late payment from the principal, though not where the creditor merely acquiesced in a payment that was late. As for performance obligations, complete discharge of a surety has followed a knowing overpayment by the employer, the payment of an instalment before it was due, and agreement by the employer to give the contractor extra time to complete. However, an employer who *unintentionally* overpays the contractor will not thereby lose the right to claim against the surety, and nor will one who fails to discover or even acquiesces in the contractor's breaches of contract.

---

[14] [1937] AC 1.
[15] (1916) SC (HL) 69.
[16] *Oval (717) Ltd v Aegon Insurance Co (UK) Ltd* (1997) 85 BLR 97.

18 Role of the contract administrator

The purpose of employing an architect, engineer, or other professional on a building project is to give the employer the benefit of that professional's skill and experience. Traditionally, the person appointed has taken responsibility for two separate functions: translating the employer's needs into drawings, specifications and the like through the processes of briefing and design, and then administering the building contract, in the sense of monitoring the contractor's work in executing that design. This is carried out in order to ensure that the work complies with the designer's intentions and meets the required standards of workmanship and quality.

Although the same person can, and frequently does, perform the functions of design and contract administration, this is not essential. Indeed, under some forms of building procurement such as construction management, it is unusual. In any event, the *functions* of a contract administrator (whether described as architect, engineer, project manager or whatever) are completely separate from those of a designer.

Design responsibilities were considered in Chapter 13. This chapter is concerned with contract administration as such. As we shall see, a contract administrator fulfils two very different roles. First, there are those duties (such as providing necessary information to the contractor), which are carried out as agent of the employer. Second, there are certain decision-making functions (such as the certification of work properly carried out) in which the contract administrator is required to act fairly between the parties and exercise independent judgement.

18.1 CONTRACT ADMINISTRATOR AS THE EMPLOYER'S AGENT

A contract administrator who is employed to supervise the carrying out of building works may in certain respects be regarded as an agent of the employer. This has a number of important consequences, notably as to the extent to which the contract administrator can bind the employer by actions and the scope of the duty of care and skill owed by the contract administrator to the employer.

18.1.1 Extent of powers

In acting for the employer, there are limits to what a contract administrator can do in terms of forming or varying contracts, instructing work to be suspended and in delegating authority to others.

Contracts and variations

An agent can of course be *expressly* authorized to do anything on behalf of a client. However, the important question of law concerns the extent of authority that will be *implied* where nothing specific is said. This is important because, where a contract administrator acts without any authority, the employer will not be bound by what is done. In such circumstances, a third party who has suffered loss may make the contract administrator personally liable for damages, under a type of legal action called breach of warranty of authority.

As a general rule, the courts take a rather restricted view of a contract administrator's implied authority. In particular, it has been consistently held that, except in very unusual circumstances, a contract administrator has no authority to create a direct contract between the employer and a sub-contractor. Thus, where an architect promised a sub-contractor that the employer would pay directly for the work, it was held that the employer was under no obligation to do so.[1]

A similarly cautious approach is seen in cases concerning the alteration of an existing contract. Unless the contract administrator acts within the terms of a variations clause, there is no power to change what was originally agreed. In *Cooper v Langdon*,[2] for instance, a contractor deviated from the plans with the architect's permission. The contractor was nonetheless held liable to the employer for breach of contract, as the architect had no authority to vary the works in this way. Likewise, in *Sharpe v San Paolo Railway*,[3] where the claimants had contracted for a lump sum to perform all the works necessary for the construction of a railway in Brazil, it was held that the engineer had no authority to order as extras any work already impliedly included in the job.

As mentioned above, a contract administrator who exceeds his or her authority risks being held personally liable to a third party with whom he or she deals. In addition, the law of agency contains another trap for the unwary. This is that any agent who signs a written contract on behalf of a client will be treated as a party to it and thus personally liable, unless the contract itself makes it clear that it is signed merely 'as agent'. In *Sika Contracts Ltd v Gill*,[4] for example, where a letter accepting a supplier's tender was signed: 'BL Gill, Chartered Civil Engineer', the engineer was held personally liable to the supplier for the price, even though the latter *knew* that he was only acting as an agent.

Suspension of work

As a general principle, neither the employer nor the contract administrator has any legal right to order the contractor to suspend work. Once the contract work has commenced, it is the contractor's right and duty to carry it through in a regular fashion, and the employer's duty to do nothing that will prevent the contractor from doing so. As a result, an unjustified order to suspend work, given by the contract administrator, will constitute a breach of contract by the employer for

---

[1] *Vigers Sons & Co Ltd v Swindell* [1939] 3 All ER 590.
[2] (1841) 9 M&W 60.
[3] (1873) LR 8 Ch App 597.
[4] (1978) 9 BLR 15.

which the contractor may claim damages. Further, if the circumstances of the suspension are such as to show an intention no longer to be bound by the contract, this may constitute a 'repudiatory breach' (discussed in Chapter 23).

This general principle is substantially modified by most of the major standard form contracts. ICE 7 clause 40, for example, gives the engineer an unfettered discretion to order the suspension of the whole or any part of the works. Depending on the reason for this order, the contractor may or may not be entitled to claim any costs incurred as a result of it. Once such a suspension has lasted for three months, the contractor is entitled to require permission to resume. If this permission is not given within 28 days, then the contractor may treat the suspended part as omitted from the contract or, if the suspension order relates to the *whole*, may treat the contract as repudiated.

JCT SBC 05 does not use the word 'suspension' in this context, but most JCT contracts empower the contract administrator to *issue instructions in regard to the postponement of any work to be executed*. If such postponement relates to the whole of the contract works, it is effectively a suspension order. In these circumstances, the contractor under JCT SBC 05 is entitled to an extension of time and to reimbursement of any direct loss and/or expense. Further, if the suspension lasts for long enough, it becomes a ground for termination of the contractor's employment.

As far as sub-contracts are concerned, these will almost invariably give the main contractor a power of suspension, which parallels the power of the employer or architect under the head contract. It is essential for the protection of the contractor that this should be so for, if it is not, the contractor may incur liabilities to the sub-contractor that cannot be passed on. This is what happened in the Canadian case of *Smith and Montgomery v Johnson Bros*,[5] which concerned a sub-contract for tunnelling work. The main contract there empowered the engineer to order work to be stopped and provided that, if this occurred, the main contractor's only entitlement would be to an extension of time. When the engineer ordered the tunnelling work to be stopped, the main contractors repeated this instruction to the sub-contractors, whereupon the latter successfully sued them for damages. The court held that the main contract terms had not been incorporated into the sub-contract, so that the suspension order was a breach by the main contractors.

Delegation of authority

As a general principle, an agent is expected to act personally for a client, and not to delegate the task to a sub-agent. However, authority to delegate may be given expressly or by implication. On this basis it has been held that, where a contract administrator is instructed to invite tenders for the contract work on the basis of bills of quantities, authority to appoint a quantity surveyor will be implied.[6] What is not clear, however, is whether this creates a direct contract between the employer and the quantity surveyor, or whether it simply means that the contract administrator can recover from the employer sums paid to the quantity surveyor. Similar authority to appoint may be implied wherever the building contract

---

[5] *Smith and Montgomery v Johnson Brothers Co Ltd* [1954] 1 DLR 392.
[6] *Waghorn v Wimbledon Local Board* (1877) HBC 4th ed, ii, 52.

provides that variations are to be measured by a quantity surveyor. In such circumstances it is clear that different parts of the 'contract administration' are to be carried out by different people.

These implications aside, the RIBA Standard Form of Agreement (SFA/99) makes express provision for the appointment of a project manager, planning supervisor, quantity surveyor, structural engineer, building services engineer and site inspector/clerk of works. These appointments are to be made by the employer after consultation with the architect, and it is clearly pointed out that some or all of these might be inapplicable. The agreement also has an express condition that the architect cannot sub-let parts of the work without the express consent of the client.

At this point mention should also be made of the possibility that an employer will wish to appoint a site inspector or clerk of works, who will be permanently on site to act as the 'eyes and ears' of the contract administrator (see for example, JCT SBC 05 clause 3.4). Since such a person is almost invariably appointed by the employer directly, the division of supervisory duties between contract administrator and clerk of works can cause problems. SFA/99 carries an express term that such a person is *appointed under the direction of the Lead Consultant* but appointed and paid under a direct agreement with the client.

The general principle is that, while matters of detail may be left to a properly briefed clerk of works, the contract administrator remains responsible for seeing that important matters of design are properly carried out. Thus, where a clerk of works, for corrupt purposes, allowed the contractor to deviate from the design in laying concrete in a way that led to dry rot in the ground floor, the architect was held liable to the employer.[7] It was said that the architect might justifiably have supervised the laying of concrete in the first building, leaving the remainder to the clerk of works; here, however, the architect had not supervised or inspected at all.

Notwithstanding this decision, it is clear that failure of proper supervision by a clerk of works is something that may rebound on the employer. Thus, for example, where a clerk of works negligently failed to notice defects in the way that artificial stone mullions were fixed, the damages which the employer was awarded against the architects for negligence were reduced by 20%, on the ground of contributory negligence by the clerk of works for which the employer was responsible.[8]

18.1.2 Functions and duties

In carrying out 'administrative' functions under a building contract, the contract administrator owes a duty of reasonable care and skill to the employer. If that duty is breached, the contract administrator will be liable in damages to the employer for any resulting loss. This is what happened in *West Faulkner Associates v Newham LBC*,[9] where a contractor was in serious breach of the obligation to maintain regular and diligent progress. The Court of Appeal held that the architect should have served a notice determining the contractor's employment, and was liable to the employer for various losses flowing from the fact that no such notice was served.

---

[7] *Leicester Board of Guardians v Trollope* (1911) 75 JP 197.
[8] *Kensington & Chelsea & Westminster AHA v Wettern Composites Ltd* [1985] 1 All ER 346.
[9] (1994) 71 BLR 1.

In considering the contract administrator's functions, two areas of special importance relate to the giving of advice and information and to the monitoring of the construction work.

Advice to the employer

The kind of matters on which it is reasonable to expect a contract administrator to give advice will naturally vary with the type of project, the professional background of the contract administrator and so on. However, some cases in which architects have been held to owe a duty of care may give some indication. In the first of these,[10] the architect greatly underestimated the extent of work required to reinstate an old property. As a result the client, having purchased the property, could not afford to refurbish it. In the second case[11] the architect, in preparing estimates of the cost of a project, completely overlooked the effects of inflation. Once again the client was forced to abandon the project. In each of these cases it was held that the architect was not entitled to any fees, since their work was worth nothing to the client.

In addition to matters of cost, a contract administrator may well advise the employer on the appointment of particular contractors or sub-contractors. This is especially likely where a tendering process has taken place. In such circumstances, a duty of care is again owed. Thus, where an architect recommended a particular contractor as 'very reliable', the architect was held liable for losses suffered by the employer when the contractor executed defective work and then became insolvent.[12]

Another matter on which employers depend upon the contract administrator is advice about rights and duties under the building contract. It is extremely important that a contract administrator explains the terms of the building contract to the employer before the contract is executed. In *William Tomkinson and Sons Ltd v Parochial Church Council of St Michael and Others*,[13] the architect under a JCT minor works contract did not advise the employer to take out insurance for the works. As a result, when damage to the works caused losses far in excess of the contract value, the employer had no insurance protection. The architect was held liable in negligence for the employer's losses. Similar liability for failing to see that a contractor had the required insurance in place has been imposed upon a project manager.[14]

A contract administrator, in the course of work, may well become aware that other consultants employed by the same client have been guilty of breaches of duty. It has been held that the contract administrator (or, for that matter, any consultant) is under an implied obligation to inform the employer of such breaches. However, the contract administrator is under no implied obligation to inform the employer of his or her own breaches.[15]

---

[10] *Ralphs v Francis Horner & Sons* (1987, unreported).
[11] *Nye Saunders and Partners v Bristow* (1987) 37 BLR 92.
[12] *Pratt v George J Hill Associates* (1987) 38 BLR 25.
[13] (1990) 6 Const LJ 319.
[14] *Pozzolanic Lytag Ltd v Bryan Hobson Associates* [1999] BLR 267.
[15] *Chesham Properties Ltd v Bucknall Austin Project Management Services Ltd* (1996) 82 BLR 92.

Instructions to the contractor

Contract administrators have no general right under building contracts to issue instructions to contractors. However, in practice standard form contracts always contain a clause permitting instructions to be issued. This is because the duration and complexity of construction projects mean that conditions are likely to change, and it is recognized that it may thus not be possible to deal in advance with every eventuality that may arise.

Contractual clauses dealing with this matter are often very widely drafted. GC/Works/1, for example, lists in clause 40 a large number of matters on which the project manager may issue instructions, before concluding with *any other matter which the PM considers necessary or expedient.* ICE 7 clause 13(1) is hardly more limited; having restated the contractor's basic obligation to build in accordance with the contract, it requires the contractor to *comply with and adhere strictly to the Engineer's instructions on any matter connected therewith (whether mentioned in the contract or not).*

The position under JCT SBC 05 is more restricted, for the contractor's duty of compliance (under clause 3.10) only applies to written instructions that the contract administrator is expressly empowered by the conditions to issue. If there is any doubt about this, the contractor can ask the contract administrator to specify in writing the provision under which the instruction is issued, following which any further dispute can be taken immediately to dispute resolution. Assuming that an instruction *is* justified, the contractor must comply with it within seven days, failing which the employer is entitled to have it carried out by someone else at the contractor's expense. On the other hand, where compliance with an instruction involves the contractor in delay, it is possible for an extension of time to be claimed. As to any extra cost involved, the contractor will normally be reimbursed for this either through the provisions for valuing variations or through those relating to loss and/or expense.

Information to the contractor

In addition to issuing instructions, the contract administrator has an important function as a source of relevant information to the contractor. Under JCT SBC 05, this is recognized by clause 2.8.2, which requires the contract administrator to furnish the contractor with one copy of the contract documents certified on behalf of the employer, two further copies of the contract drawings and (where applicable) two copies of the unpriced bills of quantities. These documents are to be furnished immediately after the execution of the contract. Clause 2.9.1 adds that the contract administrator shall provide *so soon as is possible* two copies of any further information necessary to enable the contractor to complete the works in accordance with the conditions.

Having thus started the contractor off with the basic information required, the contract administrator must keep this information up to date. The fifth recital of JCT SBC 05 identifies an Information Release Schedule (IRS), which states what information the contract administrator will produce and when it will be provided. The IRS may be modified by mutual agreement between the employer and contractor, but failure on the part of the contract administrator to keep to the IRS is one of the matters for which the contractor can claim extension of time and loss and/or expense. Moreover, Clause 2.12 provides that, quite apart from the IRS, the contract administrator shall provide such further drawings or details as are reasonably necessary either to explain the contract drawings or generally to enable the contractor to carry out the work. These are to be provided at the time it is reasonably necessary for the contractor to receive them.

One important aspect of the provision of information is dealt with separately under clause 2.10 of JCT SBC 05. This concerns the determination of any levels that may be required for the execution of the works (a level is a specified height above mean sea level, from which construction work can be set out). The contract makes this the responsibility of the contract administrator, who has further to furnish the contractor with sufficient accurately dimensioned drawings to enable the works to be set out at ground level. Provided this is duly done, the responsibility for any errors in setting out rests on the contractor. This will normally involve amending the errors, but the contract empowers the employer and contract administrator to deal with the matter instead by means of an 'appropriate' deduction from the contract sum.

An important question in this context is the extent to which a contract administrator may incur personal liability to a contractor for inaccuracies in information provided. It has long been settled that in drawing up plans, an architect or engineer does not guarantee they are practicable, nor that bills of quantities are accurate. In principle, therefore, contractors who wish to tender on the basis of such information must satisfy themselves as to its soundness. However, there are limits to this principle. For example, an architect who *fraudulently* gives inaccurate information at the pre-tender stage will be liable to the contractor. This is so even if the contract provides that the contractor must not rely on any representation contained in the plans.[16]

There seems no reason why a contract administrator should not, in similar circumstances, be held liable for a misstatement that is *negligent* rather than fraudulent. This was accepted in principle by the Court of Appeal in *J. Jarvis & Sons Ltd v Castle Wharf Developments Ltd*,[17] where a quantity surveyor was alleged to have misrepresented the planning position of a proposed scheme to one of the tendering contractors. However, the Court held that there was no duty of care on the facts, since an experienced design and build contractor could not reasonably be expected to place reliance on such statements.

Liability for a negligent misstatement was actually imposed in a New Zealand case[18] on an architect who negligently assured a contractor that he would receive full payment from the employer for certain work. The contractor relied on this assurance in completing the job and, when the employer failed to pay, the architect was held liable.

The most striking English case in this area is *Townsends v Cinema News*,[19] in which a contractor had been held liable to the employer for certain breaches of statute (installation of toilets in contravention of a by-law, and failure to serve

[16] *S Pearson & Son Ltd v Dublin Corporation* [1907] AC 351.
[17] (2001) 17 Const LJ 430.
[18] *Day v Ost* [1973] 2 NZLR 385.
[19] *Townsends (Builders) Ltd v Cinema News Property Management Ltd* [1959] 1 All ER 7.

notices). The Court of Appeal held that, since the architect had led the contractor to rely on him to serve the relevant notices and ensure compliance with the by-laws, the contractor was entitled to recover from the architect the damages that he had to pay the employer.

Inspection and monitoring of the contract work

Where a qualified architect is appointed to be contract administrator, the terms of engagement used are often those contained in the RIBA Standard Form of Agreement SFA/99 (as to this form see Section 2.4). These make clear that, while the architect will visit the site, the requirement is to make such visits (not 'inspections') *as the architect at the date of the appointment reasonably expected to be necessary*. The guide accompanying the agreement points out that the obligations of the architect as designer under SFA/99 are different from those as contract administrator, although somewhat unhelpfully gives no indication as to how an architect would become obliged to administer the contract. SFA/99, in its 1999 version, appeared to exclude altogether the architect's liability for failure to check work in progress. This is because it provided in clause 3.12 that the employer would hold the contractor or consultant *and not the architect* responsible for the proper execution of their respective work. However, the 2004 version of SFA/99 has omitted the relevant passage in this clause. Thus, it now seems to be in line with the view that a contractual clause will not override the architect's traditional duty to check general compliance,[20] or the express term in SFA/99 obliging the architect to 'co-operate with others'.

Quite apart from specific terms of appointment, the limited nature of an architect's obligation to monitor work is in any event well recognized by the English courts.[21] By contrast, an Australian court held an architect liable for the collapse of concrete that was poured between site visits, which took place on a twice-weekly basis.[22] This strict view of the architect's duties should also be compared with the case of *Clayton v Woodman*, which is discussed below.

JCT SBC 05 clearly recognizes that the contract administrator is entitled to inspect all work executed. The contractor's basic obligation, imposed by clause 2.1, is to produce the building according to the contract documents. While the contract administrator will naturally wish to check that this is indeed being done, the actual *need* to inspect relates only to those matters required to be to the contract administrator's satisfaction. In this context it may be remembered that any such stipulation means that work must be to the *reasonable* satisfaction and not the *absolute* satisfaction of the contract administrator.

As part of the function of inspecting the contractor's work, the contract administrator is given considerable powers under clause 3.17 to carry out tests of any materials and to order that work which has been covered up by other work should be opened up for inspection. The cost of making good and restoring the covering work, and any work found to be defective, depends on the status of the

---

[20] *Investors in Industry Ltd v South Bedfordshire DC* [1986] 1 All ER 787.
[21] *East Ham BC v Bernard Sunley & Sons Ltd* [1966] AC 406; *Sutcliffe v Chippendale & Edmondson* (1971) 18 BLR 149.
[22] *Florida Hotels Pty Ltd v Mayo* (1965) 113 CLR 588.

inspected work. If no defects are discovered, then the contractor must be paid for the cost of opening up and making good as if this was due to a variation. If the inspected work is defective, then this is to be made good at the contractor's expense. What is more, once defects have been discovered, clause 3.18 entitles the contract administrator to demand further tests. Provided that any such demands are in accordance with the JCT Code of Practice (which is set out in Schedule 4 of JCT SBC 05), the cost of them is borne by the contractor *whether or not any further defects are discovered.*

The contract administrator clearly owes a duty of care and skill to the employer to detect bad workmanship and defects.[23] However, due to the limited nature of the duty to inspect it cannot be said that every failure of detection will amount to negligence. Further, while it is sometimes said that a similar obligation is owed to contractors, this seems unlikely; they are responsible for their own monitoring of quality. The most that can be said is that, where the state of the work is positively *dangerous*, there may be an obligation to warn the contractor of this.[24] Of course, common sense suggests that, regardless of the strict legal position, a contract administrator who is aware of some defective work should certainly communicate this to the contractor.

Mention of danger raises the issue of whether the contract administrator has power to control the contractor's *method* of working, especially when this method proves to be dangerous. A power of this nature is undoubtedly given to the engineer by clause 13(2) of ICE 7. However, JCT contracts contain no such provision, and this has an important effect on the allocation of responsibility in case of accidents, as is shown by two contrasting cases.

In *Clayton v Woodman & Son (Builders) Ltd*,[25] the claimant bricklayer was injured by the collapse of a wall in which he was cutting a groove to take a concrete floor. The architect had insisted that this existing wall be incorporated into the new structure, rather than being demolished and replaced as the bricklayer had suggested. It was held that it was not the architect's duty under the RIBA form of contract to advise the builder on what safety precautions to take or how to conduct his operations (which specifically included shoring up and supporting walls and floors). As a result, the architect was not liable.

By contrast, in *Clay v AJ Crump Ltd*,[26] an architect agreed that demolition contractors preparing a site should leave a particular wall standing. This was at the suggestion of the demolition contractors, and the architect did not personally inspect the wall. The wall subsequently collapsed, killing two men and injuring the claimant, an employee of the main contractors. On this occasion it was held that the main contractors were entitled to assume, from the very fact that the wall had been left, that it had been inspected and declared safe. The architect was accordingly liable in negligence to the claimant.

The contract administrator may also have obligations under the Construction (Design and Management) Regulations 2007, particularly if he or she is also the planning supervisor for the purposes of the Regulations.

---

[23] *Imperial College of Science & Technology v Norman & Dawbarn* (1987) 8 Con LR 107.
[24] *Oldschool v Gleeson (Construction) Ltd* (1976) 4 BLR 103.
[25] [1962] 2 QB 533.
[26] [1964] 1 QB 533.

Quantity surveying functions

As mentioned above, certain aspects of contract administration are frequently delegated to a qualified quantity surveyor. Indeed, whether or not delegation takes place, these tasks must be performed with the skill and care that a quantity surveyor would bring to them. As to what this involves, there is little legal authority, although a quantity surveyor in an old case escaped liability for negligence where an error of £118 in a £12,000 contract was due to an arithmetical slip by a normally competent clerk.[27] On the other hand, where a local authority's quantity surveyor had accepted ridiculously high rates for work from a contractor, this was held negligent. As a result, when the contractor went into liquidation and the overpaid money was lost, the district auditor was entitled to surcharge the quantity surveyor.[28]

18.2 CONTRACT ADMINISTRATOR AS INDEPENDENT CERTIFIER

As mentioned at the beginning of this Chapter, a contract administrator has a significant part to play in exercising judgement and reaching decisions on various matters under the contract. In so doing the contract administrator acts, not as the agent of the employer, but as an independent professional.

18.2.1 Certification

The most important aspect of these decision-making powers relates to the issue of **certificates**. These have been defined by Wallace (1995), in a definition approved by the Court of Appeal,[29] as *the expression in a definite form of the exercise of the judgement, opinion or skill of the engineer, architect or surveyor in relation to some matter provided for by the terms of the contract.* However, this does not mean that every expression of opinion or decision given by the contract administrator will amount to a certificate. It will only be a certificate if it is so described in the contract, or can be so treated by implication.

The general law of building contracts does not require a certificate to be given in any particular form. Indeed, it may be given orally unless the contract provides otherwise. Interestingly, neither JCT SBC 05 nor ICE 7 specifically states that certificates are to be issued in writing. However, since both forms of contract require certificates to be sent to the employer, with a 'copy' to the contractor, the assumption is clearly that they are to be issued in writing.

Types of certificate

There are three main types of certificate found in construction contracts:

---

[27] *London School Board v Northcroft* (1889) HBC 4th ed, ii, 147.
[28] *Tyrer v District Auditor for Monmouthshire* (1973) 230 EG 973.
[29] *Token Construction Co Ltd v Charlton Estates Ltd* (1973) 1 BLR 50.

1. *Interim certificates*. These are issued at intervals as the work proceeds, and their issue entitles the contractor to be paid a certain proportion of the contract price. Under JCT SBC 05 clause 4.9.2, the period between interim certificates is whatever is stated in the contract particulars. If none is stated then it is one month, which is the usual period. As mentioned in Section 15.1.2, the regular flow of cash can be critical to a contractor's survival.
   The amount to be included in an interim certificate under JCT SBC 05 should cover the value of work done and materials delivered to date, plus the value of certain off-site materials. The contract places the responsibility for carrying out interim valuations upon the quantity surveyor. However, it is important to note that the QS should not do more than merely carry out the valuations; it is for the contract administrator to calculate what is due and to issue the interim certificates.
2. *Final certificates*. The final certificate can signify the contract administrator's satisfaction with the work, or the amount that is finally due to the contractor, or both of these things. The extent of its effect depends on the terms of the contract, but under JCT SBC 05 it clearly applies to both these issues. Under clause 4.15, the obligation is in general to issue the final certificate within two months of the end of the rectification period. In practice, it usually takes much longer than this to issue final certificates (Latham 1994). Indeed, on some projects final certificates are only issued years after completion, rather than months (Hughes 1989). There are three potential reasons for this situation. First, the last portion of the retention may be too small to worry a contractor, especially if the employer has deducted some part for any reason. Second, consultants may find that other projects, at earlier stages in their development, are more lucrative than those that can yield only the final part of the fee account. Third, on issuing the final certificate, the project will be covered by the consultant's professional indemnity insurance, and the premium may have to change to reflect this.
3. *Certificates recording an event*. In addition to confirming that a sum of money is due to the contractor, certificates may be needed to confirm that a certain event has occurred (or has not occurred). The use of this kind of certificate varies from one form of contract to another, but a good example of their use and importance may be drawn from JCT SBC 05, where the contract administrator is required to issue the following certificates:
   - *Non-Completion Certificate*. Issued under clause 2.31, this records the contractor's failure to complete the works by the completion date. It triggers the contractor's liability to pay liquidated damages.
   - *Practical Completion Certificate*. Issued under clause 2.30, this records the contract administrator's opinion that practical completion of the works has been achieved. The contractor's liability for liquidated damages ceases, half the retention money is released and the Rectification Period begins.
   - *Certificate of Making Good*. Issued under clause 2.39, this records the contract administrator's opinion that defects appearing within the defects liability period and notified to the contractor have been duly made good. The contractor is then entitled to the remainder of the retention money.

Conclusiveness of certificates

One of the most important issues concerning certification in building contracts is whether a certificate is legally binding upon the parties as to what it certifies. This question of conclusiveness depends upon the terms of the particular contract, but it would be extremely unusual to find any certificate except the final certificate treated in this way. JCT SBC 05, for example, makes it clear in clause 1.11 that nothing in any other certificate is conclusive evidence that work or materials are in accordance with the contract. This means that it is open to the employer to show that, despite the certificate, the contractor is not entitled to payment. It also means that any interim certificate can be corrected by the contract administrator when the next interim certificate is issued.

As mentioned, the issue of conclusiveness arises in practice only in relation to final certificates. The first point to make about this is that any contract clause that permits an arbitrator or adjudicator to 'open up, review and revise any certificate', inevitably destroys the conclusiveness of the certificate, at least to that extent. However, the limits of such clauses may be tightly drawn. In JCT SBC 05, for example, clause 1.10 states that the final certificate is 'conclusive evidence' in adjudication, arbitration or legal proceedings of certain matters, except where those proceedings have already commenced before the final certificate is issued, or are commenced within 28 days thereafter.

The matters on which the final certificate is 'conclusive evidence' are as follows (JCT SBC 05 clause 1.10.1):

1. That where the quality of materials or the standard of workmanship are expressly required to be to the reasonable satisfaction of the contract administrator, they are to his or her satisfaction.
2. That all appropriate additions and deductions have been made to the contract sum (though clerical or arithmetical errors can still be corrected).
3. That all extensions of time are correct.
4. That all the contractor's money claims have been properly accounted for.

As to the first of these matters, the Court of Appeal in *Crown Estate Commissioners v John Mowlem & Co Ltd*[30] interpreted the predecessor of clause 1.10 in such a way that, unless proceedings in respect of any defects were commenced within 28 days of the issue of the final certificate, the employer would lose all rights to complain of those defects. This decision was a controversial one, but it has been followed in a case where an architect, having been sued by the client, sought to hold the contractor partly responsible under the *Civil Liability (Contribution) Act 1978*. This claim failed, since it was held that the contractor could not have been liable to the client in view of a conclusive final certificate.[31]

The ruling in *Crown Estate Commissioners v Mowlem* was a much wider interpretation of clause 30.9 than the Joint Contracts Tribunal had intended, and the matter was duly dealt with by an amendment to JCT 98 which is now adopted by JCT SBC 05. This means that, unless the parties have made it clear that they intend the contract administrator to be the sole judge as to whether or not materials or workmanship comply with the contractual standards, the issue of the final

---

[30] (1994) 70 BLR 1.
[31] *Oxford University Fixed Assets Ltd v Architects Design Partnership* (1999) 64 Con LR 12.

certificate does not bar the employer from subsequently claiming that the contractor is guilty of a breach of contract through defects in the work.

An example of a conclusive final certificate may be found in the Institution of Chemical Engineers Model Form of Conditions of Contract for Process Plants (revised 1981) (the 'Red Book'). It has been held that clause 38.5 of that contract makes the issue of the final certificate conclusive evidence that the contractor has completed the works and made good all defects in accordance with his obligations under the contract; the only exception to this is where the issue of the certificate was procured by fraud.[32]

The JCT Design and Build form (JCT DB 05) makes no provision for a final certificate as such, but provides that agreement on the final account is conclusive evidence that, where it is stated in the employer's requirements that materials or workmanship are to be to the reasonable satisfaction of the employer, they are to his satisfaction. In *Barking and Dagenham LBC v Terrapin Construction Ltd*,[33] it was held by the Court of Appeal that the wording of this clause's predecessor (in JCT with Contractors Design (1981) Form of Contract) produces a similar effect to that in *Crown Estates v Mowlem* in respect of workmanship and materials but that, since it does not mention defects of design, it has no relevance to these.

Even where a final certificate is not stated by the contract to be conclusive, a question which troubled the courts for a time was how exactly that certificate was to be challenged. In the notorious *Crouch* case[34] it was held by the Court of Appeal that, where a contract gave power to an arbitrator to 'open up, review and revise' certificates, this meant that a *court* would have no equivalent power. Hence, if the strict time limits for going to arbitration were not complied with, the certificate would become in effect unchallengeable. This decision, after years of receiving heavy criticism, was finally overruled by the House of Lords in *Beaufort Developments (NI) Ltd v Gilbert-Ash NI Ltd*.[35] Their lordships held that, although the power of the courts to overturn certificates can be excluded or restricted by clear words in a contract, and also (at least in theory) by implication, the mere presence of an arbitration clause certainly does not have any such effect.

It should finally be noted that, even where a certificate is conclusive and unchallengeable *on its merits*, there are still grounds on which it can be set aside. The most important circumstances in which this will occur are:

1. Where the certificate is not issued in the correct form at the correct time by the correct person. A departure from such contractual requirements will invalidate a certificate where it is no longer clearly and unambiguously the required certificate in form, substance or intent.[36]
2. Where the contract administrator has 'certified' on something on which the contract gives no power to certify.
3. Where there is fraud or collusion between the certifier and one of the parties.
4. Where the employer has improperly pressurized or influenced the certifier.

---

[32] *Matthew Hall Ortech Ltd v Tarmac Roadstone Ltd* (1997) 87 BLR 96.
[33] [2000] BLR 479.
[34] *Northern Regional Health Authority v Derek Crouch Construction Co Ltd* [1984] QB 644.
[35] (1998) 88 BLR 1.
[36] *Cantrell v Wright & Fuller Ltd* [2003] BLR 412.

Recovery without a certificate

Is a contractor entitled to demand interim payments in the absence of an interim certificate? Is the employer entitled to deduct liquidated damages without a certificate of failure to complete the works on time? The answers to these questions depend on whether the issue of the relevant certificate by the contract administrator is a 'condition precedent'.

The general principle adopted in such cases is that, where the contract provides for payment of money following the issue of a certificate, this is indeed a condition precedent. Thus, where a certificate of non-completion is followed by the grant of an extension of time, a further certificate of non-completion must be issued before it is lawful for the employer to deduct liquidated damages.[37] Conversely, a contractor who feels that work has been undervalued on an interim certificate should either demand a further certificate from the contract administrator or seek arbitration under the contract. Such an undervaluation does not justify the contractor in leaving the site.[38]

Notwithstanding this general principle, a contract administrator's refusal or failure to issue a certificate is no more conclusive and binding upon the parties than the issue of one would be. Thus, if the circumstances are such that the *issue* of a certificate could successfully be challenged, it seems that its *non-issue* can equally be bypassed. This means that the points mentioned in the previous section on the conclusiveness of certificates are equally relevant.

In accordance with these principles, contractors have succeeded in recovering where an employer improperly ordered the architect not to certify more than a certain amount,[39] and where the certifier based a decision on whether work had been done *economically* (a question not within the certifier's remit).[40] Similarly, where an employer's claim against the contractor for defective work would have been barred by a 'conclusive' final certificate *which should have been issued*, it was held that the employer could not rely on the absence of the certificate to bring a claim.[41]

18.2.2 Other decision-making functions

Although the issue of certificates is the most important aspect of the contract administrator's 'independent' role, it is not the only one. Construction contracts may also use other forms of words, such as requiring the contract administrator to 'make decisions' or to 'give opinions'. It seems reasonable to assume that the principles outlined above as to 'conclusiveness' and 'conditions precedent' in relation to certificates would apply equally to these similar functions.

An example of 'certification by another name' is provided by JCT SBC 05 clause 2.33, which states that the employer may take possession of part of the works before practical completion is achieved. Where this occurs, the contract

---

[37] *A Bell & Son (Paddington) Ltd v CBF Residential Care & Housing Association* (1989) 46 BLR 102.
[38] *Lubenham Fidelities & Investments v S Pembrokeshire DC* (1986) 33 BLR 39.
[39] *Hickman & Co v Roberts* [1913] AC 229.
[40] *Panamena Europea Navegacion v Leyland & Co Ltd* [1947] AC 428.
[41] *Matthew Hall Ortech Ltd v Tarmac Roadstone Ltd* (1997) 87 BLR 96.

administrator is required to issue a 'written statement' identifying the part taken into possession. This statement is then treated for most purposes as if it were a certificate of practical completion for the relevant part.

An extremely important function of this kind was contained in clause 66(3) of ICE 7 before the amendment of this clause in 2004. The old version of the clause provided that any dispute whatsoever between the employer and the contractor arising out of the contract or the work must be referred in writing to the engineer for settlement. The engineer must then give a decision in writing on the dispute. Only when this has been done (or when the engineer has failed to give a decision within a prescribed period) could the matter be taken to adjudication or arbitration. This provision has caused serious problems, because the contract does not lay down any particular form which the engineer's 'decision' must take, and it can sometimes be difficult to know whether a decision has been given or not. However, according to the new version of clause 66 of ICE 7 (2004) it is no longer necessary to refer a dispute to the engineer before a referral to arbitration. Instead, the parties now have a free choice between several means of dispute resolution (see Section 24.5).

18.2.3 Liability for negligent decision-making

A question that has been the subject of important litigation in recent years is whether a contract administrator can be held liable to either of the contracting parties for any decisions.

Liability to the employer

It was for many years believed that, in issuing certificates, a contract administrator was acting in a 'quasi-judicial' capacity. By this was meant that, in exercising judgement about the quantity of work done, and making decisions about how much the contractor should be paid, the architect was acting in a manner similar to that of an arbitrator. This led to the conclusion that the contract administrator should enjoy a similar immunity from claims in negligence as an arbitrator (an immunity which is itself based on that which protects a judge in court).

The principle that a contract administrator could not be sued in negligence by the employer was laid down by the Court of Appeal in 1901.[42] However, in the important case of *Sutcliffe v Thackrah*[43] this principle was overturned by the House of Lords. In that case the architect over-valued a series of certificates and the employer duly paid the contractor on them. Unfortunately the contractor then went into liquidation before the job was completed, with the result that the employer could not recover the money that had been overpaid. It was held by the House of Lords that the architect was not acting in a quasi-judicial capacity and had no immunity from liability. The architect was accordingly liable to compensate the employer for the money lost.

---

[42] *Chambers v Goldthorpe* [1901] 1 KB 624.
[43] [1974] AC 727.

The decision of the trial judge in the *Sutcliffe* case[44] is of considerable interest in examining the practical implications of a duty of care in respect of certification. In particular, it appears that the contract administrator must notify the quantity surveyor in advance of any work regarded as not properly executed, so that it can be excluded from the quantity surveyor's valuation.

It appears that the duty of care a contract administrator owes to the employer applies, not only to certification as such, but also to the other decision-making functions. Liability may thus arise where a contract administrator, by negligently granting extensions of time to which a contractor is not in truth entitled, causes the employer to forfeit liquidated damages.[45]

Liability to the contractor

Once it had been decided that a contract administrator could be liable to the employer for negligent *over-certification*, a question which naturally arose was whether there would be an equivalent liability to the contractor for negligent *under-certification*. It was said that under-certification could cause a contractor substantial loss because, even if the certificate were ultimately corrected (for example by an arbitrator), so that the contractor recovered the correct amount from the employer, there would still be an expensive interruption of the contractor's cash flow.

The idea that a contract administrator would owe a duty of care *in tort* to the contractor was suggested by Lord Salmon in *Sutcliffe v Thackrah* itself, and this view attracted some support in subsequent cases.[46] However, it was disapproved by the Court of Appeal in the important case of *Pacific Associates Inc v Baxter*.[47] That case arose out of a dredging contract, made in a standard form (FIDIC 2nd edition), under which the claimant contractors undertook work for the ruler of Dubai. The contract provided that the contractors would be entitled to extra payment if they encountered hard material which could not have been reasonably foreseen by an experienced contractor, and made it the responsibility of the engineer to decide on any claims for extra payment on this basis. The contract also contained an arbitration clause (which meant that any decision of the engineer could be challenged) and a special condition stating that the engineer should not incur personal liability for 'acts or obligations under the contract'. In the course of dredging the contractors encountered a great deal of such material, but the engineer consistently rejected their claims (which amounted in total to some £31 million). The contractors duly settled in arbitration against the employer for £10 million and sued the engineer in negligence for the shortfall.

There is no doubt that an engineer in this situation could reasonably foresee that negligence in making a decision might cause loss to the contractors. However, the Court of Appeal held that it would not be 'just and reasonable' to impose a duty of care upon the engineer. This was because, given the contractual background, it

[44] *Sutcliffe v Chippendale & Edmondson* (1971) 18 BLR 149.
[45] *Wessex RHA v HLM Design Ltd* [1994] CILL 991.
[46] *FG Minter Ltd v Welsh Health Technical Services Organization* (1979) 11 BLR 1; *Shui On Construction Ltd v Shui Kay Co Ltd* (1985) 4 Const LJ 305; *Salliss & Co v Calil* (1988) 4 Const LJ 125.
[47] [1989] 2 All ER 159.

could not be said that the engineer had 'voluntarily assumed responsibility' to the contractors, nor that the contractors had 'relied' on him. If any extra responsibilities were to be undertaken, this could have been done by means of a collateral agreement at the time that the contract was negotiated. In the absence of any such agreement, and given that the contractors were fully aware of the contract provisions when tendering, the court would not allow the law of tort to import additional obligations into a carefully structured contractual environment.

Precisely how far this decision extends is not clear, because all three judges in the Court of Appeal stressed the importance of both the arbitration clause and the exemption condition. It *might* be therefore that the absence of one or both of these factors would lead to a contrary decision. Nonetheless, the court clearly did not favour tort claims in such circumstances, as a matter of general principle. In any event, the case has since been used by a court to justify its decision that an architect owes no duty of care whatsoever to a contractor in relation to certification duties under the contract.[48]

Quite apart from any possible liability in negligence, it appears that the contract administrator is not subject to the rules of natural justice in carrying out an independent decision-making role.[49] On the other hand, a contract administrator who colludes with the employer instead of exercising independent judgement, or who *deliberately* misapplies the contract, will probably be liable to the contractor under a tort known as wrongful interference with contract.[50]

---

[48] *Leon Engineering & Construction Co Ltd v Ka Duk Investment Co Ltd* (1989) 47 BLR 139.
[49] *AMEC Civil Engineering Ltd v Secretary of State for Transport* [2005] BLR 227.
[50] *John Mowlem & Co plc v Eagle Star Insurance Co Ltd* (1992) 62 BLR 126.

19   Sub-contracts

The UK construction industry is characterized by the prevalence of sub-contracts. Main contractors are, to an ever-increasing extent, reducing their dependence on directly employed labour. In consequence, most if not all projects involve some degree of sub-contracting, and in many cases the whole of the actual construction work will be carried out in this way.

19.1   REASONS FOR THE PREVALENCE OF SUB-CONTRACTING

Sub-contracting as a phenomenon is not unique to the construction sector. Indeed, practice in the construction sector seems to be following in the footsteps of many other non-construction businesses. There are many influences on a business that encourage sub-contracting, and it is useful to distinguish those pressures that are general (affecting all businesses), and those that are specific (affecting the unique position of construction). The general pressures for sub-contracting include the following:

- Non-wage costs of employment, such as training, pension rights, redundancy payments and sick pay.
- The increasingly diverse skill base required for the growth in complexity.
- The rising expectations of workers and a concomitant shift to free-lancing.
- The choice every firm faces between diversifying and contracting out.
- The perceived threat posed by trade unionization of permanently employed labour.
- Offsetting the risks associated with responsibility by transferring them.
- The need to employ specialists of proven reliability and repute.

The specific pressures for sub-contracting in the construction sector have combined in such a way as to make the idea of a general contractor (employing directly all of the labour) a thing of the past, until recently. These days, there is a resurgence of contractors with their own workforce and plant.

The factors affecting a contractor's decision to employ permanently or to sub-let are numerous, but are rarely directly considered as specific choices over whether to sub-let. In fact, contractors respond to the pressures on their businesses in the most appropriate way, and changes in the prevalence of sub-contracting are merely a symptom of these phenomena, each of which is dealt with below.

The nature of construction dictates that sites are geographically dispersed, meaning that some workers will be better placed for some sites than for others. Since an itinerant workforce is not common, the travelling time to the site from home will mean that employers are more inclined to employ local workers. Associated with the general pressure to specialize, mentioned above, is the fact that different types of project call for different types of skill. The pattern of required skill combinations is different for each project. Combined with the geographical constraints, it clearly makes little sense for contractors to keep permanently employed craftsmen in all of the necessary trades within each region of activity, unless they have sufficient work to keep them busy. This situation is tempered when the skill needed is specific to the firm itself, such as many middle-management roles, because in this situation the training overhead is high. Therefore, it is more economical to employ such specialists permanently, even if they are not fully utilized. Indeed, when there is rapid economic growth, labour shortages may become so acute that the difficulties of finding local labour may be more powerful than the pressures mentioned above. When labour is in short supply, it makes sense to hold on to people, even if this leads to inefficient use of their time.

In 1966, the government introduced Selective Employment Tax (SET), which was designed to tax firms on their payrolls. An immediate consequence of this tax was that, in an attempt to minimize their liability, many contracting firms sought alternatives to directly employed labour. Although SET was repealed in 1972, its effects on sub-contracting were compounded by the building industry strike in 1972, which led to a further disinclination on the part of general contractors to employ direct labour. The combined result of SET and the strike was a tremendous surge in the popularity of sub-contracting.

As a consequence of the increasing technical complexity identified in Chapter 1, there are ever-increasing numbers of specialist inputs into the building and construction processes. These take place because the more complex technologies need highly developed skills and expertise to apply them. Indeed, in many cases the increasing sophistication of the technology is a direct result of the increasing specialization of those who deal with it. Clients and designers frequently have no choice but to harness these new technologies; even if they are not particularly well disposed towards the implementation of new technologies, economic exigencies often force them into it. In this way, the construction process, which is essentially an assembly process dominated by a site-based labour force, becomes dominated by the need to specify 'high-tech' equipment at an early stage. Particularly for developers, failure to do so will result in eventual difficulty in letting the facility. This new equipment frequently takes a long time to procure, longer than the design and fabrication cycle for the whole building. One of the best ways to cope with this is to specify the installer of the equipment, leaving the main contractor no choice in the matter, so that the installer can initiate the process of getting the equipment ready by the time it is needed for installation. This obviates delays on the building site while the equipment is procured. This practice is one of the main reasons for the existence of nominated sub-contracting in particular, and for the growth of sub-contracting in general.

19.2   THE LEGAL BASIS OF SUB-CONTRACTING

It is an important principle of contract law that, where there is a contract between A and B, A may not unilaterally decide to be replaced by C. If A wishes to hand over all rights and obligations to C, and then simply drop out of the picture, there must be a **novation**. In effect, this is a new contract to replace the old one, and all three parties must agree on it. The law's view is that B has made this contract on the basis of holding A responsible on it and cannot, without consent, be deprived of the right to do so.

A novation, then, is a contract that transfers both rights and obligations from one of the original parties to a new party. An **assignment**, on the other hand, occurs where an original party transfers only contractual *rights*. For example, employers who are property developers might wish to assign to the first purchaser of a building their rights to claim against the contractor for breach of contract if any defects appear. Similarly, a contractor might wish to assign to a bank the future rights to be paid under a particular building contract, as security for a loan from the bank.

The law of contract permits assignments in principle, but this is subject to any terms in the particular contract. It should be noted that both JCT SBC 05 and ICE 7 contain restrictions. Under JCT SBC 05 clause 7.1), neither party may assign any rights under the contract without the other's written consent.[1] However, clause 7.2 provides an optional exception. If it is stated in the Contract Particulars) that this clause is to apply, then the employer can after practical completion assign the right to sue for defects to a purchaser or tenant of the building. As for the ICE conditions, clause 3 provides that neither the employer nor the contractor can assign without the other's written consent (which is not to be unreasonably withheld).

Leaving aside the issues of novation and assignment, there is the possibility that a party may wish to have its contractual obligations carried out by someone else, while remaining legally responsible for the performance of those obligations. This is known as **vicarious performance** or sub-contracting and, in principle, it is quite acceptable. However, it will *not* be permitted where the contracting party has been specifically selected on the basis of some personal qualification, skill or competence. In the building contract field, vicarious performance will be impliedly ruled out in respect of jobs calling for highly specialized skills, such as geological investigation of the site. Furthermore, while the physical work of construction is usually delegable, managerial functions (involving the co-ordination and control of the entire project) may not be, at least where an employer and a contractor have an on-going business relationship.

In any event, most modern standard form contracts place express limits on the extent to which the main contractor may satisfy its contractual obligations through the medium of sub-contractors. For example, JCT SBC 05 clause 3.7.1 provides that 'the Contractor shall not without the written consent of the Architect/Contract Administrator sub-let the whole or any part of the Works' The clause further sets out that the consent shall not be unreasonably delayed or withheld. The restrictions on sub-letting are reinforced by clause 3.9, which provides that any sub-contract must contain certain conditions that are designed to protect the employer's position. The importance attached to these matters is shown by the fact that unauthorized sub-letting is a ground for the employer to terminate the contractor's employment under clause 8.4.

---

[1] *See* Section 23.3.2.

Not surprisingly, the other contracts in the JCT range (for example DB 05, IC 05 and MW 05) deal with the basic question of sub-contracting in much the same way as does JCT SBC 05. That is to say, they require the contractor to obtain the consent of either the employer or the contract administrator and provide that such consent is not to be unreasonably delayed or withheld. By contrast, ICE 7 adopts a slightly more liberal approach. It provides in clause 4 that *the Contractor shall not sub-contract the whole of the Works without the prior written consent of the Employer*, and further that *except where otherwise provided ... the Contractor may sub-contract any part of the Works or their design*. In seeking to do this, the contractor must notify the engineer who has seven days in which to object, but only for good reason stated in writing.

A further aspect of the employer's right to control sub-contracting concerns the common practice of specifying a particular material and following this with some such phrase as 'or other approved'. It has been held by the Court of Appeal that this does *not* entitle the contractor to choose a substitute of equivalent quality.[2] The employer, through the contract administrator, has an absolute discretion to refuse any alternative, and need not even give reasons for such a refusal. It is possible that a phrase such as 'or other firm of similar quality' would give the contractor discretion, but this has not been tested in court.

19.3   THE CONTRACTUAL CHAIN

The basic position in law is that the main contract and the sub-contract (and the sub-sub-contract, if there is one) are regarded as links in a chain. The doctrine of privity of contract means that the rights and obligations contained in each contract apply only to those who are parties to it. Subject to the Contracts (Rights of Third Parties) Act 1999 (considered in Section 9.2.5), the main contract affects only the employer and the main contractor, the sub-contract affects only the main contractor and the sub-contractor, and so on. There may be other types of legal action (for example in tort) between parties who are not linked by a contract, but there can be no claim arising out of the contracts themselves.

This idea of 'chain liability' works perfectly well so long as all the links are intact. For example, where there are defects in the sub-contractor's work, the employer will have a contractual remedy against the main contractor, who will in turn take action against the sub-contractor. Similarly, the sub-contractor's right to payment will be exercised against the main contractor, who will be reimbursed by what is received from the employer. However, a chain is only as strong as its weakest link, and considerable problems arise as soon as one of the links breaks. What, for instance, is the position where the terms of the two contracts are significantly different, so that a liability may arise which cannot simply be passed down the chain? Or (an all too frequent occurrence in the construction industry) where one of the parties is insolvent and therefore unable to meet its liabilities?

The dangers of a broken chain are well illustrated by the Canadian case of *Smith and Montgomery v Johnson Bros.*[3] This concerned a sub-contract for tunnelling work, which was to be carried out 'according to the dimensions and

---

[2] *Leedsford Ltd v Bradford Corp* (1956) 24 BLR 49.
[3] [1954] 1 DLR 392.

specifications as set forth in the main contract'. A clause of that main contract empowered the engineer to order any work to be suspended and provided that, if this occurred, the main contractor would be entitled to an extension of time but not to compensation for loss and expense. Acting under this clause, the engineer ordered the tunnelling work to be stopped and the main contractor passed on this instruction to the sub-contractors. This, it was held, entitled the sub-contractors to recover damages from the main contractor for breach of contract. The sub-contract had no express term equivalent to that in the main contract permitting suspension of work, nor had the main contract term been incorporated by reference into the sub-contract.

19.4 DOMESTIC SUB-CONTRACTS

A 'domestic' sub-contractor, at least in theory, is one in whose selection and appointment the employer normally plays no part, other than simply giving consent where this is required under the terms of the main contract. The appointment of the sub-contractor is treated as being something entirely for the benefit of the main contractor, a purely 'domestic' matter. If the main contract concerns itself at all with such sub-contracting, it will usually merely insist that any sub-contract should contain certain clauses which are designed to protect the interests of the employer. These will cover such matters as termination (it is important that, if the main contract is brought to an end, the sub-contract should also fall) and the ownership of materials brought on to the site.

Until recently, the range of 'negotiated' standard-form main contracts in the construction industry was by no means matched at sub-contract level. However, JCT now publishes a number of sub-contracts for use with JCT main contracts, the Civil Engineering Contractors Association has produced a sub-contract for use with the ICE Conditions, and FIDIC has drafted a negotiated sub-contract for use with the FIDIC contract (Seppala 1995).

These examples apart, the terms of domestic sub-contracts are the result either of individual negotiation or, commonly, of imposition by one side or the other (which could well subject them to control under the Unfair Contract Terms Act 1977). As a generalization, it seems that firms who merely supply materials are more likely to have their own standard terms than those who also carry out work; the latter may well be employed on standard terms drafted by the larger main contractors. Such contracts have provoked fierce criticism from trade associations representing sub-contractors, as being one-sided to the point of unfairness. Common complaints relate to such matters as wide-ranging provisions under which the contractor may deduct from monies due in respect of cross-claims, requirements that all work be done to the main contractor's satisfaction, and the absence of any guarantee that the sub-contractor's work will be allowed to proceed in a regular and ordered manner. It should also be noted that domestic sub-contracts seldom involve the contract administrator appointed under the main contract, but leave such matters as interim payments, extensions of time and claims for loss and/or expense to be settled (and often disputed) by the contractor and sub-contractor.

In the absence of a formally drafted sub-contract (and domestic sub-contracts are frequently entered into on the most informal basis), parties sometimes attempt to incorporate terms by reference to other documents, such as a standard-form sub-contract with which they are familiar or the main contract itself. The problems caused by this latter practice, which runs the risk that a court may refuse to 'step down' all the relevant terms, have been considered in Section 10.3.

In some instances it is impossible, even with the aid of incorporation, to identify the express terms of a sub-contract in any detail. In such cases much will depend on what terms a court is prepared to imply. Some terms are readily implied, such as an undertaking by the main contractor not to cause the main contract to be terminated and thus deprive the sub-contractor of the benefits of the sub-contract.[4] Indeed, this is an area where judges, especially those of the Technology and Construction Court, have proved willing to imply quite detailed terms that will meet the normal expectations of the construction industry.[5]

However, it must always be borne in mind that there can be no implication of a term where there is an express term dealing with the matter in question. This is of particular importance as regards the timing of sub-contract works, a matter on which sub-contracts are often rather vague. For example, where a sub-contract stipulates that the work is to be carried out 'at such time or times as the contractor shall direct or require', it has been held by that there can be no implied term that the main contractor will make sufficient work available to the sub-contractors to enable them to operate in an efficient and economic manner.[6]

Before leaving the subject of 'domestic' sub-contracts, mention should be made of the procedure found in JCT SBC 05 clause 3.8. This enables an employer to gain at least some control over the identity of a sub-contractor, without accepting the risks attaching to nomination (discussed in Chapter 20). Clause 3.8 permits the contract bills to specify that certain work, although priced by the main contractor, is to be carried out by a person selected by the contractor from a list annexed to the bills. The list must contain at least three names and, if the number falls below three, either more names may be added by agreement of the parties or the main contractor may take on the work. Further, either party may in any event add names to the list, subject to the consent of the other, which is not to be unreasonably withheld. The result in practice is that a contractor may deprive the employer of control in this situation by simply adding a favourite sub-contractor to the list and then insisting on selecting that one.

If this procedure is adopted, the person selected then becomes a sub-contractor. Unfortunately, however, JCT SBC 05 fails completely to indicate what shall happen if such a sub-contractor fails to complete the work. Must there be a 're-listing', or can the main contractor simply take over the work? In either event, who is responsible for any extra costs involved? It seems highly probable that such questions (which, as we shall see in Chapter 20, have caused enormous problems in relation to nominated sub-contractors) will sooner or later also surface in relation to domestic sub-contracts under clause 3.8.

---

[4] *ER Dyer Ltd v Simon Build/Peter Lind Partnership* (1982) 23 BLR 23.
[5] See, for example, *DR Bradley (Cable Jointing) Ltd v Jefco Mechanical Services Ltd* (1988) 6-CLD-07-19.
[6] *Martin Grant & Co Ltd v Sir Lindsay Parkinson & Co Ltd* (1984) 29 BLR 31; *Kelly Pipelines Ltd v British Gas plc* (1989) 48 BLR 126.

19.5 DEFAULTS OF SUB-CONTRACTORS

A sub-contractor who is in breach of any term of the sub-contract will be liable to the main contractor, as the other party to that contract. And, of course, it should not be forgotten that the sub-contract may contain *implied* terms (as to workmanship, quality of materials and so on) equivalent to those in the main contract. In such a case the damages payable by the sub-contractor will reflect both the main contractor's own losses and also the amounts that the contractor in turn has to pay to the employer and, where appropriate, to other sub-contractors whose work is disrupted.

19.5.1 Liability of the sub-contractor to the employer

As with other relationships in construction, liability may in principle arise either in contract or in tort.

Liability in contract

The doctrine of privity of contract means that it is not possible for a sub-contractor to be made directly liable to the employer for a breach of the *main* contract. Equally the employer, who is not a party to the *sub-contract*, cannot claim damages for breach of it. This means that, unless the main contractor is liable to reimburse the employer for the loss caused by the sub-contractor, neither of them can sue the sub-contractor for that loss! The only possible exception to this would be if the sub-contractor specifically undertakes not to raise the 'no loss' defence when sued by the main contractor. Such an undertaking is given by Works Contractors operating under the JCT Management Contract (see Chapter 5). However, whether or not it is legally effective has not yet been tested in the courts.

Although the ordinary contractual structure does not provide for any direct claim by an employer against a sub-contractor, there is nothing to prevent the parties from creating an additional contractual link. This indeed is precisely what is done under those standard forms requiring a nominated sub-contractor to enter into a collateral agreement with the employer. However, while it is not unknown for an employer to insist upon direct warranties from a domestic sub-contractor, this is relatively unusual, except where the sub-contractor has an input in the design of the works.

Liability in tort

As we shall see in Chapter 20, there is at least the possibility that a **nominated sub-contractor** might incur direct liability to the employer in the tort of negligence in respect of work which is simply defective. However, even if such liability can arise, it appears very unlikely that it will extend to **domestic sub-contractors**, since their relationship with the employer is not normally of sufficient 'proximity' to carry with it a legal duty of care.

As for domestic sub-contractors who negligently cause physical damage, either to the contract works or to existing structures (in cases where the contract is not a new build), a different legal problem surfaces. It is whether something that appears to be a straightforward case of negligence may be affected by the way in which the main contract and sub-contract have chosen to allocate risks among the parties.

This problem arose in the case of *Norwich CC v Harvey*[7] where, in the course of constructing an extension to a swimming pool complex, a domestic sub-contractor negligently caused a fire that damaged both the contract works and the existing structure. The main contract (JCT 63) contained provisions relating to insurance under which all such fire damage was to be the sole risk of the employer, who was obliged to insure against that risk. The Court of Appeal held that the sub-contractors could not *directly* take the benefit of these provisions, since they were not party to the main contract in which they appeared. However, the court ruled that, against this background of contractual risk allocation, it would not be 'just and reasonable' to impose a duty of care upon the sub-contractors, who accordingly escaped liability for their negligence.

Although the allocation of risks under later versions of JCT contracts is not as clear as it was under JCT 63 (the phrase *sole risk of the employer* no longer appears), it has been held that the principle laid down in the *Norwich* case still applies, at least so as to exempt the *main contractor* from liability for damage falling within the insurance clauses.[8] However, it was held by the House of Lords in *British Telecommunications plc v James Thomson and Sons (Engineers) Ltd*[9] that, in assessing the position of a *sub-contractor*, it is necessary to consider the JCT insurance clauses (dealt with in Chapter 17) in their entirety. When this is done, it is clear that a sub-contractor who negligently causes a fire will escape liability only for damage to the contract works; damage to existing structures will be the sub-contractor's responsibility.

In *Scottish & Newcastle plc v GD Construction (St Albans) Ltd*,[10] it was held by the Court of Appeal that the position of a sub-contractor (and, for that matter, the main contractor) under IFC 84 is identical to that under JCT 98. However, decisions based on other forms of contract make it clear that the courts will be reluctant to exempt a sub-contractor from liability for negligence unless the contract wording is clear. In particular, it appears that, where a fire is caused through negligence, the wording of MW 05 provides no protection for either the main contractor[11] or a sub-contractor.[12]

19.5.2 Position of the main contractor

In theory at least, the appointment of a domestic sub-contractor is entirely for the main contractor's benefit. It is hardly surprising therefore that, as a general rule, any risks involved in such a sub-letting are to be borne by that main contractor.

---

[7] [1989] 1 All ER 1180.
[8] *Ossory Road (Skelmersdale) Ltd v Balfour Beatty Building Ltd* [1993] CILL 882.
[9] [1999] BLR 35.
[10] [2003] BLR 131.
[11] *London Borough of Barking & Dagenham v Stamford Asphalt Co Ltd* (1997) 82 BLR 25.
[12] *National Trust v Haden Young Ltd* (1994) 72 BLR 1.

The *contractual* duty of performance rests entirely on the main contractor and, if this is broken by the actions of a chosen sub-contractor, either in breaching the contract or in dropping out altogether, there can be no excuse. Thus, for example, where a contractor who was to install an up-and-over garage door for a client selected a lintel from a supplier's brochure, and this proved to be defective (it deflected some two years later), the contractor was held liable.[13] The contractor's duty to the employer was to supply suitable materials, not merely to exercise reasonable care in selecting them. Likewise, where a contractor, who was employed to carry out works including the replacement of a sprinkler system, sub-contracted the sprinkler work, the main contractor was held liable for a flood caused by the sub-contractor's negligence, in removing a sprinkler head without first shutting off the water supply.[14]

Of course, it is important not to lose sight of what the main contractor's obligations actually *are*. If, as will normally be the case under a conventional procurement method, the contractor has no responsibility in respect of design, then the contractor will not be liable for any loss or damage which results from a 'design fault' in the work of a domestic sub-contractor.

As well as incurring liability for the defaults of a domestic sub-contractor, a main contractor will in general have no claim against the employer for either time or money arising out of such defaults. However, regard must always be had to the terms of the particular contract. It was held by the House of Lords, in a case concerning a shipbuilding contract,[15] that latent defects in materials obtained for use in the work constituted 'matters beyond the contractor's control'. This meant that the contractor was entitled to be compensated by the employer under a loss and/or expense clause for losses resulting from those defects. The supplier in that case was in fact nominated by the employer, but the way the House of Lords expressed its decision would appear equally applicable to 'domestic' cases.

Interestingly, it appears that the main contractor's wide-ranging responsibility *in contract* for a sub-contractor's default has no parallel *in tort*. In *D & F Estates Ltd v Church Commissioners for England*,[16] the tenant of a flat brought an action in negligence against the main contractors who had built it, for defective plastering work carried out by a firm of sub-contractors. This claim failed on the ground that the tenant's loss was a purely financial one and that it was therefore not recoverable in tort (see Chapter 22). However, the House of Lords also ruled that, even if physical damage had occurred, the main contractors could not have been held responsible for it. This was because, not only were they not *vicariously* liable for the negligence of their sub-contractors, they did not even owe the claimants a duty of care in tort to supervise those sub-contractors. The somewhat surprising result is that, even if the main contractor is 'negligent' in failing to check the work of sub-contractors (and even if this failure amounts to a breach of the main contract), the main contractor cannot be liable for this in tort.

---

[13] *Lee v West* [1989] EGCS 160.
[14] *Raflatac Ltd v Eade* [1999] BLR 261.
[15] *Scott Lithgow Ltd v Secretary of State for Defence* 1989 SLT 236.
[16] [1988] 2 All ER 992.

19.6   RIGHTS OF SUB-CONTRACTORS

The legal rights of sub-contractors are predominantly against the main contractor, although they may also have some rights which are enforceable directly against the employer.

19.6.1 Rights against the main contractor

A sub-contractor's rights against the main contractor are concerned mainly with payment. This aspect is discussed before turning to other rights.

Payment

A straightforward application of contract law leads to the conclusion that the payment rights of a sub-contractor are to be found exclusively within the terms of the relevant sub-contract and that the main contract has no bearing on this issue. Even if the main contract expressly prohibits sub-contracting, a main contractor who disobeys that prohibition will be liable to pay the sub-contractor for work done.[17]

As we shall see in Chapter 21, a sub-contractor's demand to be paid is frequently challenged by the main contractor on the ground that the sub-contractor is guilty of delay, defective work or other breach of contract. Readers should refer to that Chapter, in view of the critical importance of this topic in practical contract administration. In the current context, we need only note that:

- An **express contractual right** to set off in this way must be exercised in strict accordance with any conditions laid down, such as a contract administrator's certificate of delay, or the service of a notice in an appropriate form.
- Any set off must also comply with the procedural requirements imposed by Part II of the Housing Grants, Construction and Regeneration Act 1996.
- The **common law right** of set off which is normally implied, and which is not subject to any conditions of this kind, will only be excluded by clear words in the contract.

In the past, many sub-contracts contained provisions to the effect that the main contractor would become liable to pay the sub-contractor, not when the relevant sum was *certified* by the contract administrator, but only when the main contractor *actually received the money* from the employer. To put it another way, these sub-contracts were based on the idea of 'pay when paid' rather than 'pay when certified'. Although none of the UK negotiated standard forms contained such terms, they were found in many of the non-standard forms of sub-contract that main contractors drafted and imposed on their sub-contractors by virtue of superior bargaining power. Statute has now greatly limited the effectiveness of such clauses in the UK (see the discussion below); however, they are still in everyday use in

---

[17] *O'Toole v Ferguson* (1912) 5 DLR 868.

standard form sub-contracts in the countries of South-east Asia, many of which are largely modelled on JCT forms.

Pay when paid clauses are inserted in sub-contracts for two main purposes. The first is to protect the main contractor's cash flow. This occurs because the main contractor will merely act as a channel of payment between the employer and the sub-contractor and will thus be in no danger of having to finance the sub-contractor's work. The second purpose, which is less obvious, is to make the sub-contractor carry the risk of the employer becoming insolvent. This will happen in circumstances where the employer's insolvency occurs after sub-contract work has been certified, but before the main contractor has been paid for it. Without a pay when paid clause, it is the main contractor who will bear the resulting loss, in the sense of having to pay the sub-contractor and then recoup whatever can be salvaged in the employer's liquidation. Under a pay when paid clause, by contrast, the main contractor need only pass on to the sub-contractor what can be recovered in the employer's liquidation, and even this will naturally be put first to meeting the main contractor's own claims.

The way in which pay when paid clauses shift risks on to sub-contractors has for a long time been criticized as unfair (Huxtable 1988). Nevertheless, courts in Hong Kong and Singapore have treated them as valid and enforceable, at least to the extent of protecting the main contractor's cash flow by merely delaying the obligation to pay the sub-contractor.[18] On the other hand, courts in both the United States[19] and New Zealand[20] have sought to interpret such clauses in such a way that they cannot make the sub-contractor bear the risk of the employer's insolvency. Those courts have effectively held that 'pay when paid' does not also mean 'pay if the main contractor is ever paid'.

The Latham Report (Latham 1994) was highly critical of pay when paid clauses and recommended that their use should be prohibited. That recommendation has been put into effect, at least in part, by section 113 of the Housing Grants, Construction and Regeneration Act 1996. This provides that any term in a construction contract that makes payment conditional on the payer receiving payment from a third party is ineffective. If this occurs, the parties must agree alternative arrangements for payment and, if they are unable to do so, the payment provisions of the statutory Scheme for Construction Contracts (see Section 15.1.2) will apply. However, this invalidation of pay when paid terms does not apply where the third party in question is insolvent. The overall effect, therefore, is that UK sub-contracts can no longer use this form of protection for the main contractor's cash flow, but can continue to make sub-contractors carry the risk of the client becoming insolvent.

---

[18] *Schindler Lifts (Hong Kong) Ltd v Shui On Construction Co Ltd* (1984) 29 BLR 95; *Nin Hing Electronic Engineering Ltd v Aoki Corporation* (1987) 40 BLR 107; *Brightside Mechanical & Electrical Services Group Ltd v Hyundai Engineering & Construction Co Ltd* (1988) 41 BLR 110.
[19] *Thomas J Dyer Co v Bishop International Engineering Co* 303 F 2d 655 (1962).
[20] *Smith & Smith Glass Ltd v Winstone Architectural Cladding Systems Ltd* [1992] 2 NZLR 473.

Other rights

The rights of a sub-contractor against the main contractor are of course dependent upon the terms of their sub-contract, so that no precise list can be compiled. It is of course possible for terms to be implied into the sub-contract, but an implied term will not be allowed to override clear express terms. For example, where a sub-contract stipulates that the work is to be carried out 'at such time or times as the contractor shall direct or require', there can be no implied term that the main contractors will make sufficient work available to the sub-contractors to enable them to operate in an efficient and economic manner.[21] On the other hand, it has been held that the express terms of DOM/2 (a sub-contract drafted for use with CD 98, the JCT design and build form of contract) are in no way inconsistent with an implied term requiring the main contractor to provide a sub-contractor with correct information concerning the works at such times as are reasonably necessary to enable the sub-contractor to fulfil all its obligations.[22]

One term which *is* likely to be implied into any sub-contract, if the matter is not covered expressly, is that the main contractor will not deprive the sub-contractor of the opportunity of carrying out the work and thus earning money. If the main contractor is guilty of a breach of the main contract that leads to the employer ejecting both the main contractor and the sub-contractor from the site, the sub-contractor will be entitled to recover damages from the main contractor.

This is what happened in *Dyer v Simon Build*[23] where, under a previous version of the ICE form of contract, the main contractors were expelled from the site after the engineer had certified that they had failed to proceed with due diligence. The main contractors argued that this entitled them to terminate their sub-contract with the claimants (made on the FCEC conditions) and to pay the claimants only what was due under those conditions, which would not include anything for the claimants' lost profits. However, it was held that the FCEC clause did not apply, since the main contract had not been 'determined' by the employer's action. Accordingly, the claimants were entitled to recover as damages an amount that included their lost profits.

19.6.2 Rights against the employer

As with rights against the contractor, payment is considered first, before turning to other rights.

Payment

As a basic principle, the lack of a direct contractual link between an employer and a sub-contractor or supplier means that the employer is not liable to pay the sub-contractor directly for work done or materials supplied. This point, which applies

---

[21] *Martin Grant & Co Ltd v Sir Lindsay Parkinson & Co Ltd* (1984) 29 BLR 31. See also *Kelly Pipelines Ltd v British Gas plc* (1989) 48 BLR 126.
[22] *Fee (J & J) Ltd v Express Lift Co Ltd* [1993] CILL 840.
[23] *ER Dyer Ltd v Simon Build/Peter Lind Partnership* (1982) 23 BLR 23.

to both domestic and nominated sub-contractors, was settled by the House of Lords in 1917. In *Hampton v Glamorgan CC*,[24] a lump sum contract to build a school in accordance with the specifications of the defendants' architect included a provisional sum for heating apparatus. The claimant sub-contractor submitted a scheme to provide this, and the architect told the main contractor to accept it. When the main contractor failed to pay the claimant the full amount due, and the claimant sued the defendants for the balance, it was held that the main contractor's overall obligation included providing the heating apparatus within the provisional sum. In employing the claimant for this purpose, the main contractor was contracting personally and not as agent for the defendants; the latter were therefore not liable to the claimant for the balance of the price.

Although the terms of the main contract may lead to the conclusion that the employer *is* personally responsible for payments to sub-contractors,[25] such an interpretation is extremely unusual. Most of the cases in which liability has been imposed have been ones in which the employer has entered into direct negotiations or dealings with a sub-contractor or supplier.[26] Even here, however, the courts are slow to reach the conclusion that any direct obligation has been assumed. In one case,[27] for example, where a main contractor was on the brink of liquidation, certain sub-contractors threatened to stop work. The employer then promised to 'ensure payment of all amounts outstanding' to the sub-contractors. However, it was held that even this did not create any legal obligation on the employer's part; it merely meant that the employer would do what it could to see that the main contractor paid the sub-contractors.

A particular aspect of the courts' attitude in this situation is their reluctance to enable a contract administrator, as agent for the employer, to bring the latter into a direct contractual relationship with a sub-contractor. This reluctance is exemplified by the case of *Vigers Sons & Co Ltd v Swindell*[28] where, under a contract which permitted direct payment of sub-contractors on the main contractor's default, the main contractor went into liquidation and the contract was taken over by one of its directors. The architect instructed sub-contractors to lay flooring and promised that the employer would pay for this. It was nonetheless held that the architect had no authority to commit the employer in this way and that, since the employer had not ratified (adopted) the architect's action, the sub-contractors could not recover payment from the employer.

Rights over materials

The general principles governing the transfer of ownership in building materials from the contractor to the employer (which we considered in Chapter 11) gain an extra dimension where a sub-contractor or supplier is involved. However, the basic position in such cases is relatively straightforward; once ownership of the materials

---

[24] [1917] AC 13.
[25] As in *Hobbs v Turner* (1902) 18 TLR 235.
[26] See, for example, *Smith v Rudhall* (1862) 3 F & F 143; *Brican Fabrications Ltd v Merchant City Developments Ltd* [2003] BLR 512.
[27] *Victorian Railway Commissioners v James L Williams Pty Ltd* (1969) 44 ALJR 32.
[28] *Vigers Sons & Co Ltd v Swindell* [1939] 3 All ER 590.

has passed from the sub-contractor to the main contractor, any lien or similar right on the sub-contractor's part comes to an end.[29] The sub-contractor accordingly loses any leverage that might otherwise have been used against the employer to ensure payment.

In an effort to improve their position, many sub-contractors and suppliers have in recent years sought to contract upon terms that incorporate a **retention of title** clause. Such clauses vary in scope, but they are generally designed to ensure that title to materials does not pass to the main contractor (and thence to the employer) unless and until the sub-contractor has been paid for them. Where a clause of this kind is validly incorporated in the relevant contract, it seems that it will in principle be legally effective.

In the leading case on this subject, *Dawber Williamson Roofing Ltd v Humberside CC*,[30] a sub-contractor who had contracted to supply and fix a roof delivered a quantity of slates to the site. The main contractor was paid by the employer for these slates, but then went into liquidation before paying the sub-contractor. It was held that a retention of title clause in the sub-contract was effective, so that the slates remained the property of the sub-contractor. Terms in the *main* contract purporting to pass ownership of unfixed materials to the employer, once their value had been included in an interim certificate, were irrelevant, since they applied only to goods the main contractor already owned. As a result, the employer was liable in damages for refusing to allow the sub-contractor to remove the slates.

It is important to appreciate that the sub-contract in the *Dawber Williamson* case was of the 'supply and fix' variety, which means that it was classified in law as a contract for 'work and materials'. Had it been a contract merely to *supply*, which would have been classified as a contract for 'sale of goods', the position might well have been different. The reason for this is section 25 of the Sale of Goods Act 1979, under which a person who has 'bought or agreed to buy goods' and who is allowed by the seller to have possession of those goods, may, by delivering them under any 'sale, pledge or other disposition', pass title to a bona fide third party.

In the Scottish case of *Archivent Sales and Developments Ltd v Strathclyde Regional Council*,[31] section 25 was applied in the context of a construction contract. The claimants in that case supplied ventilators to a firm of contractors who were working under JCT 63 on a project for the defendants. The ventilators were delivered directly to the site, their value was duly certified, and the defendants paid the contractors. However, the contractors then went into liquidation without having paid the claimants. The claimants' action to recover 'their' ventilators from the defendants failed, since it was held that the defendants had received title from the contractors by virtue of section 25. In effect, the contractors had 'agreed to buy' the ventilators from the sellers, had been allowed by the sellers to have possession of them and had delivered them to the defendants under a 'disposition'.

Although the *Archivent* case thus creates a crucial distinction between a contract to supply and install and one merely to supply, the limits of that decision

---

[29] *Pritchett and Gold and Electrical Power Storage Co Ltd v Currie* [1916] 2 Ch 515.
[30] (1979) 14 BLR 70.
[31] (1984) 27 BLR 98.

should be noted. In a more recent case,[32] where timber was supplied under a contract with a title retention clause, the main contract (unlike that in *Archivent*) contained no express terms relating to interim payments and the passing of property in materials. It was held that, although interim payments were in fact made, these did not have the effect of a 'disposition' by the contractor of any particular materials to the employer. As a result, section 25 of the Sale of Goods Act 1979 did not apply and the timber remained the property of the supplier.

This discussion clearly shows that, while the precise effect of title retention clauses may be uncertain, their presence in a sub-contract may represent a danger to the employer. In an effort to avert this danger, clause 3.9.2.1.1 of JCT SBC 05 makes the main contractor responsible for ensuring that any sub-contract contains terms that effectively rule out retention of title by the sub-contractor. Moreover, the JCT SB Sub-Contract 05 (if used) contains terms that have this effect. However, the position of a domestic supplier constitutes an interesting loophole, in that JCT SBC 05 makes no attempt to control the terms on which such a person contracts to supply materials to the main contractor. In this situation, therefore, the employer remains at risk from any title retention clause the supply contract may contain.

---

[32] *Hanson (W) (Harrow) Ltd v Rapid Civil Engineering Ltd and Usborne Developments Ltd* (1987) 38 BLR 106.

20 Employer-selected sub-contractors

Nomination is the practice by which an employer, through the contract administrator, selects persons who then enter into sub-contracts with the main contractor. This procedure is found mainly in the United Kingdom and in those countries whose standard building contracts are based on the major UK forms. However, in recent years the procedure has become much less popular with clients (partly, it appears, due to its complexity and partly because of the client's potential liability for sub-contractor defaults) and has been rather infrequently used. In recognition of this decline, JCT SBC 05 contains no provision for nomination of sub-contractors or suppliers.

20.1 REASONS FOR EMPLOYER SELECTION OF SUB-CONTRACTORS

The practice of nomination is often said to have developed in order to give the employer control over the *quality* of sub-contract work, but it also has an important bearing on *time* and *price*. As to time, it is often necessary for specialist work to be ordered well in advance, sometimes long before the main contract is let. In such circumstances, the employer will order the work and will inform the main contractor at tender stage as to who the sub-contractor is to be.

The effect of nomination on price comes about in two ways. First, where certain work is made the subject of nomination, the process of tendering by the relevant specialists for the work in question is carried out only once. Further, this process may take place before, after or at the same time as the tendering process for the main contract. If this system were not used, every main contractor would need, before submitting a tender, to obtain tenders for specialist work from each potential sub-contractor. This would be an obvious duplication of effort and a lengthy procedure. Nomination therefore reduces the time and the cost of tendering.

Second, where a **prime cost** or **provisional sum** is inserted in the main contract, the main contractor prices only for attendances and profit. This means that the final decision on the price of this work is made by the employer, rather than by the main contractor. This may be of particular importance in cases (perhaps involving design) where it is necessary to balance price against quality or long-term performance.

The foregoing explains how the nomination system benefits employers, but the system also offers some important advantages to contractors. Apart from the fact that a main contractor's responsibility for the defaults of a nominated sub-contractor is usually limited, it is common to allow the main contractor a 'cash discount' on all payments duly made to the sub-contractor. The purpose of such a provision is to encourage the main contractor not to delay in making payments, although the wording of the main contract in one case convinced the Court of Appeal that the main contractor was entitled to the discount whether or not payment was made on time.[1] Even under JCT 98, however (where timely payment is required), it is clear that the main contractor can earn the 'cash discount' of 2½% for prompt payment of the sub-contractor, without any disruption of cash flow. This is because, while the contractor must pay within 17 days of certification to qualify for the discount under terms of the relevant sub-contract, the employer must have paid the money to the contractor within 14 days of certification, under the terms of the main contract. Given the high value of work that would usually be subject to nomination, this apparently small profit on *turnover* may represent a very substantial profit in terms of *capital employed*, and is, in fact, an important element in contract profitability.

20.2   SELECTION PROCEDURES

The procedure for nomination of a sub-contractor must follow the relevant contractual provisions. Common examples are described in this Section. It will be noted that, in the absence of any provision for nomination in JCT SBC 05, we describe the procedures under JCT 98.

20.2.1 Nomination of sub-contractors under JCT 98

An employer who has decided to adopt the 'nominated sub-contractor' approach to any particular aspect of the work is, it appears, committed to it. Since the employer's power under clause 13.1 of JCT 98 cannot be used to omit work from the contract simply to give it to another contractor,[2] it seems likewise that it cannot be used to omit work that has been given to a nominated sub-contractor. What is more, it has been held that the main contractor has neither the right nor the obligation to carry out work allocated to a nominated sub-contractor, either by identification in the contract documents or by a subsequent instruction in respect of a prime cost sum.[3]

A main contractor, then, may not simply take over work designated as sub-contract work. However, it is provided in clause 35.2 of JCT 98 that the main contractor may in certain circumstances be permitted to tender for such work in competition with other 'genuine' sub-contractors. For this to take place, the work in question must be of a kind which the contractor carries out directly in the normal course of business; the main contractor must have indicated in the appendix an intention to tender for it; and the contract administrator must be willing to receive such a tender. If these conditions are satisfied, the contractor may submit a tender, but there is no obligation on the employer to accept this tender, even if it is the

---

[1] *Team Services plc v Kier Management & Design Ltd* (1993) 63 BLR 76.
[2] *Carr v JA Berriman Pty Ltd* (1953) 89 CLR 327.
[3] *North West Metropolitan Regional Hospital Board v TA Bickerton & Son Ltd* [1970] 1 All ER 1039.

lowest. If the tender *is* accepted, clause 35.2 provides that the contractor may not sublet the work without the contract administrator's consent (and there is no requirement that this consent shall not be unreasonably withheld).

The mere fact that a sub-contractor is effectively selected by the employer does not in itself make this a **nominated sub-contract** under JCT 98. That will only be so if the employer's selection comes about in one of the ways specified in clause 35.1. These are as follows:

- where the sub-contractor is named in the bills;
- where the bills contain a prime cost sum;
- where the contract administrator issues an instruction on the expenditure of a provisional sum in the bills;
- where certain variations are ordered; or
- where an agreement is made between the contract administrator and the contractor.

Where any of these is applicable, the person chosen must be nominated in accordance with the procedure laid down in clause 35. This means that the contractor and sub-contractor will enter into a specified form of sub-contract and that the sub-contractor will also be required to enter into a collateral agreement with the employer. The main purpose of this agreement is to provide a right for the employer to claim directly against sub-contractors for certain defaults, thus making it possible to free the main contractor from responsibility for those defaults. The resulting triangular relationship is extremely complex and has proved unpopular with the construction industry. In practice, clause 35 is frequently deleted from contracts, and a simpler selection system substituted. This is why the newer JCT SBC 05 has omitted the procedure altogether.

Where clause 35 applies, the procedure that it lays down requires the sub-contractor to tender on form NSC/T, which will contain as much information as possible about the main contract. Once the employer decides on the sub-contractor (who must also have completed the relevant part of the collateral agreement NSC/W), the employer completes the remainder of NSC/W and the tender is passed to the main contractor with a preliminary notice of nomination. It is then up to the main contractor and sub-contractor to settle any outstanding terms within 10 days, failing which the contract administrator may renominate. Once the main contractor and the sub-contractor have reached agreement, the contract administrator formally nominates the sub-contractor on NSC/N. When this document is issued, the main contractor and sub-contractor are automatically bound by the terms of sub-contract NSC/C.

The result of all this activity is a complex set of rights and obligations that link employer, main contractor and sub-contractor in a triangular relationship. Further complications arise from the fact that the contract administrator, who has no part to play where domestic sub-contractors are concerned, assumes a much more important position in relation to nominated sub-contractors. Apart from the important part played in the nomination process itself, the contract administrator is directly concerned with various financial provisions of the sub-contract. These include payment, variations and practical completion. The contract administrator is also responsible *indirectly* through the contractor for making decisions on applications by the sub-contractor for extensions of time, claims for loss and/or expense, and disputes between the contractor and the sub-contractor over possible termination of the sub-contract.

20.2.2 Nomination of suppliers under JCT 98

JCT 98 clause 36 contains a complex definition of a **nominated supplier**. The overall result is that a person is only a nominated supplier within the meaning of this clause where the goods or materials to be supplied are the subject of a prime cost sum. This prime cost sum may be contained in the original bills, it may arise from the conversion of a provisional sum in the original bills, or it may be included in a contract administrator's instruction on a variation. Without it, however, there is no 'nomination' under this clause, even where the intended source of supply is expressly identified or where there is only one source of supply for the specified item.

Where a prime cost sum *is* specified in one of these ways, clause 36.3 provides that the amount payable by the employer to the contractor shall be the net cost to the contractor under the supply contract. This cost includes any charges for packing, transport and delivery, and also any expense properly incurred by the contractor that would not otherwise be reimbursed under the contract. This formulation means in effect that, apart from a 5% 'discount for cash' which the nominated supplier must allow the contractor for payment within 30 days of the end of the month following delivery, a contractor who is able to negotiate the benefit of a 'trade discount' will not be able to keep it.

Although there is no compulsory form of supply contract for use with JCT 98, clause 36.4 attempts to control to some extent at least the terms on which a contractor and a nominated supplier do business. Indeed, the Joint Contracts Tribunal publishes a form of tender TNS/1 and strongly recommends its use by nominated suppliers. This document contains all the terms specified in clause 36.4 and also provides an optional warranty agreement TNS/2. When TNS/2 is used, the supplier guarantees that design, selection of materials and satisfaction of performance specifications have been and will be carried out with care and skill. The supplier also undertakes to indemnify the employer if any default entitles the contractor to an extension of time or loss and/or expense under the main contract.

Whether or not the recommended form of tender is used, clause 36.4 of JCT 98 provides that the contractor cannot be forced by the contract administrator to enter into a contract with a nominated supplier unless that contract contains certain terms. By and large, these terms merely confirm the supplier's obligations, such as to deliver in accordance with any agreed programme and to bear responsibility for defects. However, there is also mention of the contractor's 5% 'discount for cash' and, more controversially, a statement that ownership of all materials shall pass to the contractor upon *delivery*, irrespective of payment. The supply contract must also contain a term to the effect that none of its other terms shall override or modify those terms which are included to give effect to clause 36.4.

20.2.3 Nomination under ICE 7

The ICE 7 approach to nomination is different in a number of respects from that of JCT 98. Clauses 58 and 59 of ICE 6 provide a code of rules applying equally to both nominated sub-contractors and nominated suppliers. First (which is similar to JCT 98) is the contractor's right of objection to any proposed nomination. In addition to a general right to 'raise reasonable objection', clause 59(1) specifically entitles the contractor to refuse any sub-contractor who will not contract on terms which protect the contractor in various ways.

If the contractor exercises this right of objection, the situation may be dealt with in a variety of ways. Under clause 59(2) the engineer may:

- make an alternative nomination;
- vary the works;
- omit the item in question and have it carried out by a direct contractor;
- instruct the contractor to find a potential domestic sub-contractor and submit a quotation; or
- invite the contractor to carry out the work.

20.2.4 Named sub-contractors under JCT IC 05

Many users of JCT 98 compared its complex 'nomination' procedure unfavourably with the simpler situation which operated under JCT 63. As a partial response to these criticisms, the JCT Intermediate Form of Building Contract 1998 (IFC 98) introduced a system under which the employer could 'name' a sub-contractor for specific work without adopting the full nomination procedure. This mechanism has been maintained under the succeeding form JCT Intermediate Building Contract 2005 (JCT IC 05). The form of sub-contract to be used is JCT ICSub/NAM/A and C 05; the parties sign the Agreement ('A'), which incorporates the Conditions ('C'). The terms of this contract are largely the same as that drafted for domestic sub-contracts under JCT IC 05 (JCT ICSub/A and C 05), although a person named under JCT IC 05 is rather more than a domestic sub-contractor.

The JCT IC 05 system provides two alternative methods by which sub-contractors can be named, although both of these require the sub-contractor to tender on JCT Form ICSub/NAM/T and to enter into a sub-contract using JCT Form ICSub/NAM/A and C. The first method, which is dealt with in JCT IC 05 clause 3.7 and paragraphs 1-4 of Schedule 2, will be to the employer's advantage to use where both the work involved and the sub-contractor can be sufficiently identified at an early stage. Under this method, the main contract tender document names the intended sub-contractor and provides a detailed description of the work to be performed. It is then up to the main contractor to price this work when tendering. In so doing, the main contractor is in no way bound by the tender submitted by the proposed sub-contractor.

Once the main contract tender is accepted, the sub-contract must be entered into within 21 days. Should this prove impossible, because the contractor and sub-contractor cannot reach agreement, there is no power for the contract administrator simply to name an alternative sub-contractor. Paragraph 2 of Schedule 2 requires that the situation be dealt with in one of the three following ways:

1. Where the problem arises out of some 'particulars given in the contract documents', the contract administrator may change those particulars so as to remove the impediment.
2. Again, where the problem arises out of some 'particulars given in the contract documents', the contract administrator may choose to omit the work altogether.
3. Whatever the cause of the problem, the contract administrator may omit the work in question from the contract documents and substitute a provisional sum.

If either the first or the second option is chosen, the contract administrator's instruction is to be treated as a variation and valued accordingly. This means that, in respect of any subsequent delay, the contractor will be entitled to claim both an extension of time and for any direct loss and/or expense incurred (see JCT IC 05: paragraph 3 of Schedule 2). However, all three of these options place the risk of the *initial* delay on the contractor. Except in so far as the contract administrator is guilty of unreasonable delay in giving the instruction, the contractor can make no claim against the employer for an extension of time or for loss and/or expense.

The second method of naming a sub-contractor under JCT IC 05 is dealt with by Schedule 2, paragraph 5. This consists of the insertion of a provisional sum in the main contract documents, and a subsequent contract administrator's instruction as to the expenditure of that sum. Such an instruction must incorporate a full description of the work and all particulars of the tender submitted by the chosen ('named') sub-contractor. The contractor must then, unless exercising its right of objection on reasonable grounds, enter into a sub-contract with that person within 14 days. In these circumstances, while the sub-contractor is entitled to payment in accordance with the sub-contract tender, what the main contractor in turn receives from the employer is based on the valuation rules in the variations clause.

Where this method is adopted, any delay or disruption caused to the contractor may be the subject of a claim for an extension of time or for loss and/or expense. Furthermore, if it proves impossible to conclude the relevant sub-contract, then it seems that the contract administrator must issue further instructions as to the provisional sum, again at the employer's expense.

20.3   DEFAULTS OF NOMINATED SUB-CONTRACTORS

As we saw in Chapter 19, a sub-contractor who breaches any term of the sub-contract will be liable to pay damages to the main contractor. These damages will include any sums that the main contractor has in turn to pay to the employer in respect of the sub-contractor's default, and this raises an interesting point. If the terms of the main contract do not enable the employer to recover a particular loss from the main contractor, then the main contractor will have no ground on which to recover that loss from the sub-contractor who caused it. To take a simple example, where delay caused by a nominated sub-contractor entitles the main contractor to an extension of time, this means that the main contractor is not liable to pay liquidated damages to the employer during the relevant period. It follows that the main contractor has no reason to claim the amount of the liquidated damages from the sub-contractor, who therefore escapes responsibility for causing the delay. It is for this very reason that direct contractual links between an employer and a sub-contractor (such as NSC/W under JCT 98 and ICSub/NAM/E under JCT IC 05) provide that a nominated sub-contractor who causes delay in such circumstances will become personally liable to the employer for the liquidated damages which are lost.

20.3.1 Liability of the sub-contractor to the employer

A nominated sub-contractor has potential liability in contract and in tort.

Liability in contract

As we saw in Chapter 19, the normal contractual structure (consisting of a 'chain of contracts') does not provide for a direct claim by an employer against a sub-contractor. However, it is common for nomination procedures to include a requirement that the sub-contractor and the employer should enter into a direct contractual agreement. Where this is done, the agreement is likely to cover such matters as a design warranty from the sub-contractor and a direct payment undertaking by the employer. Examples of such direct agreements are NSC/W, which is used for nominated sub-contractors under JCT 98, or ICSub/NAM/E which is for use with JCT IC 05.

In addition to direct contracts that are expressly entered into, there is the possibility that a court may construct (which means 'invent') a collateral contract between the employer and a sub-contractor or supplier. This has been done in cases where an employer has been persuaded by certain assurances to nominate a particular sub-contractor or supplier, and the effect is to make those assurances legally binding. For example, in the case of *Shanklin Pier Co Ltd v Detel Products Ltd*,[4] the claimants were about to have their pier repainted. They were assured by the defendants, who were paint manufacturers, that their paint would be suitable for the job, would be impervious to rust and would last for between seven and ten years. On the basis of these assurances the claimants instructed their contractors to purchase and use the paint in question. When the defendants' claims proved to be wildly optimistic, it was held by the Court of Appeal that they formed the basis of a 'collateral contract' between the claimants and the defendants, and that the defendants were accordingly liable in damages for its breach.

In the *Shanklin Pier* case, the suppliers gave express assurances directly to the claimants, but it seems that a collateral contract may arise without this. In a later case[5] it was accepted in principle (though the claim failed on the facts) that suppliers might be made liable on this basis for making extravagant claims about their products in advertising brochures, if these were relied on by employers or their architects in deciding what materials should be specified.

---

[4] [1951] 2 KB 854.
[5] *GLC v Ryarsh Brick Co Ltd* [1985] CILL 200.

Liability in tort

Whether or not there is any direct contractual link between the employer and a sub-contractor, there exists, at least in principle, a potential claim in the tort of negligence. In so far as such a claim arises out of *physical damage* (for example, where the sub-contractor negligently causes a fire), it raises identical problems to those discussed in Chapter 19 in relation to domestic sub-contractors.

However, the fact that a nominated sub-contractor is specifically selected by the employer raises an additional possibility. This is that the reliance placed on the sub-contractor by the employer may give rise to a relationship of such 'proximity' that the sub-contractor will owe the employer a duty of care in tort not to cause **pure economic loss** through negligent performance of the sub-contract. In the case of *Junior Books Ltd v Veitchi Co Ltd*,[6] where a nominated sub-contractor laid a defective floor in a new factory, it was held on this basis by the House of Lords that the sub-contractor was directly liable in tort to the employer for the costs of relaying the floor and for the consequential loss of production while this remedial work was carried out.

The decision in the *Junior Books* case came as a shock to the entire construction industry, since there appeared no reason why it should not apply to almost any nominated sub-contract case, thus outflanking what had always been regarded as the barrier of privity of contract. However, the courts in later cases have repeatedly tried to avoid following *Junior Books*, for example by holding that a tort claim would be inconsistent with any direct warranty which exists between the employer and the nominated sub-contractor.[7] Indeed, both the Court of Appeal[8] and the High Court[9] have even stated that the decision is not binding upon them. All in all, *Junior Books* might safely be ignored altogether were it not for the fact that, in *Murphy v Brentwood DC*,[10] two members of the House of Lords regarded it with specific approval. Whether this may operate to revive the authority of *Junior Books* remains to be seen, although it appears unlikely to do so.

20.3.2 Position of the main contractor

The question: 'Who is responsible for a nominated sub-contractor?' is not one to which a simple or straightforward answer can be given. It depends upon the type of default that is in issue, the kind of loss caused, the person who suffers the loss, the subsequent actions of the parties, and (most important of all) the precise terms of the contract. With these qualifications in mind, we may turn to consider some of the more important situations that arise.

---

[6] [1983] 1 AC 520.
[7] *Greater Nottingham Co-operative Society Ltd v Cementation Piling & Foundations Ltd* [1988] 2 All ER 971.
[8] *Simaan General Contracting Co v Pilkington Glass Ltd (No 2)* [1988] 1 All ER 791.
[9] *Nitrigin Eireann Teoranta v Inco Alloys Ltd* [1992] 1 All ER 854.
[10] [1990] 2 All ER 908.

Defective work and materials

A main contractor's responsibility for a nominated sub-contractor's work is most likely to arise in respect of failure to comply with required standards of workmanship, or the quality and fitness for their purpose of any materials supplied. It has already been noted that, quite apart from any express terms in a building contract, it is the contractor's implied obligation to build in a workmanlike manner with materials that are of good quality and fit for their intended purpose. Further, while the obligation as to the *fitness* of materials only arise where the employer has relied on the 'skill and judgment' of the contractor, and will therefore be excluded where a sub-contractor is nominated, the obligation as to *quality* is strict and will therefore normally remain intact.[11] However, there may be special circumstances sufficient to displace even the implied warranty as to the quality of materials. This would occur, for example, where the main contractor has no right to object to a nomination, or where the contract administrator selects a supplier who, as the employer knows, will only contract on terms which exclude or limit the supplier's liability.[12]

Whatever the position may be in respect of workmanship and materials, it appears very unlikely that a main contractor can be held responsible for errors of *design* by a nominated sub-contractor. This is on the basis that the main contractor's obligation is merely to build in accordance with the design that is supplied. While this is logical, it should not be overlooked that all defects in *materials* are the contractor's responsibility, even those arising from errors in the design of those materials. The net result is that the contractor's legal position may be different according to whether a 'nominated sub-contractor' or a 'nominated supplier' is involved.

The 'implied term' position outlined above is by and large repeated by the express provisions of JCT 98 and its supporting sub-contract documents. Under clause 2.1 of NSC/W, the nominated sub-contractor warrants to the employer that all reasonable skill and care has been and will be exercised in design of the works, selection of materials and satisfaction of any relevant performance specification or requirement, so far as such matters are left to the sub-contractor, and this is similarly provided for in clause 1.4.1 of JCT ICSub/NAM/E. Clause 35.21 of JCT 98 then provides that, *whether or not the sub-contractor is liable under NSC/W*, the main contractor shall not be liable to the employer for anything to which it relates. However, it is made clear at all times that the basic contractual obligations of both main contractor and sub-contractor as to workmanship and materials are not affected, so that the chain of liability remains intact.

Two further provisions of JCT 98 (and NSC/C) are worth noting in this context. First, when a nominated sub-contractor is required to enter into a sub-sub-contract or a contract of supply, it may be found that the other party seeks to exclude or restrict liability. In such circumstances, the nominated sub-contractor can refuse to make the contract unless the contract administrator gives specific written approval to the exclusion or restriction. If this approval is given, then the liability of the sub-contractor to the main contractor is likewise excluded or restricted (NSC/C clause 1.7), as is that of the main contractor to the employer (JCT 98 clause 35.22).

---

[11] *Young & Marten Ltd v McManus Childs Ltd* [1969] 1 AC 454.
[12] *Gloucestershire CC v Richardson* [1969] 1 AC 480.

The second point is that the contract administrator is obliged to make provision for final payment of the nominated sub-contractor within 12 months of practical completion of the sub-contract works. This obligation, which arises under NSC/W clause 5, applies even though it may be some time before the final certificate under the main contract is issued. Where this is done, and the sub-contractor subsequently fails to rectify any defect, shrinkage or fault in the work, it is provided by JCT 98 clause 35.18 that the employer shall nominate a substitute sub-contractor to carry out the necessary remedial work. The employer must obtain the main contractor's consent (which is not to be unreasonably withheld) to this nomination, and to the price that the substitute will charge. If the employer is then unable, despite taking all reasonable steps, to recover the cost from the original sub-contractor, the main contractor must indemnify the employer against any shortfall in what is recovered.

The foregoing discussion has concentrated entirely upon those consequences of defective work by a sub-contractor, such as the cost of rectification, which fall first upon the employer and which the employer will then seek to pass along the chain of contracts to the contractor, the sub-contractor and so on. However, there may be other losses arising out of such work that do not concern the employer at all. In particular while delay arising out of a nominated sub-contractor's default may entitle the main contractor to an extension of time for completion (JCT 98 clause 25.4.7), it is most certainly not something for which claim for loss and/or expense can be made under clause 26.2. Thus, if the main contractor's work is disrupted for this reason, there can be no claim against the employer; the contractor's sole remedy lies against the sub-contractor under NSC/C clause 4.40.

A similar principle applies where the default of a nominated sub-contractor disrupts and causes loss and/or expense to another sub-contractor, either nominated or domestic. The innocent sub-contractor is entitled to claim against the main contractor (NSC/C clause 4.39 specifically makes the main contractor responsible). The main contractor may then pursue the culprit under clause 4.40 of the relevant sub-contract, and the employer is not involved at all.

Delay

A particularly controversial area of risk allocation in respect of nominated sub-contractors and suppliers is that of delay. All JCT contracts have for some time treated this as an area where the employer, as the selector of the sub-contractors, should bear responsibility for them, by providing for the main contractor to be granted an extension of time. The reason why this is controversial is that, where such an extension of time is granted, the employer is deprived of the right to claim liquidated damages *which the main contractor would otherwise have passed on to the delaying sub-contractor*. As a result, unless the employer can make a direct claim against the sub-contractor under a collateral warranty agreement, the main incentive for the sub-contractor to keep to time will simply disappear.

The underlying reasoning of such provisions was critically examined by the House of Lords in the case of *Westminster CC v Jarvis & Sons Ltd*,[13] which

---

[13] [1970] 1 All ER 942.

concerned the erection of a multi-storey car park. There a piling sub-contractor had apparently completed on time, but serious defects were subsequently discovered in the work. The sub-contractor duly returned to the site and rectified these defects, but this whole process caused considerable delay to the project. The main contractor thereupon claimed an extension of time under JCT 63 clause 23(g) for 'delay on the part of nominated sub-contractors which the contractor has taken all reasonable steps to avoid or reduce'. The House of Lords, overruling the Court of Appeal, held that no extension should be granted, since these words referred only to cases in which a nominated sub-contractor patently failed to achieve practical completion of the sub-contract work by the correct time. They did *not* cover cases where, as here, the apparent state of the sub-contract works was such as to justify the architect in accepting them, and where defects were only discovered at a later date.

Clause 23(g) was probably *intended* to cover 'any delay *caused by* a nominated sub-contractor'. The House of Lords' interpretation was therefore a very narrow one, and this is clearly because the judges regarded the clause as highly undesirable and one which ought to be restricted as far as possible. Nonetheless, despite such critical comments from the House of Lords as 'unjust and absurd', 'illogical and defective', and 'a provision under which a sub-contractor can benefit from its own default', the clause appears in a similar format in JCT 98.

The present position therefore is that, where a nominated sub-contractor is late in achieving practical completion and thus causes delay to the main contractor, the latter will be entitled to an extension of time. The employer may then take action directly against the sub-contractor under NSC/W clause 3.3 to recover the liquidated damages that have been lost. As for any losses suffered directly by the contractor (including any liability which the contractor incurs to other sub-contractors), these can be recouped from the guilty sub-contractor under NSC/C clause 4.40. The employer is not at all involved in this aspect of the loss, since delay by a nominated sub-contractor or supplier is not one of the 'relevant matters' giving rise to a claim for loss and/or expense under clause 26.2. Nor can the contractor claim extra payment for a 'variation' if, in order to help out when a supplier is guilty of late delivery, the contract administrator authorizes changes to the order of the work.[14]

One final point which may be worth noting is that, even where sub-contract works lie on a project's critical path, the fact that a nominated sub-contractor receives an extension of time does not automatically entitle the main contractor to a similar extension. However, most of the 'relevant events' listed in NSC/C are identically worded to those in JCT 98, so that parallel applications may be made. Naturally, however, the main contractor will be left unprotected in cases where the true cause of the delay is the main contractor's own default.

Repudiation

The most difficult area of nominated sub-contractor default is that which arises when the sub-contractor, without any justification, repudiates the sub-contract.

---

[14] *Kirk and Kirk Ltd v Croydon Corp* [1956] JPL 585.

Where this occurs, it leaves the parties with the question of responsibility for completing the unfinished work, the cost of which may have increased considerably due to inflation. It may also leave them with an additional trail of financial damage, including, for instance, the cost of rectifying defects (which may in effect mean paying twice for the same work), the consequent disruption of the main contractor's programme, disturbance to other sub-contractors and overall delay.

Once again, the question is whether responsibility for these losses is to lie with the main contractor, or whether the employer must bear the cost of nominating a substitute and pursue the guilty sub-contractor under any collateral agreement that exists. The answer given by English law has come about as a result of a series of cases dealing with JCT 63, the decisions in which have in turn led to drafting amendments incorporated in JCT 98.

The starting point is the decision of the House of Lords in *North West Regional Hospital Board v TA Bickerton & Son Ltd*,[15] which arose when the claimants engaged the defendants under JCT 63 to erect certain additional units at a hospital. A firm by the name of Speediwarm Ltd was nominated as sub-contractor for the installation of a heating system. The sub-contract was duly entered into, but Speediwarm soon went into liquidation. When the main contractor, who then agreed to do the relevant work on a 'without prejudice' basis, claimed that this had cost more than the original sub-contract price, the question that arose was whether the employers were liable for the increase. The House of Lords held that the employers *were* liable, on the basis that it was their responsibility, when a nominated sub-contractor dropped out, to renominate. This was because, on the wording of JCT 63, the main contractor had neither the right nor any obligation to carry out work that the contract assigned to a nominated sub-contractor. As a result, it was the duty of the employer to renominate and to pay the substitute sub-contractor's account, even if this was more than the original.

The effect of the decision in *Bickerton* was clearly that many of the costs arising out of a nominated sub-contractor repudiation would fall on the employer. However, the House of Lords did *not* treat the sub-contractor's default as if it were a breach of contract by the employer. There was thus no reason to suppose that losses suffered by the main contractor (due to disruption etc.) were recoverable from the employer.

The next case, *Percy Bilton Ltd v GLC*,[16] arose when the defendants engaged the claimants under JCT 63 to build a housing estate in South London, and nominated a firm named Lowdells as sub-contractors for mechanical services. Lowdells went into liquidation when the work was partly completed. There was then a lengthy delay (partly inevitable and partly due to the fault of the architect) in nominating a replacement, and completion of the main contract works was seriously delayed. The architect granted an extension of time to cover this delay, and the employer deducted liquidated damages for other unconnected delays which were entirely due to the contractor. The contractor then raised a technical argument to the effect that JCT 63 contained no power to extend time merely because a replacement sub-contractor could not meet the original deadline for completion, and that this delay must therefore be regarded as the employer's responsibility

---

[15] [1970] 1 All ER 1039.
[16] [1982] 2 All ER 623.

under *Bickerton*. As a result, according to the contractor, the time for completion was now at large and the liquidated damages clause therefore fell.

This argument was rejected by the House of Lords, on the ground that the employer's 'responsibility' required only a renomination within a reasonable time. The contractor would be entitled to an extension of time for any *unreasonable* delay in renominating. However, any delays arising *inevitably* were at the contractor's risk, and the employer's right to claim liquidated damages for this period (and for any other period resulting from contractor-caused delays) remained intact.

The ruling in *Bilton* (which is also expressly covered by JCT 98 clause 35.24.10), appeared at first sight to swing the legal position back in favour of the employer. However, the whole decision was effectively undermined by a dictum of Lord Fraser, who gave the only judgment in the case. In an effort to show that the decision was not unduly harsh towards contractors, Lord Fraser stated that the contractor 'could have exercised its right of 'reasonable objection' under clause 27(a) to prevent the nomination of any new sub-contractor who did not offer to complete his part of the work within the overall completion period for the contract as a whole'. If correct, this means that where, as will often be the case, a renomination will inevitably involve delay, the main contractor can create a total impasse by refusing to accept any replacement sub-contractor unless the employer undertakes to provide compensation for the effects of the first sub-contractor's default. In such circumstances, an employer who is not prepared to omit the relevant work altogether, will be forced to submit to the main contractor's demands, thereby effectively taking on virtually the whole risk.

In *Fairclough Building Ltd v Rhuddlan DC*,[17] the Court of Appeal confirmed that this was indeed the position. The defendants in that case engaged the claimants under JCT 63 to construct a leisure complex, and nominated a firm named Gunite as sub-contractors for certain specialist work on a swimming pool. When the work was partly completed, Gunite, who had received some £60,000 for work subsequently found to be worthless, repudiated the sub-contract and withdrew from the site. There was then a delay, which was held to be inevitable, of some five months before a renomination instruction was issued. At that point, the date for completion of the main contract was only some 13 weeks away, while the proposed sub-contract would give the replacement sub-contractors 27 weeks. The claimants, relying on this incompatibility of time and also on the fact that the replacement sub-contract did not include the £240,000 worth of work necessary to remedy defects in Gunite's work, refused to accept the renomination.

In ruling upon various preliminary issues, the Court of Appeal held as follows:

1. The contractors were entitled to refuse any nomination of a sub-contractor who would not undertake to complete by the stated completion date under the main contract, even if the effect of delays already accumulated would mean that *actual* completion of the main contract would not be delayed. It was also suggested by the Court of Appeal that a contractor who *did* accept such a nomination would be entitled to an extension of time until the date fixed for completion of the sub-contract work.

---

[17] (1985) 30 BLR 26.

2. According to *Bickerton*, a main contractor cannot be called upon to carry out work which is assigned to a nominated sub-contractor. It follows that the main contractor cannot be called upon to rectify any defects in such work, whether these relate to design, workmanship or materials. The contractors in this case were therefore entitled to reject any proposed renomination that did not include whatever work was necessary to remedy defects in the original sub-contractor's work. This is now expressly covered by JCT 98 clauses 35.24.7 and 35.24.8.

3. Where the defaulting sub-contractor has already been paid for work that is then shown not to be in accordance with the sub-contract, the employer is entitled as against the main contractor to be credited with the amount of the overpayment (JCT 98 clause 30.6.2.1). However, the employer must otherwise bear the full cost of renomination, including any remedial work that is necessary, except where the sub-contractor's withdrawal is itself due to the main contractor's breach (JCT 98 clause 35.24.9).

As a result of *Fairclough*, it seems that the position of the employer will in practice depend upon the relative size of the various losses caused by the withdrawal of a nominated sub-contractor. If the main contractor is already substantially in delay, the employer may find it worthwhile to omit the sub-contract work in order to preserve the right to liquidated damages. However, this may not be possible without disrupting the entire works and risking claims for loss and/or expense. In such cases, the employer may be forced to compensate the main contractor for all losses suffered, in order to induce the main contractor to accept a renomination.

The position arrived at by this series of cases has been heavily criticized (Wallace 1986) on the following grounds:

1. In cases where there is no direct agreement between employer and nominated sub-contractor, it positively *encourages* sub-contractors to repudiate a contract that has become disadvantageous or burdensome to them.

2. It gives no incentive to main contractors to keep a sinking sub-contractor afloat, although JCT 98 clause 35.25 does at least prohibit the main contractor from terminating the sub-contractor's employment under the sub-contract without first obtaining a contract administrator's instruction to this effect.

3. It involves the utter absurdity that the legal responsibility for defective work which may be borne by the main contractor and sub-contractor (on the chain liability principle) may be eliminated by the simple expedient of repudiating the sub-contract.

Other defaults

Apart from the three categories of nominated sub-contractor default considered above, the question whether 'responsibility' for nominated sub-contractors lies with the employer or the main contractor is occasionally brought into focus by other specific contractual provisions. In this context, two decisions are of particular importance.

In the first of these, *John Jarvis v Rockdale Housing Association Ltd,*[18] the defendants engaged the claimants under JCT 80 to construct 50 flats for the elderly, together with some ancillary accommodation. The firm that was nominated as piling sub-contractors started late, produced defective work and then wrongfully withdrew from the site, whereupon the architect instructed the claimants to stop work. After this suspension had lasted for more than one month, the claimants gave notice purporting to terminate their employment.

It was held by the Court of Appeal that, although such a notice could not validly be given if the instruction to suspend work was brought about 'by reason of some negligence or default of the contractor', this did not include negligence or default of a *nominated* sub-contractor. The contractor could not be held responsible for such a person, and could accordingly terminate his employment in such circumstances. The distinction thus drawn between nominated and domestic sub-contractors is now made explicit by JCT 98 clause 28.2.2.2.

The second important decision in this field was made by the House of Lords in a Scottish case[19] which arose out of a contract to build two naval submarines. Under the terms of this contract, the main contractor was entitled to claim for loss and/or expense arising out of disruption caused by any matters 'beyond the contractor's control'. The House of Lords held that this phrase included defects in cables that were obtained from a nominated supplier. What is especially surprising is that the House of Lords' reasoning was expressed in terms wide enough to apply equally to *domestic* sub-contractors and suppliers. If this is correct, it seems to undermine entirely any remaining idea that the main contractor must bear responsibility for a sub-contractor's defaults.

Other forms of nomination

The discussion of the position under JCT 98 might appear to suggest that an employer who reserves the right to nominate a sub-contractor inevitably takes on responsibility for the risks involved. Any such suggestion is simply not true; risk allocation is always a matter of choice for whoever is responsible for drafting the contract in question.

This is strikingly illustrated when one considers the second edition of GC/Works/1. Clause 31 of that contract gave the contractor the right to object on reasonable grounds to any nomination. However, it then provided that the contractor should be responsible for any sub-contractor or supplier, whether domestic or nominated and should accordingly make good any loss suffered by the employer through a sub-contractor's default. It was further spelled out by clause 38(5) that, in the event of the termination of a nominated sub-contract, it was entirely the contractor's responsibility either to find a substitute or to carry out the work personally. In either case, the main contractor would be paid only what was due on the original sub-contract.

Interestingly, subsequent editions of GC/Works/1 appear to have accepted the modern notion that the risks involved in a nominated sub-contractor's insolvency should always be borne by the employer. Accordingly, the contract provisions

---

[18] (1986) 36 BLR 48.
[19] *Scott Lithgow Ltd v Secretary of State for Defence* 1989 SLT 236.

about nomination follow those of the second edition but with one crucial exception. It is stated in clause 63A (an optional clause) that, if a nominated sub-contractor becomes insolvent, the employer must reimburse the contractor for the difference between the extra costs incurred in getting the work completed and the amount that can be extracted from the defaulting sub-contractor.

The way that ICE 7 (clauses 58 and 59) deals with nominated sub-contractors is also of considerable interest. ICE 7 begins by providing that, as a general principle, the contractor is fully responsible for any breach of contract by a nominated sub-contractor. This liability is not limited to the amount the contractor succeeds in recovering from the sub-contractor. However, it is qualified to the extent that the contractor cannot be held responsible for matters of *design*, unless this is expressly stated in both the main contract and the sub-contract.

Although the contractor is therefore fully responsible for the defaults of a nominated sub-contractor *while the sub-contract remains in operation*, things change once the sub-contract is brought to an end. Provided that termination of the sub-contract takes place with the engineer's written consent, the contractor's only obligation thereafter is to 'take all necessary steps and proceedings as are available to him' to recover from the defaulting sub-contractor all losses suffered by both the contractor and the employer. To the extent that this proves impossible, it is the employer's obligation to reimburse the contractor.

The position of a management contractor under the JCT 1987 Form of Management Contract was considered in *Copthorne Hotel (Newcastle) Ltd v Arup Associates*.[20] It was held by the Court of Appeal that, on a true construction of clause 3.21 of that contract, the employer could not hold the management contractor responsible for breaches by the works contractors of their own responsibilities. However, the clause did not protect the management contractor from liability for breach of its own obligations under the management contract (including the duty to exercise reasonable care and skill in managing, co-ordinating and supervising the works).

Defaults of nominated suppliers under JCT 98

With one exception, the general principle under JCT 98 is that the employer is not concerned with a nominated supplier's defaults. If materials are defective, in the sense that they do not comply with whatever express or implied standards are contained in the main contract, then the contractor is in breach. This will involve liability to the employer for any loss, such as the cost of replacement. The main contractor must then seek to recover this loss, plus any personal losses, from the supplier in question. The only exception to this is that delay on the part of a nominated supplier is a ground on which the main contractor may be entitled to an extension of time.

The policy of JCT 98 in this situation is clearly that a chain of liability shall be established and, as we have already noted, clause 36.4 requires the supply contract to contain certain terms designed to secure this. However, clause 36.4 is subject to contrary agreement between the contract administrator and the contractor and,

---

[20] (1997) 85 BLR 22.

moreover, there may be situations in which, although all the prescribed terms are present, the supply contract restricts the liability of the nominated supplier in other ways. Clause 36.5 deals with these possibilities by stating that, provided the contract administrator's written approval is obtained for any restriction of the supplier's liability, the liability of the contractor to the employer shall be restricted to exactly the same extent. As a result, the contractor cannot be saddled with a liability to the employer that cannot be passed on to the supplier who is responsible.

Defaults of named sub-contractors under JCT IC 05

As we saw in Section 21.2.4, the prescribed form of contract for named sub-contractors under JCT IC 05 is JCT ICSub/NAM/A 05 and JCT/ICSub/NAM/C 05 (Agreement and Conditions). Once a sub-contract is made in this form, and while it remains in operation, responsibility for any default by the named sub-contractor rests to a large extent upon the main contractor. The main contractor will be responsible to the employer (and the sub-contractor will in turn be responsible to the main contractor) for keeping to time and for rectifying defects in workmanship or materials in the sub-contract work. Further, should such defaults on the part of the sub-contractor cause problems of delay or disruption to the contractor personally, this will be of no concern to the employer, since it is not a matter for which the contractor can claim either an extension of time or compensation for loss and/or expense.

To this general principle there is one important qualification. Where the sub-contract contains any element of design, the sub-contractor may undertake direct responsibility to the employer for design, selection of materials or satisfaction of any performance specifications (as under JCT ICSub/NAM/E 05). However, whether or not such direct responsibility is undertaken, Schedule 2, paragraph 11 of JCT IC 05 makes it clear that the *main* contractor is not liable to the employer for any of these matters.

While the sub-contract remains operative, the position in respect of sub-contractor defaults is relatively straightforward. However, when it is brought to an end, a whole range of different legal situations may result, depending on *how* the sub-contract is terminated and what instructions the contract administrator gives. To take contract administrator's instructions first, Schedule 2, paragraph 7 empowers the contract administrator to name another person to execute the sub-contract work (or what is left of it); to instruct the contractor to make other arrangements for the work, in which case the contractor has the option of subletting to a domestic sub-contractor; or to omit the work altogether.

As to the question of termination, clause 7.4 of JCT ICSub/NAM/C 05 provides power for the main contractor to terminate the employment of the sub-contractor for various breaches of the sub-contract relating to such matters as suspension of the work, delay, defects and unauthorized subletting. Clause 7.5 gives a similar power on the occurrence of certain events connected with the sub-contractor's insolvency. Under each of these clauses, there are specified procedures which must be followed if the termination is to be valid.

All in all, the consequences which may arise on termination of the sub-contract are as follows:

1. Where the first sub-contractor was named in the contract documents, and the contract administrator names a replacement to complete the work, the contract sum is varied (upwards or downwards) to reflect the difference between the two sub-contract prices for that work (Schedule 2, paragraph 8.1, in connection with clause 7.1). The employer need *not* pay the contractor for the cost of rectification, nor compensate for loss and/or expense, though the contractor will be entitled to an extension of time if the main contract work is delayed.

2. Where the first sub-contractor was named in the contract documents, and the contract administrator instructs the contractor to make other arrangements for the work, this is to be valued as a variation (Schedule 2, paragraph 8.2, in connection with paragraph 7.2). Furthermore, the contractor may claim where appropriate for an extension of time and for loss and/or expense.

3. Where the first sub-contractor was named in the contract documents, and the contract administrator orders that the work shall be omitted, this is again to be valued as a variation (Schedule 2, paragraph 8.2, in connection with paragraph 7.3), and the contractor may again be entitled to an extension of time and compensation for loss and/or expense.

4. Where the first sub-contractor is named in an instruction as to the expenditure of a provisional sum, whatever the contract administrator decides is regarded as a further instruction as to this sum (Schedule 2, paragraph 9). As a result, the work will be valued under the valuation rules and, where appropriate, the contractor will be entitled to claim for an extension of time and for loss and/or expense.

In all the above cases, the employer is likely to suffer loss as a result of the termination of the sub-contract. If this does happen, it is provided by Schedule 2, paragraph 10.2 that the main contractor shall take all reasonable steps to recover such losses from the original sub-contractor, and shall then account to the employer for the amounts recovered. ('Reasonable steps' for this purpose do not include arbitration or litigation, unless the employer undertakes to cover the main contractor's legal costs.) In this connection, clause 7.7.4 of ICSub/NAM/C imposes an obligation on the sub-contractor not to argue in proceedings brought on this basis that the contractor has not personally suffered the loss in respect of which the claim is brought.

The four scenarios described above all assume that the sub-contract has been terminated in accordance with clauses 7.4–7.6 of ICSub/NAM/C. Indeed, Schedule 2, paragraph 6 of IC 05 specifically provides that the contractor shall not terminate the sub-contract in any other way, nor allow the sub-contractor to terminate it without the prior written consent of the contract administrator to such an 'acceptance' of the sub-contractor's repudiation (such consent not to be unreasonably withheld or delayed). Where termination *does* take place in some other way, the consequences for the contractor may be disastrous. The contract administrator may once again deal with the situation in any of the ways outlined above, but schedule 2, paragraph 10.1 makes it clear that this can only result in a reduction of the contract sum. Further, there can in such a case be no entitlement to any extension of time or to loss and/or expense.

20.4   RIGHTS OF NOMINATED SUB-CONTRACTORS

The rights of nominated sub-contractors are similar in most respects to those if domestic sub-contractors except for some special provisions regarding payment and retention money.

20.4.1 Payment

General

In general, the fact of nomination makes no difference to the basic principle, described in Chapter 19, that a sub-contractor's payment rights are contained in the sub-contract and are therefore exercisable only against the main contractor. However, a nominated sub-contractor's entitlement to be paid is frequently made conditional upon certification by the architect under the main contract (as is the position in NSC/C, the sub-contract conditions used under JCT 98). It should be noted that a provision of this kind does not govern the sub-contractor's rights following termination of the contracts – at that point the sub-contractor will be entitled to claim for work done, even though no certificates have been issued.[21]

Direct payment provisions

Although, as we saw in Chapter 20, the employer is not normally responsible for payment of sub-contractors, there is nothing to prevent an employer from taking on some responsibility in this matter. This is in fact often done for the benefit of nominated sub-contractors (as for example under ICE 7 clause 59). The detail of such arrangements naturally varies from one form of contract to another, but JCT 98 provides a good illustration.

While JCT 98 is careful to preserve the basic position that the employer is not directly liable to a nominated sub-contractor (clause 35.20), it sets out a scheme in clause 35.13 under which, if the main contractor fails to discharge an obligation to pay the sub-contractor, the employer may pay directly. This power is in principle to be exercised at the employer's sole discretion but, where the employer has entered into the collateral agreement NSC/W with the relevant sub-contractor, it becomes a binding obligation on the employer. It is in effect the 'price' the employer pays in return for the sub-contractor's direct warranty.

These direct payment provisions of JCT 98 require the contractor to show proof to the contract administrator that payments to nominated sub-contractors have been duly discharged. If such proof cannot be provided, then any future payment

---

[21] *Scobie & McIntosh Ltd v Clayton Bowmore Ltd* (1990) 49 BLR 119.

certified as due to the contractor is reduced by the amount owing to sub-contractors, and this amount is paid directly by the employer. It is important to note that the employer is not bound to pay any more than can be obtained in this way – it is not the employer's own money which is at risk, but only that which would otherwise be due to the main contractor. Clause 35.13 further provides that, where more than one nominated sub-contractor has been left unpaid, the employer shall apply the sums available pro rata or may adopt any other method of apportionment which appears to the employer to be fair and reasonable.

Direct payment provisions of this kind are said to be for the benefit of employers, since they encourage sub-contractors to tender for work by giving them an extra assurance of payment. However, this added security should not be too heavily relied on, since it is of doubtful value in precisely those circumstances where it is most important, namely, the main contractor's insolvency. The reasons for this lie in the general law of insolvency (bankruptcy in the case of an individual, liquidation in that of a company), which seeks to ensure that all creditors of the insolvent person are treated *pari passu* (meaning on an equal footing) and that no creditor receives an unfairly preferential share of the insolvent person's assets. If, when a main contractor defaults, one sub-contractor receives payment in full (under a direct payment scheme), it is at least arguable that this is unfair preference over other creditors, and also over other sub-contractors who do not have the benefit of such an arrangement.

The main principles by which the law of insolvency seeks to achieve its ends are worth noting. The first point to be made is that, where the main contract does *not* contain an express provision permitting direct payment of sub-contractors, an employer who makes such a payment will be at risk if the contractor then goes into liquidation. Thus, where an employer paid a supplier of materials immediately after the contractor had gone bankrupt owing money to the supplier, it was held that the employer must pay the money again to the contractor's trustee in bankruptcy, so that it would be available to *all* the contractor's creditors.[22]

Where there is an express power to make direct payments to sub-contractors, it has been held in two English cases[23] that such payments are valid against a liquidator or trustee in bankruptcy. This is apparently on the ground that what the *pari passu* principle prohibits is merely unequal dealing with the property of an insolvent person. Thus, the argument goes, since the main contractor's right to be paid is dependent upon the issue of a certificate, a contract administrator who decides not to certify in the contractor's favour is not dealing with the contractor's property at all, but rather preventing the contractor from acquiring any rights to property in the first place. This seems a reasonable suggestion, and certainly there are strong arguments of policy in favour of these clauses in the construction context. Nonetheless, doubts as to their effectiveness have been raised by a decision of the House of Lords in a quite different context, in the case of *British Eagle International Airlines Ltd v Compagnie Nationale Air France*.[24] It was there held by a bare majority of the House of Lords that, when an airline company went into liquidation, agreements it had made with other airlines for setting off mutual debts contravened insolvency law and were therefore invalid.

---

[22] *Re Holt, ex p Gray* (1888) 58 LJQB 5.
[23] *Re Wilkinson, ex p Fowler* [1905] 2 KB 713; *Re Tout and Finch Ltd* [1954] 1 All ER 127.
[24] [1975] 2 All ER 390.

This decision was reached despite the court's acknowledgment that what were being overridden were perfectly sensible business arrangements, and so it is at least possible that a court would similarly override a direct payment provision in a construction case. However, the present position remains confused: the English courts have not yet been called upon to decide the effect of *British Eagle* on direct payment provision; of the decisions from other countries with similar insolvency rules, a majority treat direct payment clauses as invalid,[25] although there is some authority to support their effectiveness.[26]

Whatever the true position may be, the uncertainties outlined above have led to the insertion of a provision in clause 35.13 of JCT 98 to the effect that the 'direct payment' procedure shall automatically cease as soon as the contractor is subject to liquidation proceedings. Further, it is provided in clause 7.2 of NSC/W that, if the employer pays a sub-contractor directly and then discovers that such proceedings had already been commenced, the money must be repaid on demand. The practical lesson for a sub-contractor is therefore clear; on becoming aware that the main contractor is in financial difficulties, the sub-contractor should inform the employer immediately. In this way, money which is ultimately intended for the sub-contractor can be prevented from reaching the main contractor, and from forming part of the main contractor's general assets on insolvency.

20.4.2 Retention money

We discussed in Chapter 15 the general rules governing the retention by an employer of a small percentage of interim payments made to the contractor, so as to create a fund from which the employer can rectify defects in the building if the contractor fails to do so. It was noted that:

- the employer is regarded as a trustee in respect of that fund, which enables the main contractor or any nominated sub-contractor to insist that the money be held in a separate bank account, so as to protect it in the event of the employer's insolvency; but
- notwithstanding this status as trustee, the employer is entitled to use retention money in satisfaction of any claim against the contractor which would give rise to a right of deduction or set off.

These general principles gain an extra dimension where nominated sub-contractors are concerned, at least under JCT 98 or similar contracts. Clauses 30.4.2 and 30.5.2.1 of JCT 98 require the contract administrator to identify separately the 'Contractor's retention' and the 'Nominated Sub-contract retention' for each such sub-contractor. The crucial question that arises is whether these provisions effectively operate to create *separate* trusts in favour of the main contractor and each nominated sub-contractor. If they do, then the employer's right to have recourse to retention money to satisfy a claim will extend only to the part

---

[25] *Re Right Time Construction Co Ltd* (1990) 52 BLR 117 (Hong Kong); *Joo Yee Construction Pty Ltd v Diethelm Industries Pty Ltd* (1990) 7 Const LJ 53 (Singapore); *Att-Gen v McMillan & Lockwood Ltd* [1991] 1 NZLR 53 (New Zealand); *B Mullan & Sons Contractors Ltd v Ross* (1996) 86 BLR 1 (Northern Ireland).
[26] *Glow Heating Ltd v Eastern Health Board* (1988) 8 Const LJ 56 (Eire).

which relates to the party in default (that is, the party in respect of whom the set off was exercised).

In one English case[27] it was held that this was indeed the position. The making of a sub-contract was treated as an assignment to the sub-contractor of the main contractor's interest in the relevant part of the retention, so as to protect it from the employer's right of set off. However, this decision was heavily criticized in a Hong Kong case,[28] where it was pointed out that *all* retention money in the employer's hands must ultimately pass to the main contractor, and that the main contractor might then be entitled to exercise rights of deduction or set off over it against the sub-contractor. As a result, it was held, the *whole* retention fund has to be treated as money 'due to' the main contractor, and thus as vulnerable to any right of recourse of the employer. The reasoning adopted by the Hong Kong court appears convincing, at least as far as JCT 98 is concerned. However, the position is very different under the JCT Management Contract. It has been held by the Court of Appeal that, on the true construction of that contract, retention deducted by the management contractor from sums due under each works contract are held by the employer on trust for the respective works contractors. As a result, the employer cannot have recourse to retention under the works contracts in order to satisfy a claim that the employer has against the management contractor.[29]

---

[27] *Re Arthur Sanders Ltd* (1981) 17 BLR 125.
[28] *Hsin Chong Construction Co Ltd v Yaton Realty Co Ltd* (1986) 40 BLR 119.
[29] *PC Harrington Contractors Ltd v Co Partnership Developments Ltd* (1998) 88 BLR 44.

21   Financial remedies for breach of contract

The primary remedy for any breach of contract is an award of damages. This remedy is always available, unlike the remedy of termination (see Chapter 24) which is only available in cases of serious breach. In addition to the normal mechanisms for claiming **general damages** for breach, some contracts include provisions for **liquidated damages**, sometimes called **LADs**. Liquidated damages are dealt with in the second part of this Chapter.

21.1   GENERAL DAMAGES

This is not the place for a detailed discussion of the rules governing damages for breach of contract, but a brief account of the more important legal principles may be useful, before we consider specific examples from the construction field.

Purpose of an award of damages

First and foremost, it should be appreciated that damages are assessed with the intention of making the innocent party's position (so far as money can do this) equivalent to what it would have been *if the contract had been properly performed*. They are *not* designed to recreate what the innocent party's position would have been *if the contract had never been made*.

Although in principle a party is entitled to be compensated for all losses flowing from a breach of contract, this is subject to a number of important qualifications.

Remoteness

An innocent party is only entitled to compensation for those losses which are not 'too remote' from the defendant's breach. To decide which losses fall within this rule, English law has adopted a complex test.[1] First, an item of damage is recoverable if, at the time the contract was made, *reasonable* parties contemplating the kind of breach that has actually occurred would have regarded that item of damage as likely to arise in the natural course of things. To put it more briefly,

---

[1] *Hadley v Baxendale* (1854) 9 Exch 341; *Victoria Laundry (Windsor) Ltd v Newman Industries Ltd* [1949] 2 KB 528.

damage that is reasonably foreseeable is not too remote. Secondly (and alternatively) an item of damage is recoverable if such damage should have been within the contemplation of the *actual* parties, at the time that they made their contract. This means that what the parties could reasonably be expected to foresee depends upon their actual knowledge at the time that they made their contract.

An example of the principle of remoteness is provided by the case of *Balfour Beatty Construction (Scotland) Ltd v Scottish Power plc.*[2] The claimants there, who were constructing a concrete aqueduct over a main road, installed a concrete batching plant and arranged for the defendants to supply electricity to it. The claimants needed to pour all the concrete in a single continuous operation and so, when the electricity supply failed, the claimants had to demolish all the work which had been done. It was held that, although the defendants were clearly in breach of contract because of the power failure, they were not liable for the extra losses involved in the demolition and reconstruction, since the claimants had not informed them that a continuous pour was essential.

Mitigation

The doctrine of mitigation of loss means that a claimant cannot recover damages for any part of a loss which the claimant could have avoided by taking reasonable steps. Thus, for example, an employer who unreasonably refuses to let a contractor return to site during a Defects Liability Period, in order to carry out necessary remedial work, will be unable to recover the extra cost of having that work carried out by another contractor.[3]

Third parties

A claimant cannot normally recover damages in respect of loss which has been suffered by another person. However, the courts have recognized an important exception to this principle in construction cases. In certain circumstances an employer who has sold or let a completed building may, when defects are subsequently discovered, recover damages from a designer or contractor on behalf of the purchaser or tenant. This matter is considered in more detail in Chapter 23.

21.1.2 Employer's damages for contractor's breach

Defective or incomplete work by a contractor normally amounts to a breach of contract. Where this occurs, an argument sometimes arises as to whether the employer's damages should consist of the cost of putting right the deficiencies, or the (usually lesser) amount by which the value of the property is diminished. As to this, it appears that the courts will usually favour **cost of repair**,[4] reverting to **diminution in value** only where it would be wholly uneconomic and/or

---

[2] [1994] CILL 925.
[3] *Pearce & High Ltd v Baxter* [1999] BLR 101.
[4] *East Ham BC v Bernard Sunley & Sons Ltd* [1966] AC 406.

unreasonable for the employer to insist on repair. Moreover, these are not the only possible measures. In *Ruxley Electronics and Construction Ltd v Forsyth*[5] a contractor installed a swimming pool at the client's house which was some 450 mm shallower than specification, but which complied with all relevant safety standards and was no less valuable than one of the correct depth. The House of Lords refused to award either the full cost of rebuilding the pool to the correct specification or nominal damages for diminution in value; instead, they identified the client's true 'loss' as being a loss of pleasure and amenity, and awarded £2500 as damages for this.

It has been held that damages based on 'cost of repair' should be assessed, not at the date of breach, but rather at the date when repairs ought reasonably to have been put in hand. This ruling can be of great significance in times of high inflation, as is shown by *Dodd Properties (Kent) Ltd v Canterbury CC*.[6] In this case (which did not involve a building contract, but a building which the defendants had negligently damaged), the claimants claimed that they were entitled to damages for the cost of repairing their premises. By the time that the case was tried, the claimants had deferred the reinstatement work for ten years. They had been waiting partly because of their financial circumstances and partly in order to be certain that the defendants, who were denying liability, would actually be held liable. It was held by the Court of Appeal that, since the claimants' decision to defer was a reasonable one, the damages should reflect the £30,000 that the repairs would now cost, and not merely the £11,000 that they would have cost ten years earlier.

In recent years, the courts have shown increasing sympathy for the physical discomfort and emotional frustration suffered by people who, for various reasons, are compelled to live in unsuitable accommodation. This is reflected, in cases concerning defective *dwellings*, by awarding such victims a fairly modest additional sum of damages by way of compensation. Such awards have become almost standard practice, but the Court of Appeal in *Hutchinson v Harris*[7] made it clear that the practice should not be allowed to extend to cases concerning *business premises*. No doubt every employer feels 'vexation' when a breach of contract occurs, but on its own this is not a sufficient basis for an award of damages.

The case last mentioned is also a good illustration of the doctrine of mitigation (mentioned above), under which claimants are required to take all reasonable steps to keep their loss to a minimum.[8] An award of damages will not include anything which is found to result from the claimant's 'failure to mitigate'. As a result, in *Hutchinson v Harris*, where an architect was guilty of negligence in supervising a conversion project, the client's claim for the loss of rental income failed. The evidence showed that the employer could easily have done what was required to render the premises lettable and then let them. Since the employer had not attempted to do this, the architect could not be held responsible for the loss.

Where a contractor's breach consists of failure to complete the contract work altogether, an additional factor comes into play. In such cases, it has been held that the damages awarded to the employer should be based on the additional cost incurred by the employer in completing the works. This will normally be what it

---

[5] [1995] 3 All ER 268.
[6] [1980] 1 All ER 928.
[7] (1978) 10 BLR 19.
[8] *British Westinghouse Electric Co Ltd v Underground Electric Railways* [1912] AC 673.

costs to have the work completed by another contractor, less the unpaid balance of the contract price.[9] If (unusually) the employer is able to complete the work for less than the outstanding balance, then the employer will only be entitled to an award of nominal damages. However, in such circumstances it may well turn out that the contractor has been overpaid for work so far carried out; if so, the employer can recover the excess under a legal doctrine known as **restitution** or **unjust enrichment**.[10]

21.1.3 Contractor's damages for employer's breach

As a general principle, where an employer is guilty of a breach of a construction contract, the contractor is entitled to damages under two headings. The first is damages for any actual loss that has been suffered, and the second is damages for any profit of which the contractor has been deprived. Examples of this principle are almost unlimited, but a good idea of the kind of losses that contractors commonly suffer can be gained from Chapter 16. That Chapter deals with contractors' claims for loss and expense under express contractual provisions, but it has long been settled that such claims are assessed in exactly the same way as damages for breach of contract.

Where the employer's breach is sufficient to justify the contractor in terminating the contract (a matter which is dealt with in Chapter 24), the contractor is entitled to damages reflecting everything which would have been received under the contract, or the proportion of it that remains outstanding at the date of termination, less what it would have cost the contractor to complete the work. It has been clearly established that the damages should include the profit element on work remaining to be done.[11] Where, however, the contract is one which the contractor had under-priced and on which the contractor would thus have made no profit, only nominal damages will be awarded for the employer's breach. This is because an award of damages should not put the claimant into a *better* position than if the contract had been performed.[12]

21.2   LIQUIDATED DAMAGES

As mentioned earlier, there is a special category of damages known as 'liquidated damages' (sometimes called 'liquidated and ascertained damages' or 'LADs'). This term applies to a predetermined sum which becomes payable by a party to a contract if certain specified breaches occur. Where it exists, this type of entitlement replaces the normal right to claim damages measured by the amount of loss actually suffered.

---

[9] *Mertens v Home Freeholds Co Ltd* [1921] 2 KB 526.
[10] *Ferguson (DO) & Associates v Sohl* (1992) 62 BLR 95.
[11] *Wraight Ltd v P H & T (Holdings) Ltd* (1968) 13 BLR 26.
[12] *C & P Haulage v Middleton* [1983] 3 All ER 94.

21.2.1 Nature and purpose of liquidated damages

A claim for 'liquidated damages' can only succeed where the contract makes express provision for it. Most construction contracts do this, by means of a clause providing that a contractor who is guilty of failure to complete the works by the contractual completion date (as extended where appropriate) shall pay or allow a certain amount of liquidated damages for every day or week of delay. Such clauses are found in JCT SBC 05 clause 2.32 (entitled 'Payment or allowance of liquidated damages'); JCT IC clause 2.23; and ICE 7 clause 47.

Liquidated damages provisions are in principle perfectly acceptable. Indeed, they are to be encouraged as they enable the parties to know from the start as much as possible about the risks they bear. They also save time and money on arbitration or litigation. However, the law recognizes that they are capable of operating rather harshly in cases where the amount to be paid or forfeited is greatly in excess of the loss caused by the breach of contract. Because of this, the courts have traditionally treated such clauses with a fair degree of suspicion. For example, the courts have always insisted that any contractual procedures are strictly adhered to and, in most cases, have interpreted any ambiguities against the employer.

It seems, however, that this strict traditional approach may have changed. In the case of *Philips Hong Kong Ltd v Attorney-General of Hong Kong*,[13] the Privy Council emphasized the desirability of upholding freedom of contract, at least between contracting parties of equal bargaining strength. It thus appears that, unless consumers are involved or there is evidence of economic duress, the courts will henceforth lean in favour of enforcing liquidated damages provisions and will not seek to strike them down on technical grounds.

The most important aspect of the courts' control of LADs is their insistence that a 'penalty clause' (which means an attempt to terrorize the contractor into completing on time) is unconscionable and wholly unenforceable. In such circumstances the employer is left to claim *unliquidated* or *general* damages in respect of whatever loss or damage can be proved.

The difference in law between liquidated damages and a penalty is thus of crucial importance. Unfortunately, this is a difference of fact and degree rather than a difference of kind, and is thus difficult to express with any great precision. However, some assistance on the matter has been forthcoming from the courts. The most notable guidelines on distinguishing between liquidated damages and penalties, which have been long accepted as the best available, were given by Lord Dunedin in 1915:[14]

1. The terms used in the contract are not conclusive, though they may be persuasive.
2. Unlike a penalty, liquidated damages represent a genuine attempt at a pre-estimate of likely damage.
3. In deciding into which category a particular contract term falls, account must be taken of circumstances at the time of making the contract, not at the time of breach.
4. The following 'tests' may be helpful, or even conclusive:

---

[13] (1993) 61 BLR 41.
[14] *Dunlop Pneumatic Tyre Co Ltd v New Garage & Motor Co Ltd* [1915] AC 79.

- If the sum stipulated is 'extravagant and unconscionable' compared with the greatest amount of loss that could be caused, it is a penalty.
- If the breach consists simply of non-payment of money, and the sum stipulated is a greater sum, it is a penalty.
- If a single sum is payable for a range of breaches of varying severity, there is a presumption (but no more) that it is a penalty.
- The fact that an accurate pre-estimation of the likely damage is almost impossible to achieve does not prevent a stipulation from being classed as liquidated damages. In fact, it is in precisely these cases – public buildings, housing association projects and other non-profit-making ventures – when a liquidated damages clause is most useful.

As Lord Dunedin's first guideline makes clear, whether or not something constitutes a penalty is a matter to be decided by the courts, not by the parties to the contract. For this reason, statements like *all sums payable by the contractor to the employer ... shall be paid as liquidated damages for delay and not as a penalty* (ICE 7 clause 47(3)) will not have their intended effect at law. Similarly, express terms stating that the amount for liquidated damages is a genuine pre-estimate of the employer's likely loss (ICE 7 clause 47(1)(a)) cannot take precedence over the guidelines.

In relation to the second guideline, it has been said that there must be a substantial discrepancy between the level of damages stipulated in the contract and the level of likely loss before a liquidated damages clause will be struck down as a penalty. However, the test of whether or not a pre-estimate is 'genuine' is an objective one and does not depend solely on the honesty of the party making it.[15]

The third guideline has encouraged contractors in some cases to argue that if, in certain hypothetical circumstances, a contract clause *could* have resulted in serious over-compensation for the employer, that clause must be regarded as a penalty. However, it has been made clear that the crucial test is what the parties might reasonably have expected to happen, not what might possibly happen.[16]

Although the vast majority of the cases in this area have concerned stipulations for the payment of a defined sum of money, it is clear that other contractual stipulations may also fall foul of the 'penalty' rules. An example of this can be seen in a civil engineering case, *Ranger v GW Railway*.[17] The contract provided that failure by the contractor to proceed regularly with the works would entitle the employer to forfeit all money due and all tools and materials. It also provided that, if this were not enough to cover the cost of completion, the contractor would be liable for the shortfall. It was held by the House of Lords that this provision was a penalty, since the value of the money and goods forfeited might far outweigh the cost to the employer of completing the work.

A similar decision was reached in *Public Works Commissioner v Hills*,[18] where a railway construction contract provided that, in case of delay, the contractor should forfeit all retention money under that contract and two other contracts. Here it was pointed out that retention money naturally increases as time goes on, and

---

[15] *Alfred McAlpine Capital Projects Ltd v Tilebox Ltd* [2006] BLR 271.
[16] *Philips Hong Kong Ltd v Attorney-General of Hong Kong* (1993) 61 BLR 41.
[17] (1854) 5 HLC 72.
[18] [1906] AC 368.

thus bears no relation to the employer's likely loss from delay – as a result, the forfeiture provision was to be treated as a penalty.

A third example of this principle is the case of *Gilbert-Ash (Northern) Ltd v Modern Engineering (Bristol) Ltd*,[19] in which a sub-contract provided that, if the sub-contractor failed to comply with any of its provisions, all payment from the main contractor could be suspended or withheld. Once again, the provision was held to be a penalty, this time because a trivial breach could lead to the retention of a wholly disproportionate sum of money.

It sometimes happens that the amount of 'liquidated damages' agreed on by the parties is likely to be *less* than the employer's actual loss. Where this can be seen from the outset, the clause in question operates as a kind of exemption clause (which are discussed in Section 10.2), since it effectively limits the contractor's liability for breach of contract. This may bring into play the very complex rules governing exemption clauses, including (in theory) the provisions of the Unfair Contract Terms Act 1977. However, the way in which that Act is drafted means that it will only apply to liquidated damages clauses where *either* the employer is a 'consumer' *or* the contract is made on the contractor's 'written standard terms of business'. In relation to JCT contracts, which are drafted following negotiations by all sides of the construction industry, it seems unlikely that a court would regard these as a contractor's written standard terms.

21.2.2 Operation and effect of liquidated damages clauses

In cases of delayed completion, the employer may seek to claim liquidated damages in accordance with the contract. To do this, the employer must show that:

1. The clause is not a penalty.
2. There is a definite date fixed by the contract from which the damages can run. As we have already seen, this date may be the completion date originally fixed or any other date that has been substituted under the provisions of an extension clause. In the latter case, the procedures for extending time must have been properly applied.
3. Any specified contractual procedures (such as a contract administrator's certificate or the giving of written notice) have been complied with.
4. The employer has not waived the right to deduct liquidated damages.

Where all these conditions are satisfied, the employer will be entitled to claim or deduct the stipulated sum, irrespective of what loss has actually been suffered or, indeed, whether there has been any loss at all. Thus in *BFI Group of Companies Ltd v DCB Integration Systems Ltd*,[20] a contractor was engaged under a JCT MW 80 form of contract to construct a warehouse. When the completion date arrived, the warehouse was almost completed, but still lacked its roller shutter doors, which had not arrived from the suppliers. The contractor allowed the employer to take possession and to fit out the warehouse and, by the time this had been done, the doors had been delivered and fitted. The employer had thus not lost any useful

---

[19] [1974] AC 689.
[20] [1987] CILL 348.

time, but was nevertheless held entitled to claim the full amount of liquidated damages provided in the contract.

A liquidated damages clause is said to be 'operative' where it applies and also where it *would* apply, but for a valid extension of time or an employer's waiver of rights. Where this is the case, it represents the sole ground of claim; it is not open to the employer to disregard it and claim unliquidated damages at common law. In *Surrey Heath BC v Lovell Construction Ltd*,[21] for example, completion of the works was delayed by a serious fire, which was alleged to be due to the negligence of the contractor. The employer granted an extension of time to cover the effects of the fire, but then claimed damages from the contractor to cover the loss of rental income from the premises resulting from the delay. This claim, it was held, must fail; the employer's losses were covered exclusively by the liquidated damages clause, which did not apply here because of the extension of time.

An even more striking example is *Temloc Ltd v Errill Properties Ltd*,[22] in which the employer had inserted '£nil' in the Appendix as the rate for liquidated damages. It was held that this precluded any claim at all for late completion; since the employer's normal right to seek unliquidated damages had been replaced by a right to seek no pounds for each week of delay! This seems a very odd result, and it may be noted that the position would almost certainly have been different if the whole liquidated damages clause had been struck out, or even if the Appendix had been left blank. In such circumstances, a court could have treated the liquidated damages provision as 'inoperative', in which case a claim for unliquidated damages would have been available.

An important practical consequence of the principle just illustrated occurs where there is a fluctuations clause in the contract, covering costs that increase at a time when the contractor is guilty of delay. In such circumstances the contractor, despite the delay, will be able to pass on these increased costs to the employer; the latter's only remedy is to claim liquidated damages.[23] However, it is of course possible for the contract to provide that a fluctuations clause shall 'freeze' when the contractor is in delay, and such provisions are indeed found in both JCT SBC 05 and ICE 7.

As stated above, an 'operative' liquidated damages clause provides the employer's sole remedy for delay. By contrast, a clause that has become 'inoperative' for some reason will leave the employer with a perfectly valid breach of contract claim for unliquidated damages. These will be assessed so as to compensate the employer for whatever losses can be proved. It is for this reason that it may be in the employer's interest, in cases where losses actually exceed the stipulated sum, to argue that the clause is in truth a 'penalty', since this would leave open the possibility of a claim for general damages. It also appears that an employer, who has rendered a liquidated damages clause inoperative (for example by causing a delay for which no extension can be granted), can resort to a common law claim.[24] However, it is doubtful whether the employer would be allowed in such circumstances to claim more than would have been received as liquidated damages. If this *were* allowed, the employer would appear to benefit from his or

---

[21] (1988) 42 BLR 25

[22] (1987) 39 BLR 30.

[23] *Peak Construction (Liverpool) Ltd v McKinney Foundations Ltd* (1970) 1 BLR 111.

[24] *Rapid Building Group Ltd v Ealing Family Housing Association Ltd* (1984) 29 BLR 5, CA.

her own breach of contract, and this is something the law does not usually allow to happen.

21.2.3 Liquidated damages clauses in the main standard forms

The operation of the principles described above may be demonstrated by reference to the clauses found in the main standard form contracts. For example, clause 2.32 of JCT SBC 05 provides that, before liquidated damages can be claimed or deducted, the contract administrator must issue a non-completion certificate (certifying that the contractor has failed to complete on time) and the employer must give the contractor written notice of intention to deduct liquidated damages. If an extension of time is subsequently granted, it is still necessary for the contract administrator, on each occasion that a new completion date is fixed, to issue a fresh non-completion certificate, but the employer's original notice of intention to deduct liquidated damages remains in force until it is specifically revoked.

The position under ICE 7 clause 47 is that liquidated damages become payable (or deductible) when the contractor has failed to complete the works within the prescribed time or any validly granted extension. Clause 47 does not specifically require the engineer to certify non-completion, but this will in effect happen anyway, since clause 44(4) requires the engineer to make a decision on any possible extension of time not later than 14 days after the due date for completion.

One further point specifically dealt with is what is to happen if an extension of time granted retrospectively means that liquidated damages should not have been paid or deducted. Both forms of contract provide that, in such circumstances, the employer must repay the money, but only ICE 7 states that interest must be paid on this sum. In fact, a Northern Ireland court has interpreted the wording used in an earlier version of the JCT contract as requiring the employer to pay interest,[25] but this decision is widely regarded as incorrect.

21.3 QUANTUM MERUIT CLAIMS

A **quantum meruit** claim is one in which the contractor seeks payment of the reasonable value of work done for the employer. Such a claim may arise in a variety of situations, not all of which involve a breach of contract by the employer. The common thread linking these situations is that there is either no contractual entitlement to payment or no contractual assessment of the amount due.

The situations in which a quantum meruit claim is most likely to arise in the construction context are as follows:

1. Where there is an express undertaking by the employer to pay a reasonable sum in return for services rendered.
2. Where professional or trade services are requested by the employer (for example under a letter of intent), but no price is agreed. Here it is implied that a reasonable sum will be payable.[26]

---

[25] *Department of the Environment for Northern Ireland v Farrans Construction Ltd* (1981) 19 BLR 1.
[26] Supply of Goods and Services Act 1982, section 15.

3. Where a price fixing clause in a contract fails to operate.
4. Where extra work is ordered which falls outside the scope of a variations clause.[27]
5. Where an apparent contract under which work is done is in fact void.

The question of how exactly a *quantum meruit* is to be assessed was considered in the case of *Laserbore Ltd v Morrison Biggs Wall Ltd.*[28] That case concerned an informal sub-contract for micro-tunnelling work, under which the defendants undertook to *reimburse the claimants' ... fair and reasonable payment*. The defendants argued that this should be assessed on a 'cost plus' basis, but the judge ruled that the correct approach was to ask: *What would be a fair commercial rate for the services provided?* A similar approach was adopted in *Costain Civil Engineering Ltd v Zanen Dredging and Contracting Co Ltd,*[29] where it was acknowledged that, in appropriate circumstances, a contractor might be entitled in effect to share in the profit generated by the work for which a *quantum meruit* was to be awarded.

One area of difficulty, which has generated considerable debate, concerns the use of a *quantum meruit* claim against an employer who is guilty of a repudiatory breach of contract. The crucial question is whether the contractor in such circumstances can simply ignore the contract and instead claim a reasonable sum for all the work done, even if this means that the contractor recovers more than would have been recovered under the contract. There is some authority to support the view that the contractor can indeed adopt this approach.[30] However, there is a strong counter-argument that the contractor should not be allowed to use this remedy to escape from a loss-making contract, and this seems the better view.

21.4   NON-PAYMENT AS A CONTRACTUAL REMEDY

A question that frequently arises under building contracts concerns the extent to which an employer may refuse to meet a contractor's claim for payment, on the ground that the employer has some cross-claim against the contractor which would reduce or even cancel out what is owed. The question seems to arise even more frequently under sub-contracts, where it concerns a main contractor's refusal on the grounds of a cross-claim to pay the sub-contractor. Such cross-claims are sometimes referred to as rights of set-off, but this is not entirely accurate; as we shall see, there are three separate defences that may arise in this situation.

21.4.1 The legal background

It is important to appreciate that, if a cross-claim does not provide a defence, the employer (or main contractor, in the case of a sub-contract) must simply make payment in accordance with the contract terms and then take the cross-claim to

---

[27] *Sir Lindsay Parkinson & Co Ltd v Commissioners of Works and Public Buildings* [1949] 2 KB 632, [1950] 1 All ER 208, CA.
[28] [1993] CILL 896.
[29] (1996) 85 BLR 77.
[30] *Lodder v Slowey* [1904] AC 442.

court or to arbitration. As a result, where the contract provides (as some do) that no such claim can be arbitrated until practical completion is achieved, the employer may have to wait a long time for reimbursement. It is thus clear, given the size of many claims that arise out of large projects, that this is a topic of immense practical significance.

It is interesting to note that the early common law provided no defences of this kind. If a builder was promised a sum of money for building a house, then the money would have to be paid as soon as the work was complete. Even if the house had already fallen down, the client would have no defence to the builder's action, but merely a right to claim damages in a separate lawsuit. Naturally, such an inconvenient rule could not possibly survive and, as we shall see, it has now been substantially eroded.

21.4.2 Counter-claim, set-off and abatement

In considering what defences may be available, much depends upon what steps a contractor takes to enforce the claim to payment. If the matter is taken to arbitration or to a full trial in court, then those proceedings should deal with all the issues between the parties. In consequence, *any* counter-claim of the employer can be set up against the contractor. It is for this reason that an application by the contractor for interim payment (see Section 25.3.2) may be defeated if the employer can raise a 'mere' **counter-claim**, since the court in such proceedings is required to consider what would happen if the action proceeded to trial.[31]

In practice, contractors often seek to enforce a claim for payment by applying for summary judgment. Where this is done, a mere counter-claim will not operate to defeat the claim.[32] However, the contractor's claim for payment *will* be defeated by a cross-claim which satisfies the more limited definition of **equitable set-off**. Whether a cross-claim will do this depends largely upon whether it satisfies the test laid down by the Court of Appeal in *Hanak v Green*.[33] This requires the claim and counter-claim to be so closely linked that it would be manifestly unjust to enforce one of them without taking the other into account.

The facts of *Hanak v Green* provide a good illustration of the doctrine of equitable set-off. The contract there was for refurbishment work in the employer's house, and it is clear that the employer and the contractor were virtually at war throughout the contract period. When the work was completed, the employer sued the contractor alleging various defects in what had been done. In answer to this action, the contractor raised claims for extra work, disruption caused by the employer's refusal to allow one of the workmen into the house, and the loss of certain tools that the employer had thrown away. It was held that all these claims could properly be raised as defences, since they arose directly under and affected the contract on which the employer relied.

In the light of *Hanak v Green*, it can safely be said that cross-claims which arise out of the same contract as the primary claim may be relied on to provide a defence of equitable set-off. In theory, the same principle may apply to claims

---

[31] *Smallman Construction Ltd v Redpath Dorman Long Ltd* (1988) 47 BLR 15.
[32] *Tubeworkers Ltd v Tilbury Construction Ltd* (1985) 30 BLR 67.
[33] [1958] 2 QB 9.

under *separate* contracts, provided that there is a sufficiently close link between them. In practice, however, the courts have proved extremely reluctant to permit a set-off in such circumstances. Thus in *Anglian Building Products Ltd v W & C French (Construction) Ltd*,[34] it was held that a contractor's claim against a supplier for defects in concrete beams used on one motorway contract could not be set off against the supplier's claim for payment for beams used on two other motorway contracts. An even more striking example is provided by *B Hargreaves Ltd v Action 2000 Ltd*,[35] where the parties had signed nine virtually identical sub-contracts (relating to different sites) on the same day. When the sub-contractors claimed payment under one of these contracts, the main contractors were not permitted to raise a set-off based on breaches of some of the other contracts. Needless to say, where the two contracts are not even between the same parties, the likelihood of successfully claiming a set off is even more remote. Thus an employer cannot avoid paying the main contractor in respect of work done by a nominated sub-contractor, on the ground that the employer has a claim against that sub-contractor under a direct warranty agreement for defective work.[36]

The *Hargreaves* case raised another point of importance in this context, by recognizing the existence of a second type of set-off. This is **common law set-off**, and it arises where both claim and cross-claim are in respect of 'liquidated debts or money demands which can be readily and without difficulty ascertained'. In such circumstances (for example where the employer raises a set-off based on liquidated damages), the two claims need not be closely linked. The defence would be available, even if they arose under separate contracts.

Until fairly recently, the term 'set-off' was used somewhat loosely to cover any complaints raised by an employer as an excuse for refusing to pay the contractor. However, the courts now recognize that, in addition to a true set-off (which consists of a cross-claim), there is a separate defence known as **abatement**. The essence of this defence is an allegation by the employer that the contractor's claim itself is unjustified, in that the work done is not worth what is being claimed. It follows from this definition that the defence of abatement may be raised only in respect of physical defects in the work; an employer's complaints about delay are the subject-matter of set-off, not abatement.[37]

Although the defence of abatement has been recognized for at least 150 years,[38] its use in the construction field raises one particular difficulty. This is where the disputed claim rests, not upon a simple assertion by the contractor that money is due, but rather on an architect's certificate or its equivalent. In such circumstances an employer seeking to defend the claim will face an uphill task, since it requires the court to be convinced of a substantial possibility that the work has been over-certified. It is established that clear evidence will be needed, and that vague allegations of defective work will not suffice. However, if the evidence is good enough, the defence of abatement will be available.[39]

---

[34] (1972) 16 BLR 1.
[35] (1992) 62 BLR 72.
[36] *George E Taylor & Co Ltd v G Percy Trentham Ltd* (1980) 16 BLR 15.
[37] *Mellowes Archital Ltd v Bell Projects Ltd* (1997) 87 BLR 26.
[38] *Mondel v Steel* (1841) 8 M&W 858.
[39] *CM Pillings & Co Ltd v Kent Investments Ltd* (1985) 30 BLR 80; *RM Douglas Construction Ltd v Bass Leisure Ltd* (1990) 53 BLR 119.

21.4.3 Certified sums

Can an employer exercise a right of set-off against sums which the contract administrator has certified as due for work properly executed? Or does the contract administrator's decision effectively override the employer's claim? This crucial question has occupied the courts from 1971.

In *Dawnays Ltd v FG Minter Ltd*,[40] the Court of Appeal held that a main contractor could not set off a claim for a sub-contractor's delay against sums certified as due to the sub-contractor. The actual decision was based on the precise wording of the 'Green Form' of sub-contract, which was used for nominated sub-contractors under JCT 63, but the court clearly regarded the maintenance of cash flow as a matter of general importance. As Lord Denning MR put it: *An interim certificate is to be regarded virtually as cash, like a bill of exchange. It must be honoured. Payment must not be withheld on account of cross-claims, whether good or bad, except so far as the contract specifically provides.*

The approach adopted in the *Dawnays* case was applied by the Court of Appeal in several subsequent cases. Most of these, like *Dawnays* itself, concerned sub-contracts, but the principle was also applied as between an employer and a contractor under a *main* contract.[41] However, this whole line of cases was heavily criticized by the House of Lords in the leading case of *Gilbert-Ash (Northern) Ltd v Modern Engineering (Bristol) Ltd*.[42] That case again concerned a sub-contract, but the terms of this one were very different. It was specifically stated that the main contractor could set off from certified sums *any bona fide contra account and/or other claims*. Not surprisingly, it was held that this clause entitled the main contractor to set off claims for delay and defective workmanship.

Because the wording of the sub-contract in the *Gilbert-Ash* case was so clear, it is possible to argue that the *Dawnays* principle still operates. However, the better view is that the courts now regard that principle as having been overruled. As a result: *In order to exclude the right to assert cross-claims admissible as equitable set-offs ... it is necessary to find some clear and express provision in the contract which has that effect.*[43] Moreover, in a case where a supplier's standard terms of business *did* purport to exclude a client's right of set-off, the Court of Appeal held that this was unreasonable and could therefore be struck down under the Unfair Contract Terms Act 1977.[44]

It should be added that, just occasionally, the right of set off may also be excluded *by implication*. This happened, for example, in *Mottram Consultants Ltd v Bernard Sunley & Sons Ltd*,[45] where a contractual clause giving a right of set off was actually deleted by the parties before the contract was signed. It was held by the House of Lords that this showed the parties' intention not to have any such right.

---

[40] [1971] 2 All ER 1389.
[41] *Frederick Mark Ltd v Schield* (1971) 1 BLR 32.
[42] [1974] AC 689.
[43] *CM Pillings & Co Ltd v Kent Investments Ltd* (1985) 30 BLR 80.
[44] *Stewart Gill Ltd v Horatio Myer & Co Ltd* [1992] 2 All ER 257.
[45] (1974) 2 BLR 28.

21.4.4 Procedural requirements

For some years, a number of standard form sub-contracts (though not main contracts) have contained provisions to the effect that a right of set-off cannot be exercised unless certain specified procedures have been followed. These commonly include the requirement to give written notice to the other party, specifying the amount of money to be withheld and the reason for it. Such provisions have in the past resulted in attempts by main contractors to exercise rights of set-off being defeated by the lack of an architect's certificate of delay,[46] failure to give the prescribed written notice,[47] and failure to quantify the set off in sufficient detail.[48]

Giving notice of an intention to withhold payment has been made a general requirement by section 111 of the Housing Grants, Construction and Regeneration Act 1996, which provides that payment of any sum due under a construction contract may not be withheld after the final date for payment (discussed in Section 15.1.2) unless an effective written notice has been given. Such a notice must specify the amount to be withheld and the ground for withholding it, and must be given not later than the prescribed period before the final date for payment. This prescribed period may be set down in the contract; if it is not, then the Scheme for Construction Contracts provides that it shall be 7 days. A sum is 'due' for this purpose if it is contained in an interim certificate, even if that certificate is subsequently challenged by the employer. It follows that an employer who disagrees with an interim certificate, but who does not issue a valid withholding notice, must pay what is stated and seek at a later stage to overturn the certificate.[49]

---

[46] *Brightside Kilpatrick Engineering Services v Mitchell Construction Ltd* (1973) 1 BLR 62.
[47] *Pillar PG Ltd v DJ Higgins Construction Ltd* (1986) 34 BLR 43.
[48] *BWP (Architectural) Ltd v Beaver Building Systems Ltd* (1988) 42 BLR 86.
[49] See *Rupert Morgan Building Services (LLC) Ltd v Jervis* [2004] BLR 18.

22 Defective buildings and subsequent owners

Where an apparently completed building is found to contain defects of design or construction, the client for whom it was originally constructed may wish to take legal action against the person or organization responsible. Such an action will normally be based on a breach of contract: the building contract if the defect is one of construction, a consultant's terms of appointment if there has been a design error. Whether or not the client's claim succeeds is likely to depend upon various factors which were discussed in earlier chapters: the terms (express and implied) of the relevant contract, when the defects were discovered, whether or not a final certificate has been issued and so on.

The question addressed in this Chapter is a different one. It relates to the legal rights, not of the original client, but of a person who has acquired the completed building from the original client. This person may be a purchaser or a tenant, and may have acquired the building directly or indirectly, after more than one change of ownership. Assuming that a defect comes to light only *after* that person has acquired the building (and has paid a price based on the assumption that there are no defects in it), the question is what remedies, if any, are available.

One point is worth making at the outset. The old legal doctrine of *caveat emptor* ('let the buyer beware') means that a contract for the sale of land and buildings, unlike one for the sale of goods, contains no implied terms relating to the quality of what is sold. Hence, a person who acquires a building and then finds that it is defective may well have no legal redress against the vendor or landlord. This is why it is so important to know whether that person can claim against whoever was responsible for creating the defect in the first place.

22.1 CLAIMS IN NEGLIGENCE

If a claim in contract against the vendor is ruled out, one might well regard the obvious next step as a claim in the tort of negligence (see Section 10.5) against any designer or constructor who is responsible for the fact that the building is defective. However, as we shall see, the law places very severe restrictions upon such a claim. These restrictions are explained in Section 22.1.2; first, however, we trace the history of negligence claims by subsequent owners, in order to show how the law developed in this area.

22.1.1 The law's development

It was laid down by the House of Lords in *Donoghue v Stevenson*[1] that a manufacturer of products would be liable for negligently causing personal injury or property damage to a 'consumer', notwithstanding the absence of any contract between them. However, while the courts in subsequent cases soon began to treat a building contractor like a manufacturer for this purpose,[2] it was held that the old immunity of an *owner-builder* (i.e. someone who constructs a building on land which he/she owns and then sells it) had not been affected by *Donoghue v Stevenson*. As a result, an owner-builder could not even be held liable in negligence for defects in the building that injured or killed people.[3]

This harsh situation was eventually rectified by the Court of Appeal in *Dutton v Bognor Regis UDC*,[4] a ruling which was endorsed by the House of Lords in *Anns v Merton LBC*.[5] The legal principles laid down by those decisions were as follows:

- A duty of care to subsequent owners was owed by anyone who participated in the design or construction of a building.
- A similar duty of care (covering approval of plans and inspection of building operations) was owed by a local authority with responsibility for building control.
- The duty of care could result in liability, not only for injury or damage to other property, but also for damage to the building itself.
- However, neither builder nor local authority was answerable in tort for *all* defects, but only for those creating a 'present or imminent danger to the health and safety' of the occupier or others. In such circumstances, the defendant was liable for the cost of making the building safe.

22.1.2 The current position

General

The principle laid down in *Anns v Merton*, which created severe problems in distinguishing between 'defects' and 'dangerous defects', soon began to provoke criticism from the courts. In the event, the protection that had been given to subsequent owners of defective buildings lasted less than ten years, before it was removed by two important decisions of the House of Lords.

The first of these cases, *D & F Estates Ltd v Church Commissioners for England*,[6] concerned a block of flats built by Wates in a joint development venture with the Church Commissioners. Some 15 years after the flats were completed, a tenant discovered that, due to the negligence of sub-contractors (who were described as 'not worth suing'), the walls and ceilings of his flat needed to be

---

[1] [1932] AC 562.
[2] *Sharpe v ET Sweeting & Son Ltd* [1963] 2 All ER 455.
[3] *Davis v Foots* [1940] 1 KB 116.
[4] [1972] 1 QB 373.
[5] [1978] AC 728.
[6] [1988] 2 All ER 992.

completely replastered. The tenant thereupon sued Wates in tort, claiming the cost of remedial work already carried out, the estimated cost of future work (some £50,000) and prospective loss of rent.

In rejecting the claimants' claims, the House of Lords held that a builder's liability in tort is limited to defects which cause either injury to persons or physical damage to *property other than the building itself* (as for instance where a defective garage roof falls on the occupier's car). Damage to the building itself is regarded as pure economic loss and therefore irrecoverable.

Although the House of Lords in *D & F Estates* was highly critical of *Anns v Merton*, it did not specifically declare the earlier case to have been wrongly decided. Two years later, however, that final step was taken. This was in *Murphy v Brentwood DC*,[7] which concerned a house-owner's action against a local authority for having negligently approved what turned out to be inadequate foundations. The House of Lords by now had clearly decided that the time had come for a thorough reform of this area of law. Their lordships acknowledged that the House's power to reverse its own previous rulings should be used sparingly (to avoid creating too much uncertainty in the law). However, the unanimous view was that this was an appropriate occasion for its use. *Anns* was declared wrong (following detailed criticism of its inherent illogicalities), and the local authority's appeal against liability was accordingly successful.

Exceptions

The House of Lords in *Murphy* emphasized once again that the loss suffered by the owner of a building who discovers defects in it is pure economic loss, which is not normally recoverable in the tort of negligence. Even so, some of their lordships suggested that, in two exceptional cases, a subsequent owner might still be permitted to recover damages (based on the cost of repair) from a negligent designer or constructor. It has to be said, with respect, that neither of these suggested exceptions is very convincing. However, *both* of them have been applied in subsequent cases, and they must therefore be treated as the law, at least for the time being.

The first exceptional case, in which a subsequent owner can still sue in negligence for damage to the building itself, is where the defect in the building creates a danger *to adjoining property* and the subsequent owner incurs expense in averting that danger. This exception was applied in *Morse v Barratt (Leeds) Ltd*,[8] where a row of houses was constructed on a sloping site, in such a way that their gardens led down to a road. When a retaining wall at the bottom of these gardens was found to be defective and in danger of falling outwards on to the road, the local authority served a dangerous structure notice on the house-owners, as a result of which they were compelled to pay for the necessary remedial work. It was held that the owners were entitled to recover this cost from the builders whose negligence had created the danger.

The second exception suggested by the House of Lords is where different parts of a building are designed or constructed by different persons, and the negligence

---

[7] [1990] 2 All ER 908.
[8] (1992) 9 Const LJ 158.

of one (in relation to the part for which he or she is responsible) causes damage to another part. It was suggested, for example, that a subsequent owner would be entitled to recover damages from a central heating sub-contractor whose negligent installation of a boiler caused an explosion damaging the rest of the building. Another example given was that of an electrical contractor or sub-contractor, whose negligent installation of a wiring circuit caused a fire.

It must be said that this exception seems likely to cause problems, if only because the division of construction work between a number of contracting organizations is normal practice in the industry. Hence, unless the exception is kept within strict limits, it could ultimately swallow up the basic rule, although this does not appear to be what the House of Lords intended.

The only reported case in which this exception has been considered is *Jacobs v Morton & Partners*,[9] which concerned a house with defective foundations. The owners of the house engaged the defendants, a firm of consulting engineers, to design a scheme to remedy the defects and, some time after the relevant work was completed, the house with its new foundations was sold to the claimant. The claimant subsequently complained that the foundations were inadequate and alleged that this was due to the defendants' negligent design. It was held that, if the defendants could indeed be shown to have been negligent, they would be liable to the claimant for the resulting damage to the rest of the house.

22.2   STATUTORY PROTECTION

The right of a subsequent owner to recover damages for negligence have thus been drastically cut back, and this has served to focus attention on whatever other legal rights may be available. We consider first a number of statutory provisions (of which the most important is the Defective Premises Act 1972), which may offer at least a partial solution to the problem.

22.2.1 Defective Premises Act 1972

Section 1(1) of this Act imposes a legal duty on any person 'taking on work for or in connection with the provision of a dwelling', a definition which includes cases where the dwelling is provided by the erection or by the conversion or enlargement of a building. This duty is owed to any person to whose order the dwelling is provided. More important for our present purposes, it is also owed to 'every person who subsequently acquires an interest, whether legal or equitable, in the dwelling'.

The nature of the duty created by section 1(1) is to see:

- that the work which has been taken on is done in a workmanlike or professional manner;
- that proper materials are used; and
- that the dwelling will be fit for habitation when completed.

---

[9] (1994) 72 BLR 92.

This wording clearly applies to every person who contributes to the process of design or construction, including the property developer or public sector equivalent.[10] All the participants are then strictly liable in respect of their respective contributions to the dwelling. It has been held that this liability attaches, not only to what a person does badly, but also to what that person fails to do.[11]

On the face of it, this provision seems to be of great potential importance, but its practical value is severely restricted by three factors:

- The Defective Premises Act 1972 applies only to 'dwellings'; there is no equivalent statutory provision relating to commercial property.
- Claims under this Act are subject to a relatively short limitation period, since they must be commenced within six years from when the dwelling is completed.
- Although the duty itself looks to have three limbs (workmanship, materials, fitness for habitation), it was held in *Thompson v Clive Alexander & Partners*[12] that this is not the correct interpretation. 'Fitness for habitation' is the *standard* by which the other two (workmanship and materials) are to be judged. If this is correct, it means that the Act will only apply to those defects of workmanship or materials serious enough to render a dwelling 'unfit for habitation'.

Although any attempt to contract out of the Defective Premises Act is void,[13] section 2 provides that the Act does not apply to any dwelling which is subject to an 'approved scheme'. This was at one time an extremely important restriction, for the scheme in question was the one administered by the National House-Building Council, which covered the vast majority of new dwellings. However, the current version of the NHBC Guarantee Scheme (known as the 'Buildmark', and in force since 1988) is not an 'approved scheme' for this purpose. As a result, the purchaser of a house or flat may now have two forms of protection. First, assuming that the house or flat is covered by the Buildmark, the builder or developer undertakes to remedy defects in the first two years and the Council provides insurance against major structural defects in the next eight years. Second, and in addition, the purchaser is also entitled to claim under the Defective Premises Act 1972.

22.2.2 Other statutory remedies

Consumer Protection Act 1987, Part I

Part I of this Act, which was based on an EU Directive, imposes a form of strict liability (i.e. liability independent of negligence) upon **manufacturers** and certain other **producers** in respect of **defective products** that cause **injury** or **damage**. There is no doubt that these provisions are capable of applying to construction, since building materials (both raw materials and manufactured components) clearly

---

[10] Section 1(4).
[11] *Andrews v Schooling* [1991] 3 All ER 723.
[12] (1992) 59 BLR 77.
[13] Section 6(3).

fall within the statutory definition of 'products', although the completed building itself does not. The way the Act applies to buildings is as follows:

- A builder or developer who constructs a building and then sells it, together with the land on which it stands, is outside the scope of the Act. Such a person 'supplies' neither the building nor its component materials.
- A building contractor who carries out work on someone else's land is within the scope of the Act in respect of all goods and materials that are supplied in the course of that work.
- In *both* these cases, the Act will apply to sub-contractors and suppliers of materials, since they undoubtedly supply products to the main contractor.

In situations to which the Act applies in principle, the following limitations should be noted:

- The Act is concerned solely with 'products' containing a 'defect'. It does not therefore apply where damage results from an error of workmanship, rather than from a defect in the materials used.
- The Act permits actions for damages to be brought in respect of death or personal injury, and also for any loss of or damage to any property (including land). However, a claim for property damage can only arise where the damaged property is of a kind ordinarily intended for private occupation or use (and actually so intended by the claimant).
- No action can be brought under the Act for damage to the defective item itself.

Building Regulations

In certain circumstances, the tort known as **breach of statutory duty** permits a claimant to recover damages on the basis that the defendant has caused loss or damage as a result of breaching some statutory rule. Whether or not such an action could be based upon a breach of the Building Regulations has never been definitively decided by the courts. It was suggested by Lord Wilberforce in *Anns v Merton LBC*[14] that such an action would be available but, while this suggestion has been applied in at least one subsequent case, judges have usually rejected it.

It may be that the position will be more clearly defined in the future, because section 38 of the Building Act 1984 makes provision for civil liability for any 'damage' caused by a breach of regulations made under that Act. For some reason, however, this provision has never been brought into force. Moreover, even if it is implemented, it should be noted that the regulations themselves impose no higher duty than that of securing reasonable standards of health and safety. It follows that a claim could only be brought in respect of those defects in a building that are positively dangerous.

---

[14] *Anns v Merton LBC* [1978] AC 728.

22.3   ALTERNATIVE FORMS OF LEGAL PROTECTION

The demise of negligence, as well as turning the spotlight on other existing legal remedies, has had another important effect upon those legal advisers who operate in the fields of construction and property. It has encouraged them to respond proactively to the requirements of purchasers and tenants and to create by way of contract rights similar to those that previously existed in tort.

There have been four main developments in this area. These are:

- The creation of a direct contractual link between a subsequent owner and each designer and constructor (by way of a **collateral warranty**).
- The transfer (by way of **assignment**) to the subsequent owner of whatever contractual rights the original client would have had against each designer and constructor.
- A claim for breach of contract against the relevant designer or constructor brought by the original owner, who then hands over the damages to the subsequent owner.
- A claim by a subsequent owner for breach of a term in the designer or constructor's contract with the original client, under the Contracts (Rights of Third Parties) Act 1999.

22.3.1 Collateral warranties

An important aspect of construction documentation in recent years has been an enormous increase in the use made of collateral warranties, as attempts are made to bring remote owners into direct contractual relationships with those against whom they would once have been able to claim in the tort of negligence. Although the contents of such warranties vary considerably, what they commonly contain is an undertaking by a consultant or a contractor to the effect that he or she has fulfilled all the obligations owed to the client. In effect, therefore, the subsequent owner of the building is given contractual rights similar to those that would have been enjoyed by the client if the building had not changed hands.

Such warranties are routinely sought from all the contractors, sub-contractors and consultants who are engaged on a construction project, and may be drafted in favour of the prospective purchasers or tenants of the completed building and/or the financial institution which is funding the development and whose money is therefore secured upon the property. The sheer number of possible relationships that may thus arise out of a single project means that the task of negotiating and drafting collateral warranties frequently involves a considerable amount of time and effort for the parties and their solicitors.

For a long time, the problem was made worse by the lack of any standard forms of collateral warranty, requiring that each one had to be negotiated on an individual basis. However, the position was improved by the publication of standard forms of warranty for consultants, which are in versions for either funding institutions or for potential purchasers and tenants. JCT have also published contractors' collateral warranties, again for funders and for purchasers and tenants (CWa/F and CWa/P&T) .

It is important to note that no contractor or consultant is automatically under an obligation to give a collateral warranty to anyone at all. It follows that the question of whether collateral warranties are to be demanded is one that should be addressed at the commencement of a project. At that stage it is possible to ensure that every consultant's terms of appointment, and every building contract, contains a term requiring that warranties be given when the client demands it.

As mentioned above, there is considerable variation in the contents of collateral warranties. Nevertheless, some general points may usefully be made. First, at least where designers are concerned, it is important to see that any liability that may arise will be covered by the designer's professional indemnity insurers. This may involve bringing the insurers into the negotiations for the warranty, which of course is likely to add to the length and complexity of those negotiations. One of the major advantages of the standard forms is that, if they have been accepted by the insurers for the participating professional institutions, they can be used by members of the relevant institutional schemes without the need for further negotiation.

The extent of liability which is undertaken by way of collateral warranty is commonly restricted in various ways. The following limits are typical:

- The warrantor's liability is limited to the cost of remedial works required because of the breach; there is thus no liability for any consequential losses (such as the cost of alternative accommodation while the work is carried out).
- In the event that the warrantor is not the only negligent party, the warrantor is liable only for the share of responsibility that a court would place on him or her if all potential defendants were sued.
- Any defence the warrantor could have raised if sued by the client may equally be used when a claim is made on the warranty.

Apart from these restrictions, three further provisions commonly found in collateral warranties are worthy of mention. First, in relation to design work, the warrantor may be under a specific obligation to take out and maintain a specified level of professional indemnity insurance cover. Second, many warranties either contain an outright prohibition on assignment, or specify the number of times that the warranty can be passed on as the building itself changes hands. Third, some warranties are expressed to be retrospective and to come into effect at the time of practical completion. The effect is that the limitation period for a claim on the warranty will be the same as that for a claim under the original contract.[15]

22.3.2 Assignment of rights to the subsequent owner

The second major device used to confer legal protection upon subsequent owners of buildings is that of assignment. What this means is that the original owner, on selling the building, transfers to the purchaser any or all of the rights of action in respect of defects that the original owner has. Such rights may include, not only those arising under the main building contract itself, but also those based on

---

[15] See *Northern Shell plc v John Laing Construction Ltd* (2003) 90 Con LR 26.

consultants' terms of engagement, sub-contractors' warranties and any other relevant contractual relationships. An assignment may also be expressed to include any rights of action *in tort* that the original owner could establish against any of these parties.

The validity of an assignment in these circumstances was considered by the House of Lords in the case of *Linden Gardens Trust Ltd v Lenesta Sludge Disposals Ltd*.[16] That case in fact consisted of two appeals (the other being *St Martins v McAlpines*) involving slightly different factual scenarios, which were heard together by the House of Lords. As a result of what was said in those cases, both in the House of Lords and in the courts below, it is now clear in principle that an assignment can effectively transfer legal rights to an assignee (the purchaser). However, it must be emphasized some important points remain uncertain, notably the nature of the assignee's rights and (most significantly) the basis on which the assignee's damages are to be assessed.

The present position on assignment may (albeit somewhat tentatively) be stated as follows:

1. The client is entitled to assign both the right to call for future performance of a building contract *and* the right to claim damages for past breaches of it. What this means is that an assignment can be effective, irrespective of whether it takes place before or after a relevant breach of contract has occurred.
2. An assignment, to be effective, need not be made at the time when the building itself is transferred. It can be carried out as a separate transaction.[17]
3. Many construction contracts either prohibit assignment by the employer altogether or place restrictions on it, for example by requiring the contractor's written consent. Such restrictions are found in JCT SBC 05 clause 7.1 and ICE 7 clause 3. An assignment in breach of such a provision will be ineffective, in the sense that the assignee will not acquire any enforceable rights.
4. In the *Linden Gardens* case at first instance,[18] it was held that a contract clause which prohibited assignment would not normally prevent the employer from assigning a right of action *in tort*. However, it should be noted that, unless and until some damage to the property has occurred (whether or not it is visible), the employer will not have acquired a cause of action in tort and there will therefore be nothing to be transferred. It should also be pointed out that at least one judge has expressed serious doubts as to whether an employer would be entitled to bring a tort claim against a contractor.[19]
5. Where an assignment is valid, the assignee is entitled to be placed in as good a position as the assignor. However, an assignee is not supposed to be in a *better* position than the assignor. This principle has led contractors to claim that, since an assignor who sells a building to an assignee at a price which assumes no defects has suffered no loss, the assignee (who of course

---

[16] (1993) 63 BLR 1.
[17] *GUS Property Management Ltd v Littlewoods Mail Order Stores Ltd* 1982 SLT 533.
[18] (1990) 52 BLR 93.
[19] *Nitrigin Eireann Teoranta v Inco Alloys Ltd* [1992] 1 All ER 854; see Chapter 10.

has suffered a real loss) will be limited to recovering nominal damages! Thankfully, the courts have simply rejected this technical argument which, if accepted, would mean that the 'loss' resulting from the defective building would somehow disappear into a black hole.[20]

6. Although the courts have thus rejected the 'no loss' argument, they have nonetheless stated that an assignee 'cannot enforce any claims, let alone under new heads of damage, which would not have been available to his assignor'.[21] Precisely how this limitation will operate in practice remains obscure, for there have as yet been no reported cases in which it has been applied. It might mean that, where a subsequent owner suffers far greater business disruption costs during remedial works than would have been suffered by the original owner (for example because the subsequent owner has specialist requirements for which there is no suitable alternative accommodation), those extra losses cannot be recovered as damages.

22.3.3 Claims by original owner on subsequent owner's behalf

As mentioned above, the House of Lords in the *Linden Gardens* case held that a purported assignment which contravened a contractual prohibition was of no effect. This ruling was uncontroversial, but the House of Lords then astonished both the construction industry and the legal profession by the second part of their decision. This was that the *employer* could still sue the contractor for breach of contract, notwithstanding that the employer (having sold the building before any defects appeared) had suffered no loss. Their lordships further held that, if an employer took this course of action, the damages recovered (which would be assessed on the basis of the *purchaser's* loss) would then have to be handed over by the employer to the purchaser.

In reaching this decision, the House of Lords acknowledged that, as a general principle of contract law, a party who has not suffered loss cannot recover substantial damages, and cannot recover damages on behalf of a third party. However, their lordships pointed out that there were certain exceptions to this principle, and identified one that they said would apply in the present case. This is that where the parties to a building contract would expect the completed building to be transferred to a third party, who could not acquire any rights under the contract itself, the parties might well intend that the original employer should be able to enforce the contract on behalf of the third party. And, if that is what the parties intended, then their contract would contain an implied term permitting such enforcement.

The precise extent of this new legal development is extremely difficult to determine. In *Linden Gardens* the House of Lords stated that their decision would provide 'a remedy where no other would be available to a person sustaining loss that under a rational legal system ought to be compensated by the person who has caused it'. This emphasis on the lack of any other remedy has led the House of Lords to conclude that a direct claim by the original employer is not possible where

---

[20] *GUS Property Management Ltd v Littlewoods Mail Order Stores Ltd* 1982 SLT 533; *Darlington BC v Wiltshier Northern Ltd* (1994) 69 BLR 1.
[21] *Dawson v Great Northern & City Railway Co* [1905] 1 KB 260.

the purchaser has been given an alternative remedy against the contractor, under a duty of care deed or a collateral warranty (irrespective of the extent of this alternative remedy).[22] It has been suggested that the decision would only apply where the contract in question prohibits assignment (and thus prevents the purchaser from acquiring a remedy by that route). However, this argument was rejected by the Court of Appeal in *Darlington BC v Wiltshier Northern Ltd.*[23]

The *Darlington* case also offered a solution to another practical problem caused by this 'third party' claim, namely, what happens if the employer refuses to sue on the purchaser's behalf? Construction lawyers have attempted to deal with this possibility by extracting a promise from the employer, as part of the contract for the sale of the building, to join in any future legal action that the purchaser wishes to bring (in return for an indemnity against any costs incurred). However, the Court of Appeal's decision suggests that this may not be necessary. According to their lordships, the employer in a case of this kind is to be regarded as a constructive trustee of the right to claim damages for the purchaser. And, as a trustee, the employer is bound to enforce those rights and can be liable to the beneficiary (the purchaser) for failure to do so.

It must be said, in conclusion, that the law on both assignment and claims by third party remains very obscure. This is regrettable, since the question of effective remedies for purchasers is one of considerable practical importance, and something moreover which impacts on the marketability of new buildings.

22.3.4 Direct claim by subsequent owner

Under the Contracts (Rights of Third Parties) Act 1999, the subsequent owner of a defective building may be able to enforce relevant terms in the contract between the original employer and whichever contractor or designer is to blame for the defects. The circumstances in which such third party actions (which were formerly ruled out under the privity of contract doctrine) may be brought are discussed in Section 9.2.5; to recap briefly, the contract in question must have expressly provided for third party enforcement or, alternatively, must be seen as purporting to confer a benefit on the third party. In either case, the third party (here the subsequent owner), must have been identified in the contract (by name, description or by class). If this statutory provision were to be widely used, it could take over this area of law and effectively render superfluous the devices described above. It is too early to judge whether this will happen but the fact that, while JCT 98 explicitly ruled out such third party rights, JCT SBC 05 makes express provision for them, suggests that it may well do so.

---

[22] *Alfred McAlpine Construction Ltd v Panatown Ltd* [2000] BLR 331.
[23] (1994) 69 BLR 1.

23 Suspension and termination of contracts

As noted in Chapter 21, the legal remedy of damages can be awarded for any breach of contract, whether major or trivial. This Chapter deals with other remedies that may be available for serious breaches of contract, or in other situations where events fundamentally change the contractual environment. In particular, we are concerned with the circumstances in which the contract work may lawfully be stopped, either temporarily or permanently. This requires consideration of the remedies of suspension of work, termination of the contract for breach and termination of the contractor's employment. There is also the possibility that the contract may be brought to an end under the doctrine of frustration.

23.1 SUSPENSION OF WORK

At common law (unlike French law), one contracting party (A) has no right to suspend performance of contractual obligations on a temporary basis, on the ground that the other party (B) is in breach of contract.[1] Unless B's breach is sufficiently serious to justify A in terminating the contract altogether, A's only remedy is to claim damages, in the meantime continuing with the contract. This means that, unless the contract specifically provides for suspension of work as a remedy for non-payment (as do JCT forms of both main contract and sub-contract), a contractor who suspends work on the ground of not having been paid will be guilty of a breach of contract in failing to maintain regular and diligent progress.

As far as construction contracts are concerned, the common law position has been reversed by statute. Section 112 of the Housing Grants, Construction and Regeneration Act 1996 provides that where any sum due under a construction contract has not been paid in full by the final date for payment (see Section 15.1.2), and no effective notice to withhold payment has been given (see Section 22.4.4), the payee has the right to suspend work until full payment is made. This right may not be exercised without giving the other party at least seven days notice of the intention to suspend, stating the grounds on which this action is based.

The Act provides that any time during which work is suspended in this way is to be disregarded in working out the length of time taken to complete the contract work, and any other work which is directly or indirectly affected (for example under a sub-contract). This means, in effect, that the contractor and any affected sub-contractor is given an extension of time for the period of suspension. However,

---

[1] *Channel Tunnel Group Ltd v Balfour Beatty Construction Ltd* [1992] 2 All ER 609.

it is worth noting that nothing in the Act itself gives either the contractor or any sub-contractor the right to claim for loss and/or expense in respect of costs arising out of the delay. Not surprisingly, many contracts provide specifically that the contractor is entitled to recover such loss and/or expense see (for example JCT SBC 05, clause 4.24).

23.2   TERMINATION FOR BREACH AT COMMON LAW

Termination of a contract for breach refers to the situation where the misconduct of one of the parties is so serious that the law gives the other party the option to bring the contract to an end. This Section explores the issue of termination for breach, contrasting it with termination of the contractor's employment, and examines the circumstances in which termination commonly happens in construction contracts. It may be noted at this stage that, while modern contracts such as JCT SBC 05 and ICE 7 speak of 'termination' of the contractor's employment, many older contracts (and therefore cases arising out of them) used the word 'determination'. In this Chapter we use the word: 'termination' throughout.

23.2.1 Termination for breach and termination of employment distinguished

It is important at the outset to understand the distinction between the two concepts of **termination for breach** and **termination of the contractor's employment**. The common law right to terminate or 'repudiate' a contract can arise in either of two situations. First, one party may make clear that it has no intention of performing its side of the bargain. Secondly, that party may be guilty of such a serious breach of contract that it will be treated as having no intention of performing. A breach of either kind is known as a 'repudiatory breach'. In both cases, the innocent party has a choice; either to 'affirm' the contract and hold the other party to its obligations (while claiming damages as appropriate for the breach), or to bring the contract to an end. If repudiation is opted for, then *both* parties are released from any further contractual obligation to perform. However, the terms of the contract remain relevant to such matters as establishing liability (for example where there is an exemption clause), assessing damages (including liquidated damages), and resolving disputes (for example where there is an arbitration clause).

By contrast, many building contracts make provision for 'termination of the contractor's employment' in specified circumstances. Not all of these circumstances amount to sufficiently serious breaches of contract to justify termination; indeed, some of them are not breaches at all. Such clauses normally lay down procedures (the giving of notice and so on), which must be followed if the termination is to be effective. They also deal with the consequences, financial and otherwise, of the termination. Most importantly, while common law termination brings the contract to an end altogether (subject to what was said in the previous paragraph), the contractual remedy of termination only terminates the contractor's right and obligation to carry out the contract works, but does not release the parties from any further obligation.

Some events would justify termination of the contract at common law as well as triggering a termination clause in the contract. Where this is so, it was for many years assumed that the innocent party would have a completely free choice as to which remedy to pursue. However, it was held by the Court of Appeal in *Lockland Builders Ltd v Rickwood*[2] that, where a contract lays down a clear procedure for termination, this may by implication exclude any common law right of termination, except where the other party demonstrates a clear intention not to be bound by the contract (that is, where the guilty party commits the first type of 'repudiatory breach' identified above). While this decision has not been overruled, it is somewhat difficult to reconcile with the subsequent decision of the House of Lords in *Beaufort Developments (NI) Ltd v Gilbert-Ash NI Ltd*,[3] and may therefore not be followed in future. In any event, JCT SBC 05 makes clear in clause 8.3 that the contractual provisions for termination of the contractor's employment are in addition to any other legal rights or remedies of the parties.

Where the *Lockland* decision does not apply (so that both remedies are in principle available), the innocent party must elect for one remedy or the other, and cannot combine the best elements of the two remedies in some complex way. Thus, for example, a party who seeks the favourable remedies provided by a contractual termination clause must follow the specified procedures. Failure to do so may result in the innocent party having to rely on its common law rights instead, but this can only be done if the contractor's breach is a sufficiently serious one.[4] Likewise, where a clause of the contract provides for termination in circumstances where the common law does not, the innocent party can *only* take whatever remedy the contract offers. This situation is exemplified by a case[5] in which a local authority contract provided that, in the event of unauthorized sub-contracting, the authority could either terminate the contractor's employment or claim £100 liquidated damages. When the contractor breached this provision, the local authority gave notice of termination, engaged another contractor to complete the work and then claimed against the original contractor for the extra cost incurred (a total of some £21,000). However, it was held that, since the contractor's breach would not have justified termination of the contract at common law, the local authority was limited to those remedies specifically given by the contract. These did not include unliquidated damages, and so the local authority's claim failed.

23.2.2 Nature and effect of repudiation

As mentioned above, repudiatory conduct by a contracting party does not of itself bring that contract to an end. This will come about only when the innocent party chooses to exercise its right to terminate the contract and notifies the guilty party to that effect. Such notification, once made, is irrevocable. Indeed, notification by the innocent party of intention to 'affirm' the contract is also irrevocable. Since intentions may be implied from conduct as well as expressed in words, it is very important for the innocent party to make clear decisions and to act promptly upon

---

[2] (1995) 77 BLR 38.
[3] (1998) 88 BLR 1.
[4] *Architectural Installation Services Ltd v James Gibbons Windows Ltd* (1989) 46 BLR 91.
[5] *Thomas Feather & Co (Bradford) Ltd v Keighley Corporation* (1953) 53 LGR 30.

them. In the Canadian case of *Pigott Construction v WJ Crowe Ltd*,[6] for instance, plastering sub-contractors were told in September that they would soon be required. However, due to delays arising because the main contractor failed (in breach of the main contract) to provide temporary heating, they were not called upon until the following April. It was held that, even if this delay could have been treated as sufficiently serious to justify termination, the sub-contractors' failure to complain about it (despite inspecting the site in March) deprived them of the right to terminate the sub-contract. As a result, the sub-contractors were not in a position to demand a renegotiation of their prices.

An even stronger case is *Felton v Wharrie*[7] where, under a 42-day demolition contract with a liquidated damages provision, the contractor delayed beyond the completion date and, on being asked when he expected to finish, replied that he could not say. Thirteen days later, the employer without warning entered the site and refused to allow the contractor to continue with the work. It was held that this purported exercise of the right to terminate the contract came too late.

One particular problem which the law has not yet satisfactorily resolved is whether an innocent party who 'affirms' a contract can insist on carrying out that contract in the face of the other party's clear (though wrongful) wish to terminate it. The innocent party's right to act in this way was upheld by a bare majority of the House of Lords in the non-construction case of *White & Carter (Councils) Ltd v McGregor*,[8] albeit with the proviso that the innocent party must have a 'legitimate interest' in performing the contract instead of claiming damages.

The particular difficulties that this raises in the construction context were recognized in *Hounslow LBC v Twickenham Garden Developments Ltd*,[9] where a contractor claimed to be entitled to continue with a JCT 63 contract that the employer had purported to terminate. Megarry J was firmly of the opinion that the *White & Carter* principle did not apply where performance of the contract required the co-operation of the other party, especially where it consisted of work done on the other party's property. If this view is correct, it means that the *White & Carter* principle will seldom apply in the context of a building contract.

The *Hounslow* case raised another issue of enormous practical importance, and unfortunately dealt with it in a way that has attracted considerable criticism. It was there held that, where an employer purports either to terminate a contract for breach or to operate a contractual provision for termination, but the contractor disputes the validity of the employer's act, the employer will not be entitled to an interim injunction compelling the contractor to leave the site. If this is correct, it means that the contractor can effectively prevent the employer from getting in another contractor to complete the work, even if it is later found that the employer's actions were perfectly valid. Such a conclusion creates an intolerable situation in practical terms, and it is significant that, while the decision has not been overruled, courts in both England[10] and New Zealand[11] have refused to follow it.

---

[6] (1961) 27 DLR (2d) 258.
[7] HBC 4th ed, Vol 2, 398.
[8] [1962] AC 413.
[9] [1971] Ch 233.
[10] *Tara Civil Engineering Ltd v Moorfield Developments Ltd* (1989) 46 BLR 72.
[11] *Mayfield Holdings Ltd v Moana Reef Ltd* [1973] 1 NZLR 309.

23.2.3 Breach of contract justifying termination

It sometimes happens that one contracting party ('A') is in breach of contract and the other party ('B') treats this as a repudiatory breach, but it is later held that A's breach was not sufficiently serious to justify this. The question which then arises is whether this mistake means that B, who clearly intended no longer to be bound by the contract, is now guilty of a repudiatory breach, so that A is entitled to terminate the contract. There are good reasons for suggesting that this *should* be the result, since it enables the parties to know where they stand. However, the courts have shown some reluctance to punish a party for an honest but mistaken belief as to the rights and remedies available, and it has thus been held by a majority of the House of Lords that to invoke a contractual remedy in error, unless it is totally abusive or in bad faith, does not in itself constitute a repudiation of the contract.[12]

Employer's breach

Various acts by the employer can result in a repudiatory breach and thus entitle the contractor to terminate the contract. These are listed and explained below:

- **Failure to give possession of the site:** While minor interference by the employer with the contractor's possession of the site is not a repudiatory breach,[13] an outright refusal to give possession in the first place will be so. Similarly, wrongful ejection of the contractor from the site is such a breach.[14] As to mere delay in giving possession, the crucial question (and it is one of degree) is whether the employer's conduct indicates an intention no longer to be bound by the contract. For example, where an employer delayed giving possession of the site for two months despite repeated requests from the contractor, and also announced that part of the contract work was to be omitted and given to another contractor, it was held that these two breaches, taken together, amounted to a repudiatory breach of contract.[15]
- **Non-payment of sums due:** As a general principle of law, failure to pay on time what is due under a contract will not normally be treated as a sufficient breach to justify the other party in terminating that contract.[16] Failure to pay on time what is owed on *another* contract is even less likely to be a repudiatory breach.[17] This is why contracts commonly provide express rights of termination for non-payment. However, while late payment (or short payment) is not in itself repudiatory, a continued refusal to pay may become so.

    The practical problems which may arise in this area are exemplified by the case of *DR Bradley (Cable Jointing) Ltd v Jefco Mechanical Services*

---

[12] *Woodar Investment Development Ltd v Wimpey Construction UK Ltd* [1980] 1 All ER 571.
[13] *Earth & General Contracts Ltd v Manchester Corporation* (1958) 108 LJ 665.
[14] *Roberts v Bury Commissioners* (1870) LR 4 CP 755.
[15] *Carr v JA Berriman Pty Ltd* (1953) 89 CLR 327.
[16] *Mersey Steel & Iron Co v Naylor, Benzon & Co* (1884) 9 App Cas 434.
[17] *Small & Sons Ltd v Middlesex Real Estates Ltd* [1921] WN 245.

Ltd.[18] That case concerned a £113,000 sub-contract for electrical work which was made by word of mouth, with no express agreement as to time or methods of payment (although it was accepted on all sides that the parties intended interim payments to be made). After a succession of five payments which were both late and short, the claimant sub-contractors withdrew from the site, whereupon the defendant contractors threatened that after seven days they would treat this as a repudiatory breach and would then employ a substitute at the claimants' expense. The claimants then submitted a claim for £31,000, to which the defendants offered a mere £5,000.

On these facts it was held that the claimants were not entitled to terminate the contract for the original failures to pay, and that they would thus have been guilty themselves of a repudiatory breach, had not the defendants waived this by giving them seven days to return to work. However, the defendants' subsequent derisory offer was sufficient to shatter the claimants' confidence in ever getting paid, and it thus justified them in bringing the contract to an end at that stage.

- **Withholding of certificates:** If the contract administrator refuses to certify at the appropriate time, or negligently under-certifies, this may well constitute a breach of contract on the employer's part. It certainly will do so if the contract administrator's conduct is due to positive interference by the employer. Such events will undoubtedly enable the contractor to claim damages, or possibly to recover what is due without the necessity of a certificate.[19] Whether they will justify termination of the contract will once again depend on whether the breach is sufficiently serious to be regarded as repudiatory.

- **Hindrance of the contractor:** A breach by the employer of the duties of non-hindrance and positive co-operation may be so serious as to indicate an intention not to be bound. This has been held to be the case where an employer wrongfully ordered the contractor not to complete the work,[20] where an employer failed to provide the necessary drawings as required by the contract[21] and where an employer, concerned that work was falling behind schedule, employed other workmen to carry out part of the contractor's work.[22]

Contractor's breach

Acts by a contractor that may constitute a repudiatory breach fall into the following groups:

- **Abandonment or suspension of the work:** Perhaps the most obvious example of 'repudiatory breach' by a contractor is total abandonment of the work in circumstances where this is unjustified.[23] Whether less extreme

---

[18] (1988) 6-CLD-07-19.
[19] *Perini Corporation v Commonwealth of Australia* (1969) 12 BLR 82.
[20] *Cort v Ambergate, Nottingham, Boston & Eastern Junction Railway Co* (1851) 17 QB 127.
[21] *Kingdom v Cox* (1848) 5 CB 522.
[22] *Sweatfield Ltd v Hathaway Roofing Ltd* [1997] CILL 1235.
[23] *Marshall v Mackintosh* (1898) 78 LT 750.

action by the contractor will have this effect is, as usual, a question of fact and degree. For instance, contractors who were complaining of late payment retaliated by withdrawing their labour and most of their plant from the site and thus slowed down progress considerably. However, they retained a presence on site through their supervisory staff, and they did nothing to discourage sub-contractors from continuing with their work. It was held by the Court of Appeal that the contractors' tactics, though clearly in breach of their duty to maintain regular and diligent progress, were not sufficient to justify the employers in terminating the contract.[24]

- **Defective work:** As a general principle, defects in the work do not entitle the employer to terminate the contract and refuse payment altogether. The employer's remedy is to claim damages for the cost of rectification.[25] However, very serious defects may justify the conclusion that there has not been 'substantial performance' by the contractor. Where this can be established, the employer need pay nothing. For example, the installation of a central heating system at an inclusive price of £560 was defectively carried out. The system was only 90% efficient (70% in some rooms) and gave off fumes in the living room. The Court of Appeal held that the claimant was entitled to nothing for this work; the defendant was not limited to setting off the £174, which it would cost to put right the defects.[26] It is even possible that an accumulation of lesser defects may amount to a repudiatory breach of contract, even though none of them would be sufficient on its own. It was thus held in one case[27] that the contractors' 'manifest inability to comply with the completion date requirements, the nature and number of complaints from sub-contractors and their own admission that ... the quality of work was deteriorating and the number of defects was multiplying' entitled the employer to terminate the contract and to order the contractors to leave the site. The employer had justifiably concluded that the contractors had neither the ability, competence nor the will to complete the work in accordance with the contract.

- **Delay:** At common law, there are only three situations in which delay by a contractor will justify the employer in terminating the contract. First, the delay may be so great as to demonstrate the contractor's intention not to be bound by the contract. Second, the contract itself may make clear that the time of completion is of fundamental importance (the legal phrase is that 'time is of the essence'). This can be done by express words; it can also be done by implication, although this is unlikely to be the case in a construction contract. Third, once delay has occurred, the non-delaying party may make time of the essence (where it is not already so) by giving reasonable notice to the other party.

- **Miscellaneous:** Whether or not a contractor's breach of contract is sufficiently serious to justify the employer in terminating is always a question of fact and degree. Thus, where a contractor sub-contracted in contravention of an express term in the contract, it was held that this did not

---

[24] *JM Hill & Sons Ltd v Camden LBC* (1980) 18 BLR 31.
[25] *Hoenig v Isaacs* [1952] 2 All ER 176.
[26] *Bolton v Mahadeva* [1972] 2 All ER 1322.
[27] *Sutcliffe v Chippendale & Edmondson* (1971) 18 BLR 149.

amount to a repudiatory breach.[28] However, in a case where a contractor failed to obtain a performance bond as required by the contract, a court came to the opposite conclusion and decided that this was repudiatory.[29]

23.3   TERMINATION UNDER JCT CONTRACTS

As explained above, building contracts commonly provide for the contract to be brought to an end in various circumstances, some but not all of which concern a breach of contract by one of the parties. We now consider the 'termination' provisions of JCT SBC 05 and its related sub-contract JCT SBCSub 05.

23.3.1 Termination by the employer under JCT SBC 05 clauses 8.4 to 8.8

Grounds for termination

Clause 8.4.1 entitles the employer to terminate the contractor's employment for certain defaults, namely:

1. Suspension of work without reasonable cause.
2. Failure to proceed regularly and diligently.
3. Failure to comply with a contract administrator's instruction to remove defective work, so that the works are materially affected.
4. Unauthorized sub-contracting or assignment.
5. Failure to comply with contractual obligations regarding CDM Regulations.

Clause 8.6 entitles the employer to terminate the contractor's employment if the latter (directly or via employees) is guilty of corrupt practices relating to the obtaining or execution of contracts.

A contractor who becomes insolvent (as defined in clause 8.1) is obliged to inform the employer about it (clause 8.5.2). The insolvency entitles the employer to terminate the contractor's employment by notice to the contractor (clause 8.5.1). However, JCT SBC 05 has abandoned the feature of JCT 98 according to which the contractor's employment was automatically terminated in certain types of insolvency. Whether or not the employer terminates the contractor's employment, there are important consequences which follow from the contractor's insolvency (clauses 8.5.3.1–3): particular payment provisions apply, the obligations of the contractor to carry out and complete the works are suspended, and the employer is entitled to secure the site.

Procedures

Where the contractor is guilty of any of the defaults specified in clause 8.4.1, the contract administrator is to issue a written notice specifying the default. If the

---

[28] *Thomas Feather & Co (Bradford) Ltd v Keighley Corporation* (1953) 53 LGR 30.
[29] *Swartz & Son (Pty) Ltd v Wolmaranstadt Town Council* 1960 (2) SA 1.

default is then continued for 14 days, the employer may within 10 days of the continuance terminate the contractor's employment by issuing a notice to this effect. Furthermore, if termination does not take place on this occasion, any subsequent repetition of a specified default gives the employer the right to terminate immediately; there is no need (and indeed no power) to issue a second default notice.[30] It may be noted that this 'two-notice' procedure is not required in cases of termination on insolvency or corruption.

It is expressly provided that a notice of termination is not to be given 'unreasonably or vexatiously' (clause 8.2.1). This would presumably prevent the employer from using a technical breach of contract that is causing no loss (such as unauthorized sub-contracting of some unimportant part of the work) to escape from a contract that had proved to be disadvantageous.

Any notice under clause 8 is supposed to be actually delivered to the contractor, or to be sent by registered post or recorded delivery (in which case delivery is presumed on the second business day after the date of posting). It has been held by courts in the Commonwealth[31] that conditions governing the method of sending notices must be strictly complied with, if the notices are to be valid. However, it seems that the English courts take a more liberal approach and will uphold a notice that actually arrives, irrespective of the method used.[32] Furthermore, a notice that fails to satisfy the contractual requirements may still be a valid notice of termination at common law (provided, of course, that the contractor is guilty of a sufficient breach to justify the employer in taking this step).

A good example of the last point is provided by a case in which main contractors sent a letter to labour-only sub-contractors, requiring them to comply with a term of the contract that dealt with the length of the working day. Eleven months later and without further warning, the main contractors sent a telex purporting to terminate the sub-contract for breach of this term. It was held that this telex did not comply with the contractual provisions for termination after due warning, since a reasonable person would have seen no connection between the two notices. However, it was a valid termination of the contract at common law.[33]

Consequences of termination of employment

Clause 8.7 sets out the rights and duties of the parties following a termination of the contractor's employment by the employer. The position, briefly, is as follows:

1. The employer may employ others to complete the works, and may for this purpose make use of the contractor's temporary buildings, plant, tools, equipment and materials.
2. All of the contractor's equipment, etc. must be removed from the site, but only when the contract administrator has given the instruction so to do.
3. The contractor is required to provide the employer with two copies of all contractor's design documents (if there is a contractor's designed portion).

---

[30] *Robin Ellis Ltd v Vinexsa International Ltd* [2003] BLR 373.
[31] *Eriksson v Whalley* [1971] 1 NSWLR 397 (Australia); *Central Provident Fund Board v Ho Bock Kee* (1981) 17 BLR 21 (Singapore).
[32] *Goodwin v Fawcett* (1965) 175 EG 27; *JM Hill & Sons Ltd v Camden LBC* (1980) 18 BLR 31.
[33] *Architectural Installation Services Ltd v James Gibbons Windows Ltd* (1989) 46 BLR 91.

4. The employer may insist on taking an assignment from the contractor of all sub-contracts and all contracts for the supply of materials.
5. The contractor is not entitled to any further payment (including, it appears, payments that the employer should already have made[34]) until the works are completed by another contractor or until six months after the employer decides not to complete them (clauses 8.7.4 and 8.8 respectively). The contractor is then entitled to the difference, if any, between what would have been earned by completing the contract and what the breach has cost the employer (the expenses incurred in completion plus any direct loss and/or damage). If, as is likely, the employer's losses exceed what would have been due to the contractor, the latter is liable for the difference.

23.3.2 Termination by the contractor under JCT SBC 05 clauses 8.9 and 8.10

Grounds for termination

Clause 8.9 entitles the contractor to terminate his or her employment on various grounds which may be loosely described as 'defaults' of either the employer or persons acting on behalf of the employer (other than the contractor). These are:

1. Failure by the employer to pay amounts properly due within the time limits.
2. Interference by the employer with the issue of any certificate.
3. Unauthorized assignment by the employer.
4. Failure by the employer to comply with contractual obligations regarding CDM Regulations.
5. Suspension of the whole or substantially the whole of the works for a continuous period stated in the Contract Particulars (where a default provision of two months is provided) by reason of one of the following specified events:
   - Contract administrator's instructions dealing with discrepancies in contract documents, variations or postponement of work, unless these are caused by negligence or default of any statutory undertaker or of the contractor.
   - Any impediment, prevention or default by the employer, contract administrator, quantity surveyor or any other person employed by the employer, unless caused by negligence or default of the contractor.

Clause 8.10 relates to the employer's insolvency. The clause is an equivalent to the provisions relating to contractor insolvency under clause 8.5. The employer has to inform the contractor about any of the different forms of insolvency set out in clause 8.1, and the contractor is entitled to terminate his or her employment.

---

[34] *Melville Dundas Ltd v George Wimpey UK Ltd* [2007] 1 WLR 1136.

Procedures

On all the above grounds, contractors may terminate their employment by a written notice delivered to either the employer or the contract administrator, or sent to them by registered post or recorded delivery. As with termination by the employer, the contractor must first (except in the case of the employer's insolvency) give a 'warning shot'.

The general requirement that notices shall not be given 'unreasonably or vexatiously' (clause 8.2.1) applies to termination notices of either party. In this connection, it is worth noting that all the 'suspension of work' grounds are also 'relevant events' in respect of which the contractor would be entitled to an extension of time (clause 2.29) and 'relevant matters' in respect of which the contractor is entitled to compensation for direct loss and/or expense (clause 4.24). The availability of these alternative methods of safeguarding the contractor's interests might well make it seem unreasonable to terminate in any but the most serious of cases.

Consequences of termination

Clause 8.12 requires the contractor to remove all equipment and materials, including that of sub-contractors, from the site. This must be done with all reasonable dispatch, but with all necessary safety precautions. Moreover, the contractor must provide the employer with two copies of all contractor's design documents (if there is a contractor's designed portion).

As to payment, the contractor is entitled to:

1. The value of all work completed at the date of termination.
2. The value of all work begun but not completed at that date.
3. Any direct loss and/or expense arising out of other matters for which the contractor has a claim.
4. The contractor's costs of removal from the site.
5. The cost of materials ordered by the contractor for the works *which then become the employer's property.*
6. Any direct loss and/or damage caused to the contractor by the termination.

The contractor has to prepare an account with all the sums listed above. The amount properly due (after taking account of all amounts previously paid) has to be paid within 28 days after the submission of the account to the employer, without deduction of any retention money.

23.3.3 Termination on neutral grounds under JCT SBC 05

There are circumstances beyond the control of both parties that can give rise to termination of the contractor's employment. Surprisingly, perhaps, JCT SBC 05 has not collected all of these into a single clause.

Clause 8.11

Clause 8.11 of JCT SBC 05 deals with the situation where the contract work is suspended by certain causes that could be regarded as 'neutral', in the sense that they are the fault of neither the employer nor the contractor. In certain circumstances, this clause entitles either party to terminate the contractor's employment by written notice which, like all termination notices, is not to be given unreasonably or vexatiously. The relevant circumstances are described in clause 8.11.1, and the clause applies where the whole or substantially the whole of the contract works has to be suspended for a continuous period stated in the Contract Particulars or, if none is stated, two months.

The relevant circumstances are:

1. *Force majeure.*
2. Contract administrator's instructions dealing with discrepancies in contract documents, variations or postponement of work, each of them as result of the negligence or default by a statutory undertaker (a term which is defined in clause 1.1 of JCT SBC 05, for example a local authority).
3. Loss or damage to the works caused by a 'Specified Peril'.
4. Civil commotion, the use or threat of terrorism or the thereby triggered activities of authorities.
5. The exercise of any statutory power by the UK Government which directly affects the execution of the works.

According to clause 8.11.2, the third ground (specified peril) cannot be relied upon if the event has arisen out of the contractor's own negligence or default. This includes negligence or default of those for whom the contractor is responsible (including sub-contractors and sub-sub-contractors). As to the procedure two notices are required: after the expiry of the relevant period of suspension, either party can give notice in writing that she or he may terminate the contractor's employment under the contract. Then, after seven days, a notice of termination can be issued (clause 8.11.1).

The consequences of termination under clause 8.11 are very similar to those under clauses 8.9 and 8.10 (termination by contractor). However, the contractor is not entitled to reimbursement for any direct loss and/or expense arising out of the termination, except where it is caused by damage from a 'Specified Peril' which is itself the result of negligence on the part of the employer (clause 8.12.4.2).

Insurance Option C – Paragraph C.4.4 of Schedule 3

Where the contract consists of work on an existing structure and insurance option C is stated in the Contract Particulars (against clause 6.7) to apply, an additional ground of termination exists which overlaps rather awkwardly with clause 8.11. This ground is set out in paragraph C.4.4 of Schedule 3: if loss or damage to any part of the work is caused by a specified peril (whether or not due to negligence on the part of the contractor), then either the employer or the contractor may within 28 days serve written notice to terminate the contractor's employment. The effectiveness of such a notice depends upon the very vague criterion of whether termination would be 'just and equitable', and the other party is given seven days to take this question to arbitration or court (whichever is stated in the contract to be the preferred dispute resolution procedure).

If the contractor's employment is indeed terminated under this clause, the financial consequences from clause 8.12 apply. If it is not (either because neither party serves notice, or because a notice is successfully challenged), the contractor is entitled to be paid for necessary works of restoration and repair as if for a variation.

23.3.4 Termination under sub-contracts

In general terms, the termination provisions of the JCT Standard Building Sub-Contract (SBCSub 05) follow, 'one step down', those of JCT SBC 05, as regards both the grounds for termination and the consequences. However, there are some significant differences between main contract and sub-contract termination, the most important of which are as follows.

Grounds for termination

Termination of the sub-contractor's employment by the main contractor is provided for by clauses 7.4 to 7.6 of JCT SBCSub 05. The grounds are virtually identical to those on which the employer may terminate the main contract, the only difference being that it is the main contractor rather than the contract administrator who issues the first notice.

Clause 7.8 of JCT SBCSub 05 entitles the sub-contractor to terminate their employment if the main contractor is guilty of one of the following defaults:

1. Suspension, without reasonable cause, either wholly or substantially, of the carrying out of the main contract works.
2. Failure to proceed, without reasonable cause, with the main contract works so that reasonable progress of the sub-contract is seriously affected.
3. Failure to make payment in accordance with the sub-contract.
4. Failure to comply with the CDM regulations.

Two more grounds on which the sub-contractor may terminate are contained in clauses 7.9 and 7.10 of JCT SBCSub 05. The former provides for automatic termination of the sub-contractor's employment if the main contractor's employment is itself terminated. The latter sets out a right to terminate where the main contractor is insolvent.

Procedures and consequences

While the procedural provisions of JCT SBC 05 are followed to a great extent, there are minor differences (for example as to the period which must elapse between the two notices required to terminate). As to the financial and other consequences of termination, once again the JCT SBC 05 pattern is largely adhered to. Sub-contractors are well protected since they can claim any direct loss and/or expense caused by the termination (JCT SBCSub 05: 7.11.3.5). Clause 7.11.5 of the sub-contract sets out that (parallel to clause 8.12.5 of JCT SBC 05) the amount properly due has to be paid to the sub-contractor within 28 days after the submission of the appropriate account, without deduction of any retention money.

23.4   TERMINATION OF CONTRACT BY FRUSTRATION

This Chapter would be incomplete without at least some mention of a legal doctrine which may bring about the termination of a contract on 'neutral' grounds. This doctrine, known as 'frustration of contract', applies where, due to some external event, performance of a contract becomes impossible, illegal or radically different from what was originally envisaged. If this happens through the fault of neither party, and the contract itself makes no sufficient provision for what has occurred, it is possible that the law may treat the contract as terminated. In such a case both parties are freed from any further obligations under the contract. As for any losses already incurred, these will be allocated between the parties in accordance with principles contained in the Law Reform (Frustrated Contracts) Act 1943.

It is important to appreciate that this doctrine is very limited in its application. If the general law terminates a contract in this way, it interferes with the balance of risks between the parties. In the building and civil engineering fields in particular, the courts recognize that the kind of risks involved in such cases often fall naturally on one party or the other, and that to give the risk-bearing party an escape route would unfairly distort this balance. What is more, most standard form contracts make express provision for many of the eventualities that might lead a party to claim that a contract has been frustrated, and the doctrine cannot be used to override clear contract terms.

Two decisions of the House of Lords may be used to illustrate the doctrine of frustration and give an idea of its limits. In the first of these, *Davis Contractors Ltd v Fareham UDC*,[35] the parties in 1946 entered into a fixed price contract for the construction of 78 houses in 8 months. The work in fact took 22 months to complete and cost the contractors far more than the contract price. This was due partly to bad weather and partly to labour shortages caused by the slow demobilization of World War II troops, both of which were unforeseeable. The contractors sought to argue that the contract was frustrated, and thus to claim a reasonable sum for the value of the work. However, it was held that what had happened was squarely within the risk assumed by the contractors, so that no relief could be granted to them.

The earlier case of *Metropolitan Water Board v Dick, Kerr & Co Ltd*[36] arose out of a fixed-price contract for the construction of a reservoir. The contract, which was entered into on the eve of World War I, provided that the work was to be completed within six years, but gave the engineer very wide powers to order

---

[35] [1956] AC 696.
[36] [1918] AC 119.

extensions of time. After some 18 months, the government ordered the contractors to stop the work and to sell all their plant. This development, it was held, was sufficient to bring the contract to an end. It went far beyond the extension of time provisions in the contract, and made the project fundamentally different from what had been envisaged.

24 Non-adversarial dispute resolution

Contractual disputes in construction arise because of a series of factors that combine in various ways to produce arguments, disagreements and, ultimately, disputes. Some of these factors are basic to all disputes between humans, such as the motivating factors of individuals, human behaviour, organizational behaviour, culture, etc. What makes construction contract disputes different is the nature of the dispute. This, again, depends upon a variety of things such as the terms of the contract, the technological issues of the site and the building, the character of the project personnel, the amount of time and money available, the realism of peoples' expectations, project environmental factors, the legal basis of the argument, the magnitude of the issue, and so on. The third group of variables in this area concerns the choice of methods for resolving disputes once they have arisen. What are the options available in the contract, and what other options are there? These are the topics confronted in this Chapter.

24.1 BACKGROUND TO DISPUTES

In trying to understand why disputes occur on building contracts, the key clearly lies with the fact that *people* are interacting in some way. Although many disputes are based upon arguments about technical or legal points, disagreement escalates when people become intransigent. It is important to be clear about the basic concepts. In particular, conflict should be distinguished from dispute. These words are often interchanged in common parlance but for our purposes **conflict** occurs when objectives are incompatible. This is to be expected in construction. In fact, one of the reasons for appointing a team of professionals to contribute advice to a client is to engage in this kind of conflict. Each specialist will bring a particular agenda and skill to the problems at hand. The ensuing conflict of objectives is a central part of the project development process and should be expected. On the other hand, **disputes** arise when a conflict becomes an altercation; perhaps when one or both of the parties becomes intransigent (from a behavioural point of view), but definitely when the argument revolves around rights and is justiciable. These are more likely to arise when a project is on site because the differences of opinion are more contractually based. Pride or ignorance may also play a significant role in influencing the outcome of a disagreement, perhaps more than the nature of the dispute. Disputes arise, not just because people are entering into building contracts but because of a wide variety of interactions between diverse people.

24.1.1 Motivating factors of individuals and organizations

It is often assumed that those who work on building projects all have a common objective, but it has been shown, and indeed it is fairly self-evident upon reflection, that most people bring with them very different sets of objectives. These personal objectives can be very difficult indeed to predict, partly because people are from a wide range of backgrounds. The fact that people are from different organizations will also affect their interest in a project, because organizations (i.e. companies) have their own objectives. Moreover, not all the people from any one organization will have similar objectives. Objectives are influenced by upbringing, education, family, friends and even the media. To make matters even more complex, objectives change for an individual as time passes. The complexity of objectives is thus due to the fact that they can be personal and/or organizational, as well as dynamic. This inherent flux and likelihood of mismatch means that conflict can develop and escalate very quickly. However, of itself, conflict is not a problem, as demonstrated in Section 2.1.5.

24.1.2 Preconceptions about roles

Another factor influencing the development and escalation of conflict is the fact that most participants in a building project bring with them preconceived notions about what their role ought to be. An architect will expect to be doing certain things and taking certain decisions. In the same way an engineer, a quantity surveyor and a builder will all have very definite ideas not only about their own roles but also about the roles of others. These expectations are often modified by personal objectives, and can thus be yet another source of conflict. In addition, the definite and entrenched views which construction professionals have about their expected roles make it difficult to introduce new systems of building procurement.

Of course, one participant who may well have no preconceptions of this kind is the client, whether a corporate client or an individual. Such people are frequently new to the building process and find it full of surprises! An extremely important function for an architect or project manager is to explain to the client at the earliest opportunity what the conditions of contract actually mean. The client needs to be told in plain English the extent and nature of what is being promised, and by whom it is being promised (see Section 18.1.2). If the client understands this, many disputes can be avoided.

24.1.3 Project success or failure

In the light of these issues, it is not at all surprising that conflict on building projects leads to disputes which can very quickly become acrimonious. These disputes, if not dealt with swiftly and equitably, can ruin everybody's chances of 'success', by which we mean the satisfying of one or more of their objectives. In this connection it is as well to reflect that, since objectives are so diverse, it is just as easy for everyone to be satisfied as it is for everybody to be dissatisfied. In other words, each individual's potential for success within a building project is not necessarily at the expense of the others' potential. This is contrary to the belief that avarice and conflict are the only routes to success in construction! The concept that all parties to a project can be equally satisfied has gained much support in recent years with the publication and wide dissemination of strategic reports (especially Latham 1994, Egan 1998) and by governmental initiatives such as 'Achieving Excellence in Construction' (Office of Government Commerce 2007). However, it is not always clear that procurement methods under the label of 'partnering' truly allow all parties a reasonable profit margin and abstain from overburdening the weaker party with unbearable risks (see, for example, Bresnen and Marshall 2000, Gruneberg and Hughes 2005).

24.1.4 The roots of contractual dispute

In order to identify the type of dispute, it is essential to examine the position of the parties in terms of the amount of time and money available, project environmental factors, the magnitude of the issue and so on. Of particular importance are the duties of the parties to the contract. Contractual disputes tend to arise when one party alleges that the other party has not kept to the bargain. Since the performance or non-performance of obligations is at the root of any contractual dispute, it follows that the contracting parties must have a very clear understanding of what they are undertaking.

24.1.5 Business relations

A further point affecting the origin of disputes is the preservation of good business relations. There are often circumstances where the parties to the contract are from the same business environment and will probably be contracting with each other again in the future. In these circumstances they are likely to understand both the rules and each other's needs and requirements. For the sake of future business a party that feels aggrieved by the other party will often seek effective and quick resolution of points of disagreement, even if that implies giving up a claim that would have good chances to succeed in court. The practicalities of business are often more crucial for the outcome of disputes than questions of liability. This is particularly marked in civil engineering, where the client is usually the government or some other public agency. Contractors are very keen to preserve a good relationship with such clients.

On the other hand, there are circumstances where one party is deliberately obstructive and seeks to exploit every possible opportunity to its limit. This can happen where the parties are not likely to work together in the future. In the building industry, as opposed to civil engineering, most work is done in the private sector, and the number of disputes is much larger: perhaps these two factors are connected. The growth of partnering and the importance of developing long-term business relationships are, at least in part, reactions to the problems of doing business with strangers.

This need to preserve business relationships is particularly marked where the construction market is closed, or where access to the market is limited in some way. For example, for many years the Japanese construction market was characterized by a strong ethos of trading only with people already known. While there is now considerably more competition, the tradition creates an atmosphere where it is more usual to find commercial disputes settled by private negotiation rather than public litigation. It must also be remembered that a contractor's risk of losing long-term clients can easily change from being an inducement to seek a negotiated settlement of an argument into a threat by an unscrupulous client.

24.2   THE NATURE OF CONSTRUCTION DISPUTES

The first factor defining the nature of a construction dispute is the terms of the contract. Basically, a contract is an enforceable promise. And the subject of this enforceable promise is the production of a unique, technical artefact, using temporary management systems.

24.2.1 Enforceable promises

Building contracts, like any other contract, are concerned with making promises, with the expectation that one can be forced to carry them out. A person who has no intention of doing a thing should not sign a contract recording such an intention!

Of course, it can happen that people enter into contracts that they did not completely intend. A shared mistake is no real problem, as the parties can rectify it by mutual consent. However, what sometimes happens is that one party claims, due to oversight or mistake, to have signed a contract that does not accurately reflect his or her intentions. If there is a difference of interpretation, then the type of contract will be important in terms of the way in which it will be interpreted. If it is not a standard-form contract, or if it is an organization's own standard-form contract, the principle of 'contra proferentem' will prevail. This means that any ambiguity in the contract will be construed against the party who seeks to benefit by exclusions or limitations in it. This will usually, although not always, be the party which put it forward (also see Section 10.2).

It is during disagreements about the intentions of contracting parties that such details as notes of telephone conversations, minutes of meetings, correspondence and the like may become relevant. These seek to provide evidence of the parties' intentions. However, it will in most cases be too late for the dissenting party to alter the contract. Building contracts, as we have seen, are very comprehensive and specific about what is expected of each party, and it is difficult to claim that the obligations arising from entering into such a contract were not properly understood at the time it was made.

24.2.2 Technical matters

Disagreements often arise over technical questions. The technology involved in construction is idiosyncratic, difficult to understand and subject to change. Added to this is any change that may be associated with the technology of the client organization. The use of different and/or unfamiliar techniques is often the cause of arguments and disagreements.

For example, the nature of the site is often a source of contention. While the site itself is clearly visible at ground level, it can hold many surprises once excavation starts. It is not enough merely to look at a site in order to ascertain the site conditions. The part where the building is going to sit is actually several metres below the part that is visible. Adequate site investigation is a constant source of problems in the industry. Whose responsibility is it? In order to answer that question, one must look at the clauses in the contract. Do they represent what is intended? It is depressing to realize how many clients are shocked to discover the extent of their liability for site conditions once problems arise.

Disputes that escalate to arbitration or litigation often hinge on an intricate understanding of some particular technical matter. An example concerns the failure of the Emley Moor television transmitter.[1] A large tower fell down in a storm, on a moor, in the middle of the night. Nobody actually saw it fall down. How did it fall down? What caused it? Was it the wind, the rain, the frost, creep, metal fatigue, foundation failure or something else? It is a purely technical question to ascertain the cause of such a problem. Once the cause is identified, it is a fairly straight-forward matter to allocate blame and with it legal responsibility.

24.2.3 Legal matters

Some disputes are technically simple, and turn on what is the law on a specific point. The law is not infinite! There are many day-to-day occurrences that have not previously been decided upon by the courts. There are many spheres of activity not covered by statute. The resolution of a dispute may hinge upon the ascertainment of the law in a previously undefined area. Also, even where there are statutes or precedents, they may be inappropriate in the particular case. No two cases are identical, so there are often tremendous difficulties to overcome in interpreting and applying the law.

One particular legal problem area arises from the inconsistencies between various contracts. A consequence of the fragmentation of the industry mentioned in Section 2.1.3 is that each participant's involvement with the project is formalized with a contract. A major problem mentioned by Latham (1994) was the inconsistencies and gaps between the various consultant appointments and the building contract. As a result, most of the bodies who draft contracts now seek to produce integrated packages of contracts, rather than a standard form for just one of the relationships.

24.2.4 Entitlement and magnitude

The rough division of disputes into 'technical' and 'legal' is often reflected in two aspects of a claim. The first aspect is that, for any claim to succeed, legal

---

[1] *Independent Broadcasting Authority v EMI Electronics Ltd and BICC Construction Ltd* (1978) 14 BLR 1, HL.

entitlement to the money must first be proven. After this, the magnitude of the claim must be established. Entitlement arises from the legal interpretation of the contract and associated documents. Magnitude then follows as a factual ascertainment of technical data. In consequence, most disputes contain elements of both types of dispute.

24.3   THE ROLE OF THE CONTRACT ADMINISTRATOR

One of the most distinctive features of construction contracts in the UK is the part played by the contract administrator (whether described as architect, engineer, supervising officer, project manager or whatever). In fact, the contract administrator plays not one part but two in a construction project: first, as agent for the employer in such matters as the issue of instructions, and second, as an impartial decision maker in such matters as certification and the valuation of variations and other claims. The latter role is obviously of considerable relevance to the topics now under discussion – decisions of the contract administrator may avoid or resolve potential disputes, although it must be said that such decisions are often the *cause* of disputes!

The legal principles governing a contract administrator's certification and decision-making functions are considered in Chapter 18.

24.4   METHODS OF DISPUTE RESOLUTION

It is sometimes, perhaps too often, thought that any dispute arising from a construction contract must be resolved by court action, arbitration or adjudication. This is simply not true. First of all, the parties have a statutory right to refer their disputes to adjudication (dealt with in Section 25.1). In addition, whether or not there is an arbitration agreement in the contract, it must always be remembered that contracting parties can alter the terms of their contract (and any dispute about its interpretation) *at any time, by mutual consent.* This fact was not stressed, or even hinted at, anywhere in the clauses of earlier versions of JCT building contracts. However, JCT SBC 05 (and also the other major building contracts of JCT's 2005 contract suite) now expressly mentions mediation (see Section 24.5 below). This mode of dispute resolution, in effect, is based on the assumption that the parties can resolve their disputes by agreement. Similar to the JCT forms, the ICE 7 form offers the parties several modes of Alternative Dispute Resolution (ADR), in particular conciliation and mediation) (see Section 24.5 below). In any event, whatever the contract says, if a dispute arises, the parties are *not* compelled to resort to the courts for the settling of their differences. They can choose instead to attempt to settle their differences amicably. Of course, if this should happen, it is extremely important to record exactly what has been agreed, and to have it signed by both parties, in case of any future disagreement about what was agreed.

The use of the courts for settling disputes is expensive, uncertain and time-consuming. It is also very public. The development of ADR techniques arises from dissatisfaction with and alienation from the legal system (Brown and Marriott 1999). Indeed, arbitration was originally devised as an alternative to litigation but is prone to many of the problems that beset litigation, not least because of the Arbitration Act (see Chapter 25).

Several techniques are available to help contracting parties to come to some form of settlement without resorting to adjudication, arbitration or litigation. These procedures are informal, and the terminology used to describe them is rather loose and vague. Their chief features are that they are cheap and investigatorial rather than accusatorial. Although they are informal, there is an emerging trend to formalize them (see Section 24.5).

The terms most commonly found to describe these procedures are Conciliation, Quasi-conciliation, Mediation, Private Enquiry and Mini-trial. However, these terms are often used interchangeably, and sometimes inconsistently. Collectively, they may be referred to as reconciliation. Similar to the widespread term ADR, reconciliation is a generic term that indicates private, non-adversarial methods of resolving disagreement.

These techniques can be seen as an intermediate step between having an argument or disagreement, and referring to the courts (or to an adjudicator or arbitrator). Since reconciliation is voluntary, either party may pull out at any time and refer the matter to the courts instead, if satisfactory progress is not being made. What is more, the parties are not bound to accept the decisions of the person deciding the issue. As Brown and Marriott (1993) point out, ADR is based on a philosophy of empowering the disputants, putting them back in control of their own dispute. Much of the dissatisfaction with traditional dispute resolution procedures is because of the lawyers' professionalization of a dispute, which leads to a legal dispute having a life of its own, almost displacing the disputants from the process. ADR techniques tend to place the disputants at the centre of the process and seek to help them to find their own way out of an impasse. This requires a will on both sides, without which ADR procedures are probably doomed.

Finally, before describing the distinct modes of ADR, it is useful to consider a relatively recent development, namely the active support of the courts in favour of ADR. The leading case is *Dunnett v Railtrack*.[2] Here, the Court of Appeal refused to make an order as to the costs with the reasoning that the defendants (who won the case) had refused *to contemplate alternative dispute resolution at a stage before the costs of this appeal started to flow*. Hence, the party who won the case could not recover its costs (i.e. especially the fees of its lawyers) from the losing party. This approach can also be gleaned from the Pre-Action Protocol for construction and engineering disputes (which is part of the Civil Procedure Rules[3]), that have to be satisfied by the claimant before bringing a case to court. Section 5.3 states: *In respect of... the dispute ..., the parties should consider whether some form of alternative dispute resolution procedure would be more suitable than litigation, and if so, endeavour to agree which form to adopt.*

24.4.1 Adjudication, arbitration and litigation

These three methods of dispute resolution, discussed in detail in Chapter 25, are adversarial in nature. Contrary stances have to be taken, and this frequently results

---

[2] [2002] 2 All ER 850.
[3] Civil Procedure Rules (SI 1998 No 3132).

in people becoming entrenched in their views. Often in adjudication, arbitration and litigation, the parties are represented by counsel whose skill and ability lies in the art of arguing and scoring points over each other. They 'play' the adversarial roles with much skill. Unfortunately, this is not conducive to amicable settlement. However, good lawyers will not encourage their clients to take legal action if they do not see a good chance of winning the case. The expense of litigious actions is so enormous that, in the past (before the introduction of adjudication as a statutory right), most cases were settled out of court shortly before they get there, after the parties had done all their preparation work and had become aware of each other's stance. However, the avenue of adjudication (considered in detail in Section 25.1) has revolutionized the landscape of dispute resolution. Most cases which would have been dealt with by the courts or by arbitration in the past are nowadays dealt with by adjudication. Although the adjudicator's decision is not necessarily binding (see Section 25.1), it is believed that well over 80% of adjudication decisions are accepted by the parties (Gaitskell 2005: 11), other estimations even propose the figure of 'well over 90%' (Uff 2005: 63). Of an estimated 15,000 adjudication decisions which occurred between the coming into force of the Construction Act on 1st May 1998 and 2005, only about 300 are believed to have reached the courts (Gaitskell 2005: 11). Exact figures about the shift from litigation/arbitration to adjudication are difficult to obtain since it is impossible to establish if an adjudication case would have gone to court (or arbitration) if this statutory right of adjudication had not existed. Surely, many adjudication decisions would have been settled by the parties without going to court. What is clear, though, is that since the Construction Act came into force, the number of construction disputes coming before the courts has reduced to about a third of what they were.

24.4.2 Conciliation

A 'conciliator' must be absolutely independent of the parties to the contract. Impartiality is essential, since the purpose of this process is to precipitate an agreement by persuasion and suggestion. Conciliators do not take sides, take decisions or make judgments. They talk to each party in private, and must be sure not to reveal anything to the other party. Confidentiality is essential in order for discussions to be frank and meaningful. The conciliator may bring the parties together after a while for an open discussion, which he or she chairs and leads.

The conciliator will be seeking to establish common ground, ascertaining the facts that are in dispute. In order to undertake this function effectively, the conciliator needs considerable knowledge of construction disputes. There may be previous judicial precedents which are appropriate to refer to, and the conciliator must be able to advise the parties of these, if necessary.

Where conciliation is adopted, it is ultimately up to the parties themselves to reach an agreement, and to decide upon the precise terms of that agreement.

24.4.3 Quasi-conciliation

Although it is a variation on the procedure described above, quasi-conciliation starts in a very different way. It comes about when one of the parties unilaterally appoints an expert professional to advise on a dispute, perhaps in order to obtain a second opinion. The purpose of the appointment is not simply to maximize the return on a claim, as would be the case where a claims consultant is engaged, but rather to help the party to overcome some technical or contractual difficulty. The investigator may be appointed to discover the facts and to make a recommendation to the party about how to proceed. It is often the case that this quasi-conciliator will need to talk to the other party and find out exactly what is at issue. If the other party also appoints such a professional, then the two of them may get together and compare their findings and their conclusions. In theory at least, they ought to do this impartially.

This type of process is often likely to happen in the public sector, because it is very difficult for a public sector agency simply to appoint such a person without first seeking relevant approvals from various committees. There are many factors to consider when spending money because of the notion of accountability. However, a private contractor, working for a public sector client, has the freedom to appoint someone unilaterally. In this case it is not unheard of for the public sector agency to listen to an expert second opinion.

Once the quasi-conciliator has reached a decision, a report is made to the client, which can then be used as a negotiating instrument. In addition, if the dispute continues and goes to court, the report may be used as evidence.

24.4.4 Mediation

Mediation is the most widely used and therefore the most important mode of ADR. It is like an extended version of conciliation. The initial stages will probably follow a very similar process, often referred to as shuttle diplomacy, as the mediator consults first with one party and then with the other. However, the end of this process is very different from conciliation in that, if no negotiated settlement results from the process, the mediator will make recommendations to settle based on his or her findings. This process accordingly retains the flexibility of conciliation, while encouraging a slightly more active role for the mediator. As a result, it tends to be less open-ended. According to anecdotal evidence, over 70% of cases referred to a mediator result in settlements. This is probably why recourse to arbitration seldom follows mediation. The Joint Contracts Tribunal has recognised the growing importance of mediation by referring to it in clause 9.1 of JCT SBC 05.

24.4.5 Private enquiry

This procedure involves the appointment of an independent professional to investigate some aspect of the project. It is commonly used for highly technical disputes, but is also valuable where the issues to be resolved are sensitive. Because the uses to which such enquiries are put are so wide-ranging, there is no fixed procedure for such an enquiry; it has to be created each time to suit the occasion. On the basis of the report produced, the parties are in a much better position to negotiate and reach a settlement.

It is very important that the appointee is given precise terms of reference in order that he or she can identify and carry out the intended task. Where this is done, it is usually found that private enquiries discover technical facts much more quickly than would a judicial enquiry. This is partly at least because arbitrators and judges are prohibited from using their own experience, and must reach a decision purely on the basis of upon what is put before them. This is by contrast with a private enquiry, where the person conducting the enquiry can make use of his own knowledge and professional expertise in arriving at a conclusion.

One of the greatest benefits of a private enquiry is its speed. It is interesting to note that the private enquiries set up to investigate the Lockerbie air disaster and the King's Cross fire came to their conclusions after a few months had elapsed. This is by stark contrast to litigation. It is not untypical for litigated cases to take years, perhaps decades.

24.4.6 Mini-trial

This procedure actually requires the disputing parties to present their cases to a board consisting of themselves! To be more precise, it means that representatives of the employer's and contractor's organizations will conduct something like a trial in front of a panel of senior executives from those organizations. It is important that this group should have the necessary authority to reach and implement decisions, and also that its members should not have been personally involved in the dispute up to this point.

Strictly speaking, this does not really fit the definition of Reconciliation given above, because it is adversarial in nature. The two parties are expected to take opposite stances and to argue their cases in front of the panel. Having heard the evidence, the panel can then negotiate their respective positions until they reach agreement.

24.4.7 Dispute Resolution Board

The term Dispute Resolution Board (DRB) is a generic term that covers, in particular, Dispute Adjudication Boards (DABs), Dispute Review Boards and Dispute Conciliation Boards (Institution of Civil Engineers 2005a: 1). Such boards are very common in major civil engineering projects, especially at international level. They have replaced the engineer in its role of being the first instance of dispute resolution in such projects (Draper 2007). The setting-up of a DRB requires an agreement between the parties which will often be part of a standard-from contract. The board commonly consists of three members (more rarely: only one member) and it is usually established at the commencement of a project, i.e. not only when a dispute actually arises. From the project's beginning, the board members conduct routine visits on the site on a regular basis. Their presence on site serves to maintain a good working relationship between the employer, the contractors and the engineer; often they are used as sounding board by the parties: problems are being discussed before they develop into disputes. Another positive effect of the board members' regular site visits is their familiarity with the project. As a consequence, when a conflict arises, the need for an evidential and hearing process can be kept to a minimum. Hence, although considerable costs may accrue already before any dispute arises, this mode of dispute resolution can be cost-effective if it avoids the more expensive modes of arbitration and litigation. This result, however, requires that the parties accept the decision of the board. It should also be noted that the costs inherent to the early establishment of DABs make this less appropriate in construction projects of a smaller scale (Draper 2007).

When a dispute arises, either party may refer it to the DRB which in turn will look into it and render its decision within a short period of time. To prepare its decision the board may, among other things, request information and hear witness statements. If one of the parties is dissatisfied with the DRB's decision, it can refer the dispute to arbitration or, if there is no arbitration agreement, to litigation. The arbitrator (or judge) does not review the board's decision since it is the dispute and not the board's decision that is referred to arbitration or litigation. However, according to anecdotal evidence, it is very rare that a decision of the DRB is not accepted by the parties.

24.5 REFERENCES TO ADR PROCEDURES IN STANDARD FORMS

The techniques discussed above are consensual (meaning that a party cannot be forced to agree upon a mode of ADR) and thus they are often agreed upon on a 'one off' basis when a construction dispute actually arises. However, modern standard forms encourage the parties to resolve their disputes amicably, i.e. without adjudication, arbitration or litigation. This encouragement takes place in form of explicit references to various modes of ADR and, in some cases, to particular ADR procedures.

In terms of JCT standard-form contracts, most forms of the 2005 suite contain the following reference in their conditions: *The parties may by agreement seek to resolve any dispute or difference arising under the Contract through mediation* (JCT SBC 05 clause 9.1, JCT IC 05 clause 9.1, JCT MW 05 clause 7.1, JCT DB 05 clause 9.1, JCT PCC 06 clause 8.1, JCT FA 05 clause 24, JCT RM 06 clause 7.1). A particular 'mediation procedure' is not specified. This is because the Joint Contracts Tribunal is of the opinion that this decision should be better taken by the parties when a dispute arises (Joint Contracts Tribunal 2005b: 14). However, the parties can use the mediation agreement (which contains simple procedural rules) published by the JCT in its 1995 practice note 25 (Joint Contracts Tribunal 1995).

Regarding the ICE 7 form, major changes were introduced in 2004, namely by an amendment to ICE 7. (For a concise description of this amendment see Shearer 2004.) In particular, the former two-step approach (engineer's decision, arbitration) has been replaced by a free choice between various amicable modes of ADR (of which conciliation and mediation are particularly mentioned), adjudication and arbitration (clauses 66A, 66B, 66C of ICE 7, as amended in 2004). This has scaled down the role of the engineer to a great extent: it is no longer necessary to refer a matter to the engineer, and to await the engineer's decision, before referring it to arbitration. In legal terms, the engineer's decision has ceased to be a 'condition precedent' to the referral to arbitration. Regarding the new dispute resolution provisions of the form ICE 7 (amendment 2004), the following points are important.

Clause 66(2) provides an advance warning procedure. As soon as possible, either party should inform the other party *of any matter which if not resolved might become a dispute*. It is not necessary that a 'dispute' has already occurred; the warning shall be served when a matter arises that might lead to a dispute. Following such a warning, a copy of which shall be issued to the engineer, the parties shall meet within seven days to attempt to resolve the matter. If any parts of the matter remain unresolved after the consultations have taken place for a reasonable period of time, these parts shall be defined in writing by the parties. According to the guidance notes, this procedure is a matter of voluntariness (Conditions of Contract Standing Joint Committee 2004: 2) although the word 'shall' (which is used in clause 66(2): 'shall advise;' 'shall meet;' 'shall define') generally has a mandatory meaning when used in standard-form contracts. Thus the procedure may be seen as a management tool (Shearer 2004). However, the guidance notes also explain that, if the matter leads to a dispute, any failure to operate the advance warning system may be taken into account by a subsequent tribunal in its decision about the costs.

Clause 66A(1) concerns the 'notice of dispute'. This notice may be served at any time by either party on the other side, with a copy to the engineer. However, the notice may only be served if the person issuing it has first taken all steps or procedures described elsewhere in the contract which may have overcome the dispute, and allowed sufficient time for reasonable steps to have been taken. According to the guidance notes this notice serves as a basis for defining when a dispute comes into existence and it provides a starting point for any further action (CCSJC 2004: 2).

Clause 66A(2) makes reference to different modes of 'amicable dispute resolution'. The clause offers negotiation and *other means including conciliation or mediation*. The desired option is initiated by a written notice seeking the other party's agreement to the requested mode of ADR. According to clause 66A(2)(a), for the case that the parties agree on conciliation or mediation, the respective ICE procedure shall apply (i.e. the ICE Conciliation Procedure 1999 or the ICE Construction Mediation Procedure 2002; in the respective version in force at the date of the notice).

Clauses 66B and 66C relate to adjudication and arbitration. These are means of adversarial dispute resolution which are discussed in Chapter 25.

The standard form NEC 3 deals with dispute resolution in its options W1 and W2. These contain two different adjudication procedures, one compliant with the Housing Grants, Construction and Regeneration Act 1996 (option W2) and one non-compliant with the act (option W1). The latter is drafted for use in international contracts to which the Act does not apply. Most surprisingly, the form NEC 3 does not contain a reference to mediation or any other form of ADR. One would have expected such a reference because the NEC standard forms have always been at the forefront of modern procedures. In this regard, as Eggleston (2006: 317) puts it, *NEC 3 has not kept pace with the growing trend for fully structured dispute resolution procedures* such as *negotiations at various levels, expert determination, conciliation and/or mediation, dispute resolution boards, adjudication and, only finally, arbitration* or litigation. However, nothing prevents the parties to an NEC 3 contract from resolving their disputes amicably, i.e. by means of mediation or other modes of ADR.

The standard form FIDIC Construction Contract (1st ed. 1999), the successor of the 'Red Book' 4th ed. 1987, provides for a Dispute Adjudication Board (DAB), i.e. a form of a DRB (dealt with in Section 24.4.7 above). According to the FIDIC contract, the decision of the board becomes final and binding if no notice of dissatisfaction has been issued within a period of 28 days. If such a notice has been served by either party, the contract sets out that the parties have to attempt to resolve their dispute amicably before the commencement of arbitration. Hence, arbitration cannot be commenced before the 56th day following the day on which notice of dissatisfaction was served. From a legal perspective, since one cannot impose an obligation on the parties to attempt to reach a settlement, this period simply prolongs the time before which arbitration may be commenced. However, there is nothing to stop the parties from attempting to reach agreement without going to arbitration. In fact, the clause only serves as a timely reminder to the parties that they really ought to try.

25 Adversarial dispute resolution

Where a dispute arises under a construction contract, the first method of resolving that dispute in a binding manner is often through the contract administrator. As seen in Chapter 18, the contract administrator usually has extensive decision-making powers, and the terms of the contract will govern the extent to which a decision, whether or not it is formally embodied in a certificate, will be legally binding upon the parties, at least for the time being. However, unless the contract makes a particular certificate 'conclusive' (in the sense discussed in Section 18.2.1), its effect can be challenged and, if appropriate, overturned in later proceedings of the kinds described below.

Until 1996, a party wishing to pursue a dispute beyond the contract administrator (or to challenge the contract administrator's decision on an existing dispute) had two options: either to go to arbitration, if the contract made provision for this, or to begin proceedings in court. However, an important statutory development means that this is no longer the case. Under section 108(1) of the Housing Grants, Construction and Regeneration Act 1996 (referred to throughout this Chapter as the 'Construction Act'), a party to every construction contract to which the Act applies (see Section 9.1) has the right to refer any dispute arising under that contract to an independent third party for a process called 'adjudication'. The intention of the lawmakers, following recommendations of Latham (1994), was that this should be a speedy and inexpensive procedure and that it should result in a decision that would be legally enforceable by both parties, at least until such time as the dispute might be finally settled at arbitration or in court.

This chapter describes the process of adjudication and also the more traditional mechanisms of arbitration and litigation, considering in particular the relative merits of the latter two as means of achieving a final solution to a construction dispute. In this connection it may be noted that the law of arbitration has been substantially reformed by the Arbitration Act 1996, with the intention of providing a more 'user friendly' form of dispute resolution procedure. However, many of the procedural reforms are dependent on the will of the parties to implement them, and the construction industry appears to have been rather slow to make use of them.

25.1 ADJUDICATION

Section 108(1) of the Construction Act creates a statutory right to adjudication, designed to produce a decision which is, at least temporarily, binding on the parties.

25.1.1 The nature of adjudication

Prior to the Construction Act, a number of standard form construction contracts (mainly sub-contracts) already incorporated a form of adjudication procedure for the temporary resolution of disputes. These frequently provided that the adjudicator should act as 'an independent expert and not as an arbitrator' in reaching a decision. However, it is clear that an adjudicator under the Construction Act, while not an arbitrator, is more than an independent expert. He or she must accept and consider any information properly submitted by the parties and must make any information submitted by one party, which the adjudicator intends to take into account, available to the other party. An independent expert would not be required to do either of these things.

This quasi-judicial aspect of adjudication has led to suggestions that it constitutes 'legal proceedings' for the purposes of the Human Rights Act 1998. However, in *Austin Hall Building Ltd v Buckland Securities Ltd*[1] it was held that this is not the case; the adjudicator's decision is not directly enforceable and so the adjudicator does not satisfy the definition of a 'court or tribunal'. The court in that case also held that, even if adjudication did fall within the Human Rights Act, the adjudicator would not have to hold a public hearing. This is because the adjudicator's decision can only be enforced by a court order (considered in Section 25.1.4), and so the 'hearing' required to satisfy the Human Rights Act would take place in court. Nor could an adjudicator's decision be regarded as contravening human rights by virtue of any restrictions imposed by the Construction Act, such as the very short time allowed for making a decision.

Where a construction contract is made with a residential occupier, it has been held that an adjudication clause is potentially an unusual and onerous provision which, if it has not been individually negotiated, may be invalidated by the Unfair Terms in Consumer Contracts Regulations 1999.[2] However, where it is the employer or their professional adviser who is responsible for the inclusion of the adjudication clause, it will be enforceable.[3]

25.1.2 The right to adjudication

As mentioned above, section 108(1) of the Construction Act 1996 provides that every party to a construction contract (the definition of which is discussed in Chapter 9) has the right to refer any dispute arising under the contract for adjudication under a procedure complying with the Act. This will be most useful while the contract work is being carried out (since it avoids undue delay); however, the right to adjudication may be exercised at any time, even after the contract has been determined.[4] Indeed, there is nothing in the statutory provisions to prevent adjudication from taking place at the same time that the parties are engaged in arbitration or litigation.[5]

The remainder of section 108 lays down certain minimum provisions about adjudication which the contract itself must contain. If it does not do so, either explicitly or by incorporating an appropriate set of rules to govern adjudication, the

---

[1] [2001] BLR 272.
[2] *Picardi v Cuniberti* [2003] BLR 487.
[3] *Bryen & Langley Ltd v Boston* [2005] BLR 508.
[4] *A & D Maintenance and Construction Ltd v Pagehurst Construction Services Ltd* (1999) 16 Const LJ 199.
[5] *Herschel Engineering Ltd v Breen Property Ltd* (2000) 70 Con LR 1.

parties will be subject to the Scheme for Construction Contracts, which was issued by the government as a statutory instrument made under the Act.[6] It should be noted that, even if the contract only fails to provide a part of what is required, the whole of the Scheme is then incorporated into the contract.

In order to satisfy the Act (and hence avoid the intervention of the Scheme), the contract must:

- Enable a party to give notice at any time of the intention to refer a dispute to adjudication.
- Provide a timetable with the object of appointing an adjudicator and referring the dispute within seven days of the notice.
- Require the adjudicator to reach a decision within 28 days (which the adjudicator may extend by up to 14 days if the referring party agrees), or such longer period as both parties agree to.
- Impose a duty on the adjudicator to act impartially.
- Enable the adjudicator to take the initiative in ascertaining the facts and the law.
- Provide that the adjudicator's decision is binding unless and until the dispute is finally determined by legal proceedings, arbitration or agreement of the parties (though the parties may agree that the adjudicator's decision is finally binding).
- Provide that the adjudicator is not liable for anything done or omitted unless the act or omission is in bad faith.

A number of organizations have drafted sets of adjudication rules which satisfy these minimum conditions and which parties may choose to incorporate into their contracts. Of these, the most important are:

- Construction Industry Council (CIC) Model Adjudication Procedure.
- Institution of Civil Engineers (ICE) Adjudication Procedure.
- Technology and Construction Solicitors' Association (TeCSA) Adjudication Rules.
- Centre for Effective Dispute Resolution (CEDR) Rules for Adjudication.

JCT 98 contains its own statute-compliant rules in clause 41A, which also provides that no-one shall be appointed as an adjudicator who is not prepared to enter into the 'JCT Adjudication Agreement', published in 1998 and updated in 2005. In JCT SBC 05, however, the whole adjudication procedure has been omitted. Clause 9.2 of the contract states that the statutory adjudication scheme shall apply, subject only to certain provisions which relate to the nomination of adjudicators and to cases of opening up and testing.

ICE 7 (as amended in July 2004) also contains statute-compliant rules for adjudication in clause 66B and in the ICE Adjudication Procedure (which is incorporated by reference). In accordance to clause 66B(1), either party has the right to refer any matter in dispute at any time to adjudication. This referral has to be done by a 'notice of adjudication' to the other side. A very important rule is set out in clause 66B(3): the adjudicator's decision becomes final and binding if

---

[6] SI 1998 No 649, Scheme for Construction Contracts (England and Wales) Regulations.

neither party refers the dispute (i.e. not the adjudicator's decision) to arbitration within three months of the decision.

25.1.3 Adjudication under the statutory scheme

Where the Scheme for Construction Contracts applies, it lays down detailed rules to govern both the setting up and the conduct of the adjudication process (though the parties may agree to waive any of these rules). The process begins when the party seeking adjudication (the referring party) serves written notice on every other party to the contract, briefly setting out the nature of the dispute and the relief sought. A copy of this notice is sent to the adjudicator named in the contract, if there is one, or to one of the bodies (such as professional institutions) that offer an adjudicator nominating service. Such a body must nominate an adjudicator within five days, and the person nominated has a further two days to confirm his or her willingness and availability. The Scheme provides that this person must not be an employee of a party to the dispute, but does not specifically prohibit the appointment of the contract administrator as adjudicator. However, such an appointment would not normally be advisable, since it could easily result in a conflict of interest if the adjudication involved the challenge of a contract administrator's decision.

Once the adjudicator is appointed, and no more than seven days from the original notice of adjudication, the referring party must serve a referral notice on the adjudicator, accompanied by copies of all the documents on which that party intends to rely. Copies of this notice and the documents are sent at the same time to all other parties to the dispute. Thereafter it is for the adjudicator to decide on the appropriate procedure, but it is clearly intended to be one in which the adjudicator takes the initiative. The Scheme accordingly empowers the adjudicator to demand documents from the parties, appoint expert advisers, make site visits, meet and question any party to the contract and issue timetables and deadlines. If a party fails to comply with any direction of the adjudicator, the latter may draw whatever inferences he or she feels are justified. If the adjudicator decides that an oral hearing would be helpful, the Scheme limits each party to one representative unless the adjudicator specifically directs otherwise.

The adjudicator must reach a decision on all matters referred, impartially and in accordance with all relevant terms of the contract, within 28 days of the referral notice. However, this period may be extended by up to 14 days with the consent of the referring party, or indefinitely if both parties agree. The adjudicator is not automatically obliged to give reasons for the decision, but any party is entitled to require this to be done. In addition to directing that something should be done or not done or, more commonly, that a sum of money be paid by one party to another, the adjudicator has power to award interest. Interestingly, however, the Scheme does not specifically empower the adjudicator to award costs, stating merely that the parties are jointly responsible for the adjudicator's reasonable fees and expenses.

25.1.4 Enforcement of the adjudicator's decision

Exactly what form an adjudicator's decision takes will depend on the nature of the dispute. The statutory provisions are not very helpful, the Scheme for Construction Contracts merely stating that 'the adjudicator shall decide the matters in dispute'. However, the Scheme goes on to say that the adjudicator may 'open up, revise and review' any certificates issued under the contract, other than those which are stated to be conclusive, and may decide that a party is liable to pay money to another (specifying the due date and final date for payment).

The intention underlying the Construction Act is clearly that an adjudicator's decision should be binding and enforceable, at least until such time as it may be overturned in legal proceedings or arbitration. The Scheme for Construction Contracts (following the pattern of section 42 of the Arbitration Act 1996), accordingly requires the adjudicator to issue the decision in the form of a (peremptory) order, following which the party who gained an award has to apply to the court for a second order (a mandatory injunction), instructing the other party to comply with the adjudicator's order.

In practice, this route is seldom if ever pursued, at least not in respect of adjudication decisions on payment. Instead, as pointed out by the TCC judge in *Macob Civil Engineering Ltd v Morrison Construction Ltd*[7], the usual remedy to enforce an adjudicator's order for the payment of a sum of money will be to issue proceedings claiming the some due, followed by an application for summary judgment. This procedural remedy (which is not mentioned in the Scheme) is considered in more detail in Section 25.3.2 below.

Three further points are worth mentioning. First, actions to enforce adjudicators' decisions are processed very quickly by the courts. For example, the judgment in the *Macob* case (12 February 1999) was delivered just five weeks after publication of the adjudicator's decision (6 January 1999). Second, the route to apply to the courts for a mandatory injunction is the appropriate route to enforce other types of order, for example an order to provide access or to open up work.[8] Third, it appears that, where an adjudicator makes an order for payment, this must be complied with; the paying party may not rely on a defence, such as a right of set-off, which it did not raise at the adjudication.[9]

25.1.5 Challenges to the adjudicator's decision

There is no mechanism by which an adjudicator's decision can be appealed, since it is inherently binding only temporarily. In subsequent proceedings, whether by arbitration or litigation, the dispute is dealt with afresh. Nevertheless, there are certain grounds on which the enforcement of an adjudicator's decision may be resisted, notably lack of jurisdiction, error of law and procedural unfairness. These will be briefly considered, although it must be said that the courts have so far

---

[7] [1999] BLR 93; confirmed by *Outwing Construction Ltd v H Randell & Son Ltd* [1999] CILL 1482.
[8] *Macob Civil Engineering Ltd v Morrison Construction Ltd* [1999] BLR 93.
[9] *Levolux AT Ltd v Ferson Contractors Ltd* (2003) 86 Con LR 98.

appeared reluctant to allow a party to evade an adjudicator's decision by raising what are in truth technical objections.[10]

Lack of jurisdiction

An adjudicator has no jurisdiction to act unless certain conditions are fulfilled. There must be an existing dispute between contracting parties, arising under a construction contract, made in writing, which falls within the Housing Grants, Construction and Regeneration Act 1996. The dispute must be referred, in accordance with the relevant contractual provisions, to an adjudicator properly appointed under the contract.

If it is alleged that one of these conditions has not been fulfilled, and the court finds that there is sufficient doubt to make this a triable issue, it will not award summary judgment in a claim to enforce the adjudicator's decision.[11]

Error of law

Although the legal position is not entirely clear, it has been accepted by both the Technology and Construction Court and the Court of Appeal that an error of law is not in itself sufficient to render an adjudicator's decision invalid, so long as the effect of the error is merely that the adjudicator has given the 'wrong answer to the right question'.[12] If, however, the error is such that the adjudicator has answered the 'wrong question', the decision is outside the adjudicator's jurisdiction and is therefore invalid.

Procedural irregularities

All judicial and quasi-judicial proceedings are subject to the principles of natural justice. The most important of these principles involve:

- Giving each party a reasonable opportunity of learning what allegations are made by the other party and of putting forward a case in answer to those allegations;
- Not acting with bias, or in a way that would lead a fair-minded observer to conclude that there was a real possibility of bias.

There is no doubt that an adjudicator must act in accordance with these principles, or as fairly as the statutory rules permit (given, for example, the very short time limits and the fact that this is not an adversarial procedure). Any failure

---

[10] See, for example, *Macob Civil Engineering Ltd v Morrison Construction Ltd* [1999] BLR 93; *Carillion Construction Ltd v Devonport Royal Dockyard Ltd* [2006] BLR 15.
[11] *The Project Consultancy Group v Trustees of the Gray Trust* (1999) 65 Con LR 146.
[12] *Bouygues UK Ltd v Dahl-Jensen UK Ltd* [2000] BLR 49; 522.

to do so, at least where the breach is a relevant and substantial one, will mean that the adjudicator's decision will not be enforced.[13]

Where procedural irregularities do not raise questions of natural justice, the courts have appeared reluctant to allow them to invalidate an adjudicator's decision. In one case, for example, a court upheld an adjudicator's decision which, although made within the contractual time limit, was not communicated to the parties until shortly after the time limit had expired.[14]

25.2 ARBITRATION

Arbitration is a procedure for the resolution of disputes which is, for the most part, under the control of the parties.

25.2.1 The nature of arbitration

In any consideration of arbitration, it is important to understand at the outset that, whereas there is normally an automatic right to take a dispute to court, the right to go to arbitration is by its nature more restricted. Arbitration can only arise where certain conditions are met. In particular, there must be:

1. A genuine dispute or difference between the parties, of a kind 'justiciable at law'. This means that the arbitrator's award must be capable of enforcement as if it were a judgment of a court.
2. A binding contractual agreement to submit that dispute to arbitration by a third party chosen either by the parties or in accordance with their instructions. Such an agreement may be made when a dispute has arisen; however, it is more common for a contract to provide from the outset that any dispute arising between the parties shall be finally determined by arbitration (as, for example, in JCT 98, Article 7A and clause 41B; ICE 7, as amended in 2004, clause 66C(1)(a); as to JCT SBC 05, see below). In either case, the arbitration agreement itself need not name the arbitrator; the parties may and frequently do provide for a professional body to nominate a suitable person.
3. A reference to arbitration in accordance with whatever procedure is laid down in the agreement. Thus, for example, ICE 7, as amended in 2004, clause 66C(1) requires a 'notice to refer' (to arbitration), which must be served no later than three months after the decision of an adjudicator.

It should be noted that, while it is perfectly possible to make a valid arbitration agreement by word of mouth, the Arbitration Act 1996 only applies to arbitration agreements which are made in writing. This is important, because the Act lays down detailed rules about the way in which arbitrations are to be conducted, including details of how they are controlled by the courts in terms of hearing appeals, removing arbitrators for misconduct and so on. Fortunately, section 5 of

---

[13] *Discain Project Services Ltd v Opecprime Development Ltd* [2000] BLR 402; *Discain Project Services Ltd v Opecprime Development Ltd (No 2)* [2001] BLR 285.
[14] *Barnes & Elliott Ltd v Taylor Woodrow Holdings Ltd* [2004] BLR 111.

the Act provides a very wide definition of what is meant by 'in writing' for this purpose. It will include not only a written agreement (whether or not it is signed), but also an agreement made by exchange of letters and any agreement evidenced in writing. Moreover, in a provision that changes the previous legal position, section 5 provides that an oral agreement to be bound by a set of written terms (for example those contained in a standard form building contract) is to be treated as an agreement in writing.

Arbitration has traditionally been the construction industry's preferred means of settling disputes, and most standard form contracts have therefore contained arbitration clauses (for example JCT 98 Article 7A; ICE 7 (amendment 2004) clause 66C). However, the most recent review of JCT contracts has seen an important change of attitude towards this issue. Aware that most disputes are now resolved by adjudication, and that issues which are not so resolved are frequently concerned with points of law, such as the extent of the adjudicator's powers, the drafters of JCT SBC 05 (and, indeed, of the 2005 suite of JCT contracts) have decided to make litigation the default dispute resolution procedure. According to JCT SBC 05 Articles 8 and 9, therefore, disputes will be settled in court unless the parties specifically show their preference for arbitration by making an appropriate entry in the Contract Particulars.

25.2.2 The arbitrator's jurisdiction

The jurisdiction of the courts to hear cases arising out of disputes between contracting parties is virtually unlimited (although a contract may provide that it cannot be exercised until after the parties have been to arbitration). By contrast, an arbitrator is empowered to decide only those disputes which are specifically referred, and which fall within the scope of the arbitration agreement under which the appointment is made. It may therefore be that, for certain kinds of dispute, litigation is the only available solution.

It was formerly the case that an arbitrator did not have power to decide any question on which his or her own jurisdiction depended, such as whether the contract containing the arbitration agreement was void. Hence, if a dispute arose over such a matter, it could only be dealt with by a court. However, this has been altered by section 30 of the Arbitration Act 1996, which provides that, unless the parties agree otherwise, the arbitrator may decide whether there is a valid arbitration agreement, whether the arbitration tribunal is properly constituted and whether matters have been properly referred to arbitration. This does not completely rule out a reference to the court, but section 32 provides that such a reference can only be made with the written consent of all parties, or by permission of the arbitrator in circumstances where the court is satisfied that it is likely to save costs and that there is good reason for it. In any event, section 31 provides that a party who wishes to claim that the arbitrator lacks jurisdiction from the outset must do so before taking any steps to answer the case on its merits. Section 31 also provides that, if the question of jurisdiction arises during the course of arbitration proceedings, a party who wishes to object on that ground must do so as soon as possible.

Whether or not a dispute falls within an arbitration agreement depends on the wording of that agreement. In this connection it appears that the phrase 'arising under' is narrower than either 'arising out of' or 'arising in connection with' a contract, and that it does not include claims in tort for misrepresentation or attempts to establish a collateral contract.[15] Where the wider form of words is used, the arbitrator has power to rectify the contract on the ground of mistake or misrepresentation.[16] Such power is given expressly under JCT 98 clause 41B.2 and (where the parties specifically choose arbitration) under JCT SBC 05 clause 9.5. The wider form of words has also been held to cover ordering a sub-contractor to contribute towards the damages payable by a main contractor in a fatal accident claim, and to decide whether damages should be reduced on the ground of contributory negligence.[17]

The arbitration clauses found in most standard form construction contracts are widely drafted. For example, JCT 98, Article 7A, includes 'any dispute or difference as to any matter or thing of whatsoever nature arising under this contract or in connection therewith...'. JCT SBC 05, Article 8 is couched in even wider terms, the crucial passage stating 'any dispute or difference between the Parties of any kind whatsoever arising out of or in connection with this Contract'. ICE 7 (as amended in 2004) clause 66C states 'all disputes arising under or in connection with the Contract'. However, all these forms of contract specifically exclude from arbitration disputes concerning the enforcement of an adjudicator's decision.

It has been decided by the House of Lords that formulations of this kind are sufficient to cover a dispute as to whether the contract has been terminated by repudiatory breach or frustration.[18]

In former centuries arbitrators were expected to decide disputes, not according to legal principles, but on the basis of what they regarded as just and equitable. This has remained the position in many countries but, until recently, English law insisted that an arbitrator's award must be based on strict legal principles, irrespective of the parties' wishes or instructions. That rigid rule has been altered by section 46 of the Arbitration Act 1996, which provides that, if the parties agree, their dispute can be decided in accordance with whatever considerations are agreed by them or determined by the arbitrator. Notwithstanding this increased flexibility, however, an arbitrator can only award those remedies arising under the general law (such as damages) or for which the contract in question makes provision. Thus, for example, a contractor cannot be ordered to do remedial work as an alternative to paying liquidated damages.[19]

25.2.3 Arbitration procedure under the 1996 Act

The intention underpinning the Arbitration Act 1996 is that the parties should be encouraged to take control of the dispute resolution procedure that they have chosen and to use its inherent flexibility so as to obtain the maximum possible

---

[15] *Fillite (Runcorn) Ltd v Aqua-Lift* (1989) 45 BLR 27.
[16] *Ashville Investments Ltd v Elmer Contractors Ltd* [1988] 2 All ER 577.
[17] *Wealands v CLC Contractors Ltd* [1999] BLR 401.
[18] *Heyman v Darwins Ltd* [1942] AC 356.
[19] *BFI Group of Companies Ltd v DCB Integration Systems Ltd* [1987] CILL 348.

benefit. This intention, and indeed the general philosophy of the Act, may be seen in three specific provisions. First, section 1 sets out three basic principles, in accordance with which the rest of the Act is to be construed. These are:

- that the object of arbitration is to obtain the fair resolution of disputes by an impartial tribunal without unnecessary delay or expense.
- that the parties should be free to agree how their disputes are resolved, subject only to such safeguards as are necessary in the public interest.
- that the court should not intervene except where the Act itself provides.

Section 33 then sets out the general duty of the arbitration tribunal. This is:

- to act fairly and impartially as between the parties, giving each party a reasonable opportunity of putting its case and of dealing with that of its opponent; and
- to adopt suitable procedures so as to provide a fair means for the resolution of the matters to be determined, while avoiding unnecessary delay or expense.

Lastly, section 40 imposes a general duty on the parties to do all things necessary for the proper and expeditious conduct of the proceedings, including complying without delay with any decision of the arbitrator as to procedural or evidential matters, or with any other order or directions of the arbitrator.

Procedure in general

Sections 33 to 41 of the Arbitration Act 1996 make it very clear that the procedure to be adopted at arbitration is, first and foremost, for the parties to agree (which they may do for instance by incorporating a set of model rules in their contract). Anything not specified by the parties is for the arbitrator to decide. There is clearly no presumption that the proceedings will follow closely along the lines of an action in court; the Act provides, for example, that (subject to what the parties may agree) the arbitrator may decide whether or not to enforce the strict rules of evidence, and may take the initiative in ascertaining the facts and the law. However, it is also provided that, in the absence of agreement to the contrary, a party to arbitration is entitled to be represented, either by a lawyer or by any other person.

The control exercised by the arbitrator is considerable, and it is backed by far-reaching powers to deal with a party who fails to 'do something necessary for the proper and expeditious conduct of the arbitration'. Thus, where a claimant is guilty of 'inordinate and inexcusable delay' in pursuing a claim, leading to a substantial risk that a fair resolution of the dispute will not be possible or causing serious prejudice to the respondent, the arbitrator may dismiss the claim altogether. If a party without good reason fails to attend a hearing or to make a written submission when required, the arbitrator may continue with the proceedings on the basis of such evidence as is available (which will of course normally consist of the evidence given by the other party). And, if a party fails without good reason to comply with any of the arbitrator's orders or directions, the arbitrator may make a 'peremptory order' specifying a time for compliance and this, in turn, may be enforced by a court.

Whether or not the procedure at arbitration follows that of a court, the result that it produces, namely the arbitrator's award, will closely resemble a court judgment. This resemblance is strengthened by the fact that the Arbitration Act 1996 empowers the arbitrator (provided the parties agree to this) to grant a range of remedies similar to those obtainable in court (including injunctions, orders for specific performance of a contract and declarations). The arbitrator may also make provisional awards, add interest to any sum due and make orders as to liability for costs.

It has long been the practice, in industries where arbitration is a common method for the resolution of disputes, to have standard sets of rules to govern arbitration procedures. The arbitration clauses in the contracts used by those industries can then simply state that any arbitration shall be conducted in accordance with the appropriate set of rules.

In the construction industry, the passing of the Arbitration Act 1996 has brought about the drafting of two sets of rules of particular importance. The current versions of these are the Construction Industry Model Arbitration Rules 2005 (CIMAR) and the Institution of Civil Engineers' (ICE) Arbitration Procedure 2006.

CIMAR Arbitration Rules 2005

CIMAR not only follows the 1996 Act, but actually reproduces much of the statutory text, which it then amplifies and extends. These Rules do not remove any of the powers which the arbitrator would have under the Act and they also provide that, once an arbitrator is appointed, the parties will not make any agreement which would have the effect of amending or conflicting with the Rules. CIMAR further contains important provisions dealing with the consolidation of arbitrations under different contracts, although the effectiveness of these will inevitably depend upon the agreement of all the parties involved, and empowers the arbitrator to grant provisional relief.

CIMAR provides a choice of three procedures, each of which may be suitable for different types of dispute. At one extreme is the 'full procedure' (rule 9), which would be expected to follow fairly closely the litigation model. This is regarded as appropriate for large and complex cases, where disputes over the facts are such that witnesses ought to give evidence orally and be subjected to cross-examination. By contrast, the 'documents only' procedure (rule 8) is one in which the arbitrator reaches a decision on the basis of written statements, documents such as specifications, letters, site minutes and experts' reports and (if necessary) a site inspection. This procedure is appropriate where the issues involved do not require oral evidence or the sums in dispute do not warrant the cost of a hearing. This might be, for example, where the dispute relates to the valuation of work, investigation into the cause of defects or interpretation of the contract. An award following a 'documents only' arbitration is to be made within one month.

The 'short hearing' procedure (rule 7) is specifically designed for disputes which are to be settled primarily by the arbitrator inspecting work or materials. The expectation is that, in addition to making a site visit, the arbitrator will conduct a hearing of not more than one day (which the parties may agree to extend) at which each party will have a reasonable opportunity to state its case. Expert evidence may be given, but the cost of this is only recoverable if the arbitrator rules that it was necessary in order to reach his or her decision. Once again, the arbitrator is required to make an award within one month.

As to which of the three procedures is to be adopted in any given case, CIMAR (rule 6.2) makes clear that this is for the arbitrator, rather than the parties, to decide. However, the Rules provide that the parties are to inform the arbitrator of their views as to the which procedure would be appropriate (and as to the probable length of any required hearing) and, if the parties are in agreement as to which procedure they wish to follow, it seems unlikely that an arbitrator would insist on a different one.

Where the parties to a JCT contract opt for arbitration as their dispute resolution procedure, a special version of CIMAR, called the 'JCT 2005 edition' is used (see JCT SBC 05 Article 8 and clause 9.3). This contains, in addition to the basic rules, a section called 'JCT Supplementary and Advisory Procedures', of which Part A lists a number of mandatory provisions and Part B a number which the parties may expressly agree to include. Of the ways in which these change the basic CIMAR rules, the most important is probably to include a presumption in favour of the 'documents only' procedure, subject to any contrary agreement by the parties or a considered decision by the arbitrator to use the full procedure.

ICE Arbitration Procedure

The ICE arbitration procedure is drafted so as to comply with the Arbitration Act 1996, but to be quite independent of it. As a consequence, this procedure is thought to be suitable, not only for arbitrations within the UK, but also for international contracts, which may not be subject to UK statutes. Notwithstanding this difference in approach, however, the effect of the ICE Procedure is largely similar to CIMAR, including the powers on consolidation and provisional relief.

The ICE Procedure provides that, if the total of the sums claimed does not exceed £50,000 or if the parties so agree, the arbitrator shall adopt a 'Short Procedure' (Part F), which may include a brief hearing or may be conducted on a 'documents only' basis. Alternatively (again with the agreement of the parties), the arbitrator may use the 'Special Procedure for Experts' (Part H), which is very similar to the Short Procedure but with the arbitrator meeting only the experts appointed by each party.

25.2.4 Control by the courts

The Arbitration Act 1996 contains a number of provisions under which the courts exercise a measure of control over arbitrations, while they are taking place. Two of the most important of these are the court's power, in appropriate circumstances, to remove an arbitrator and to give a ruling on a particular point of law.

Removal of arbitrator

Section 24 of the Arbitration Act 1996 (a provision which cannot be excluded) provides that any party may, on giving notice to the other party and to the arbitrator, apply to the court to remove the arbitrator from office on certain specified grounds. If such an application is made, the arbitrator is entitled to appear before and be heard by the court before it makes its decision. Moreover, the fact that an application has been made does not prevent the arbitrator from continuing with the arbitration and even making an award.

Earlier legislation gave power to remove an arbitrator for 'misconduct', a term which was widely interpreted so as to include procedural irregularity as well as any intentional wrongdoing. The 1996 Act is clearly intended to be somewhat narrower in scope, as may be seen from the limited grounds on which an application to remove the arbitrator may be made. These are:

- That there are justifiable doubts as to the arbitrator's impartiality.
- That the arbitrator lacks the qualifications required by the arbitration agreement.
- That there are justifiable doubts as to the arbitrator's physical or mental capacity to conduct the arbitration.
- That the arbitrator has refused or failed to conduct the proceedings or to use all reasonable despatch and that this has caused substantial injustice to the applicant.

The first of these grounds is arguably the most important, since it is essential that an arbitrator should be free from any form of bias in favour of one party. In cases decided on similar provisions in earlier legislation, it has been held that the crucial test is whether there is a 'real danger' of bias. The fact that an architect or engineer named as arbitrator is associated with, or even an employee of, the client has never been regarded as an *automatic* reason for disqualification.[20] However, where the connection is too close it may give rise to reasonable suspicion of bias.[21] Moreover, where a contract administrator is appointed as arbitrator (at one time a common practice, though rare today), they will be disqualified from dealing with a dispute in which they are the real defendant.[22] In other cases, arbitrators have been removed for showing bias against a witness on racial grounds, or simply for using such sarcastic and hostile language as to convince one party that their minds are already made up and that a fair hearing is impossible.[23]

The last of the grounds for removal might seem at first sight to be potentially very wide. However, it is clear from the government committee on whose report the 1996 Act was based that this is intended only to apply to those rare cases where the arbitrator's conduct is in effect frustrating the purpose of arbitration as stated in section 1 of the Act.

---

[20] *Eckersley v Mersey Docks and Harbour Board* [1894] 2 QB 667.
[21] *Verital Shipping Corporation v Anglo-Canadian Cement Ltd* [1966] 1 Lloyd's Rep 76.
[22] *Nuttall v Manchester Corporation* (1892) 9 TLR 513.
[23] *Turner (East Asia) PTE Ltd v Builders Federal (Hong Kong)* (1988) 42 BLR 122.

Determination of preliminary point of law

Section 45 of the Arbitration Act 1996 provides that, unless the arbitration agreement provides otherwise, a party may apply to the court for a ruling on any question of law arising in the proceedings. A party making such an application must first give notice to the other party, and the court can only give its ruling if it is satisfied that the point in question is one which 'substantially affects the rights of one or more of the parties'.

This procedure is subject to complex procedural rules. In particular, it requires the consent *either* of the arbitrator *or* of all the parties to the arbitration. Furthermore, where it is made without the consent of all the parties, the court must be satisfied that determining the question of law is likely to produce substantial savings in costs and that the application was made without delay.

25.2.5 Challenging an arbitrator's award

Appeals on a point of law

Section 69 of the Arbitration Act 1996 provides for a right of appeal to the High Court (normally, in construction cases, the Technology and Construction Court) on any question of law arising out of an arbitrator's award. Such an appeal cannot be brought unless any process of appeal or review provided by the arbitration agreement itself has first been exhausted. Furthermore, an appeal must normally be brought within 28 days of the date of the award (or the end of an arbitral appeal or review process).

It should be noted that the procedural rules governing this right of appeal are extremely complex. Similar wording in previous legislation has given rise to a great deal of litigation and has been interpreted by the courts in a highly restrictive way. Briefly, the position is that an appeal against an award requires *either* the consent of all parties to the arbitration *or* the leave of the court. Such leave is only to be granted where the court is satisfied that four requirements have all been met. These are that:

- determination of the question of law will substantially affect the parties' rights;
- the question is one that the arbitrator was asked to determine;
- either the arbitrator's decision is obviously wrong or the question is one of general public importance (a test which is very difficult to satisfy in relation to a 'one off' contract) and the arbitrator's decision is open to serious doubt; and
- despite the agreement of the parties to resolve the matter by arbitration, it is just and proper in the circumstances for the court to determine the question.

It is also worth noting that the conditions on which it is possible to appeal from the court's decision (either on the point of law itself or on whether or not to grant leave to appeal) are even more restrictive.

Lack of jurisdiction

Section 67 of the Arbitration Act 1996 provides that in two circumstances a party may, after giving notice to the other party and to the arbitrator, challenge an arbitration award on the ground of lack of jurisdiction. In both cases, a challenge of this kind, like an appeal on a point of law, cannot be brought unless any process of appeal or review provided by the arbitration agreement itself has first been exhausted. And, again like an appeal on a point of law, a challenge must normally be brought within 28 days of the date of the award or the end of an arbitral appeal or review process.

The first type of challenge arises where the arbitrator has made a decision as to his or her own jurisdiction. As we saw in Section 25.2.2, this is a power that arbitrators formerly did not have, but which was given to them by section 30 of the 1996 Act. It is provided that, while a challenge of this kind is under way, the arbitrator may continue with the proceedings.

The second circumstance mentioned above is where the arbitrator has made an award that the applicant believes there was no jurisdiction to make. The Act provides that, where such a challenge is made, the court is empowered to confirm the award, vary it, or set it aside in whole or in part.

Serious irregularity

Section 68 of the Arbitration Act 1996 provides that a party may, after giving notice to the other party and to the arbitrator, challenge an arbitration award on the ground of 'serious irregularity affecting the tribunal, the proceedings or the award'. Once again, a challenge of this kind cannot be brought unless any process of appeal or review provided by the arbitration agreement itself has first been exhausted, and it is subject to the 28-day time limit.

Section 68 contains an exhaustive list of the things that may constitute 'serious irregularities' for this purpose. They range from the arbitrator's failure to act impartially, or acting in excess of his or her powers, through procedural errors in conducting the arbitration, to awards that are uncertain, ambiguous, or contrary to public policy. However, an overriding requirement is that, in order to take action, the court must consider that the irregularity in question has caused or will cause substantial injustice to the applicant.

Where a court decides that there has been a serious irregularity of the kind described, it will in most cases remit the award to the arbitrator for reconsideration. However, where the court is satisfied that this would be inappropriate, it may instead set aside all or part of the award, or declare all or part of it to be of no effect.

25.3   LITIGATION

As a general principle, any dispute arising between the parties to a contract may be settled by an action in court. As to which court is appropriate, this will depend upon a number of factors including the size of the dispute, its nature and, in some cases, its location. The size of the dispute is relevant because claims up to a certain financial limit tend to go to a county court. Its nature is important because that will affect which division of the High Court is selected. Thirdly, the location may influence the choice of court since it may be tried locally in the relevant county court or by a circuit judge hearing High Court business.

In practice, building and civil engineering cases of significant size are tried in the Technology and Construction Court. This is a specialist sub-division of the Queen's Bench Division of the High Court, which sits in separate premises in London (though some cases are tried at provincial locations). It is staffed by circuit judges, formerly known as 'Official Referees', who are almost all appointed from barristers who have specialized in construction disputes. Since some 80% of the business of the Technology and Construction Court is construction-based, it effectively constitutes a specialist court for the construction industry.

A litigant who is dissatisfied with a decision in the Technology and Construction Court may appeal to the Court of Appeal on a point of either law or fact. However, an appeal on fact is only possible if either the trial judge or the Court of Appeal gives leave. It seems that, in the interests of finality and certainty, leave should not be granted unless the court believes there is a real prospect of success.[24]

25.3.1 Litigation procedure

Since 26 April 1999, all civil litigation has been governed by the Civil Procedure Rules (CPR) (SI 1998 No 3132). These rules are specifically intended to be a complete break from the old system, and the courts have confirmed that the old rules and case law are not to be relied upon, even as a guide. Instead, the CPR are to be developed by judges from the fundamental principles stated at Part 1.1:

1. These Rules are a new procedural code with the overriding objective of enabling the court to deal with cases justly.
2. Dealing with a case justly includes, so far as practicable:
   a. ensuring that the parties are on an equal footing;
   b. saving expense;
   c. dealing with the case in ways which are proportionate:
      i. to the amount of money involved;
      ii. to the importance of the case;
      iii. to the complexity of the issues; and
      iv. to the financial position of each party.
   d. ensuring that it is dealt with expeditiously and fairly; and
   e. allotting to it an appropriate share of the court's resources, while taking into account the need to allot resources to other cases.

---

[24] *Hoskisson v Moody Homes Ltd* (1989) CSW, 25 May, 69.

25.3.2 Useful remedies in construction cases

An important difference between arbitration and litigation is that parties whose dispute is heard in court may be able to invoke two very powerful legal remedies, which are not readily available at arbitration. First, in very clear-cut cases, there may be a power to obtain 'summary judgment' under Part 24 of the CPR. Where this is so, it means that the whole of a claim, or a particular issue, can be decided once and for all on the basis of affidavit evidence (written statements made on oath). This avoids the expense and delay of a full-scale trial, which can save up to 80% of the costs and up to two years of waiting.

Whether or not a claimant seeks summary judgment under Part 24 of the CPR, it may be appropriate to apply for 'interim payment' under Part 25. This is a procedure under which a claimant may, by satisfying the court that he or she is likely to be awarded a substantial sum of money, become entitled to payment in advance of a reasonable proportion of that money. However, this procedure does not operate as a final decision on the merits of the case and so, if the later proceedings do not confirm the award, then the money must be repaid.

Summary judgment under CPR Part 24

Summary judgment enables a court, either of its own volition or on the application of a party, to dispose of a claim, a defence or a particular issue without a full trial. This procedure, which can save a great deal of time and money, is available where two conditions are satisfied:

- the court considers either that the claimant has no real prospect of success on the claim or issue, or that the defendant has no real prospect of success on the defence; and
- there is no other reason why the case or issue should go to a full trial.

Since summary judgment effectively denies to a defendant the opportunity to argue the case in full, it follows that a court must feel very convinced that there is no arguable defence before making such an award. In cases decided on an earlier version of this procedure, the courts used such phrases as 'indisputably due', 'as plain as could be' and 'beyond reasonable doubt', which indicate that the onus of proof on the claimant is a heavy one. In the construction context, a contract administrator's certificate has usually been regarded as sufficient evidence that a sum is due, except where the defendant has been able to challenge the certificate in some way. Where no such challenge has been possible, then vague allegations of defects or delay, without good supporting evidence, have not been enough to prevent the contractor from obtaining summary judgment.

Part 24 is intended to provide a speedy method of disposing of disputes to which in truth there can be only one answer. In consequence, the court should not allow such proceedings to escalate into a full-scale trial. However, if the defendant raises a point of law, which can be disposed of after a brief argument, the court can deal with it.

Interim payment under CPR Part 25

A claimant may apply for an order for interim payment on account of damages or other sums that a defendant may subsequently be held liable to pay. Such an order may only be made in three circumstances:

- where the defendant has admitted liability;
- where the defendant has already been held liable; or
- where the court is satisfied that, if the matter went to trial, the claimant would be awarded a significant amount of damages.

If the court is satisfied on this, it may order the defendant to make an interim payment of such amount as it thinks just, not exceeding a reasonable proportion of the ultimate liability and taking into account any contributory negligence, set-off or counter-claim.

25.4   ARBITRATION OR LITIGATION?

The parties to a dispute are in principle free to choose a method of resolving it. Arbitration and litigation each offer certain advantages. Even where a contract contains a clause stating that disputes shall be settled by arbitration, the parties may agree to ignore this and instead go to court. However, the presence of an arbitration clause in the contract will normally mean that, unless both parties agree otherwise, disputes *must* be settled by arbitration. Of course, such a clause affects neither the statutory right of adjudication nor the option of finding a settlement through a form of amicable dispute resolution.

25.4.1 Relative advantages and disadvantages

As to which procedure (arbitration or litigation) would be a more appropriate means of resolving a dispute, no definitive answer can be given, but there are various matters that may be taken into account in any particular case. In briefly noting these factors, one point should be borne in mind. While 'litigation' will follow much the same pattern wherever it takes place, 'arbitration' covers a much wider range of possible procedures. Thus, any comparison between litigation and arbitration inevitably raises the question of what kind of arbitration is being considered. The relative merits of the two forms of dispute resolution, which are described below, are largely based on the assumption of a large-scale construction arbitration involving complex arguments, legal representation and procedures similar to those in court.

Advantages of arbitration

The advantages most commonly claimed for arbitration are that it is cheap, quick, suitable for matters of technical complexity, convenient, private and commercially expedient. These points are discussed below:

- **Cost:** It is often said that arbitration is cheaper than litigation. Unfortunately, while this *can* be true in simple cases, where a short informal procedure (for example without legal representation) is used, it tends not to be so in complex construction disputes. Indeed, in such cases, where the procedures adopted are similar to those of a court, arbitration is likely to be the more expensive option. This is because the parties must pay for the arbitrator, the venue, and such other items as a transcript of the proceedings.
- **Speed:** Again, while arbitration in simple cases is likely to be much quicker than litigation, this is dependent upon the parties' willingness to adopt a suitable procedure. If what is required is effectively a trial, then litigation is likely to be quicker, since judges tend to be more ruthless than arbitrators in enforcing the prescribed time limits for the various procedural stages, and in refusing to give extensions of time. This is not to say that arbitrators lack the necessary powers, rather that they do not always have the confidence to apply them.
- **Technical complexity:** It may be that, where the legal issues in a case are relatively straightforward, but the factual questions are complicated (for example as to detailed matters of design loading), it is better to have the dispute heard by an arbitrator possessing relevant technical qualifications. However, where 'complexity' means only 'construction industry practice', this is well understood by the judges of the Technology and Construction Court, who spend most of their time hearing such cases.
- **Convenience:** Arbitration can be arranged to suit the parties (and the arbitrator) and, for example, it may be possible to hold hearings in the evening or at a weekend. Once again, however, this is a less realistic option in a complex case where lawyers and experts are involved.
- **Privacy:** It is not easy to maintain complete confidentiality in large construction arbitrations, if only because so many prominent people within the construction industry and professions will be involved. However, what may quite legitimately be said is that litigation is officially in the public domain; the fact that one company is suing another will be known as soon as the writ is issued, and the general public has a right of access to court proceedings. Thus, there is likely to be more publicity attached to litigation than arbitration.
- **Commercial expediency:** One other matter, which might influence a party in favour of arbitration, is its rather less confrontational nature. This may be important where the parties' contractual relationship is continuing.

Advantages of litigation

The perceived advantages of litigation include the ability to join third parties in the action, the availability of legal aid, the ability to deal with legal complexities and a more decisive approach by the decision-maker. These are expanded below:

- **Third parties:** The right to take a dispute to arbitration is conferred, not by law, but by the terms of a contract. In consequence, only the parties to that contract are bound. This means that where, as is often the case in construction disputes, more than two parties are involved, they can only be brought into the same arbitration proceedings if either they all agree or provision is made for this in all the relevant contracts. JCT SBC 05 and the accompanying sub-contract are drafted with the intention of enabling proceedings to be joined. However, the drafting is complex, and it is by no means certain that it achieves its object.

- **Legal aid:** A private individual involved in a dispute may qualify for legal aid. If so, that person will almost certainly prefer litigation to arbitration, since legal aid is not available for the latter.
- **Legal complexity:** Where a dispute essentially turns on a point of law (which includes the meaning to be given to a term of the contract) it is probably better to have it decided by a judge rather than by an arbitrator without legal qualifications. Indeed, where a dispute can be narrowed down to being only on a point of law, it may be resolved by a judge on a 'construction summons', a quick and simple procedure not involving a full scale court case.
- **Decisiveness:** Whether it is justified or not, there is a belief among lawyers specializing in construction cases that arbitrators are instinctively reluctant to rule wholly in favour of one party or the other but prefer, to some degree at least, to 'split the difference'. As a result, it is felt that a contractor who makes a wholly spurious claim against the employer is likely to come away from arbitration with at least something, whereas a judge would have less hesitation in denying the claim altogether. Consequently, a party who genuinely believes that a claim is 100% justified may, given the choice, prefer to litigate rather than to arbitrate.

25.4.2 Staying proceedings

In some standard form contracts, such as those relating to insurance, it is common to find a clause which makes an arbitrator's award a condition precedent to the right to bring an action in court. These are called *Scott v Avery* clauses, after the case in which their effectiveness was first recognized.[25] Such clauses are rarely found in modern construction contracts. However, where a contract contains an arbitration clause, there is a procedure whereby one party may seek to prevent the other from outflanking that clause by starting proceedings in court. Under section 9 of the Arbitration Act 1996, a defendant in such circumstances may apply to have a legal action barred ('stayed' is the technical term) so that arbitration can take place.

An application under section 9 must be made by the defendant before taking any step in the proceedings to answer the substantive claim made by the other party. Under previous statutory wording, defendants sometimes inadvertently lost their right to a stay by taking some fairly trivial procedural step in response to the issue of legal proceedings against them. The wording used in the 1996 Act means that this is much less of a danger and the Court of Appeal, in interpreting section 9, has shown itself very unwilling to deprive a defendant of the right to a stay on technical grounds.[26]

---

[25] *Scott v Avery* (1856) 5 HLC 811.
[26] *Patel v Patel* [1999] BLR 227.

Under previous legislation (section 4 of the Arbitration Act 1950), the court had complete discretion over whether or not it would grant a stay of proceedings, and there was a substantial body of case law concerning the principles on which that discretion would be exercised. For example, it was normal for a stay to be refused in cases involving more than two parties, since such disputes are much more easily handled in court than by an arbitrator. However, the 1996 Act has removed the court's discretion in this respect, by providing that the court *must* order a stay of proceedings unless satisfied that the arbitration agreement is null and void, inoperative or incapable of being performed.

The way in which the courts have interpreted the 1996 Act has brought about another crucial change in this area. Under previous legislation it had been held that, where a claim was sufficiently 'indisputable' for a court to give summary judgment (see Section 25.3.2), there was by definition no 'dispute' to be arbitrated. Hence, the defendant would not be entitled to have the proceedings for summary judgment stayed. This line of argument is, it appears, no longer valid; however clear-cut a dispute may be, it is still a dispute and, if it falls within a valid arbitration clause, the defendant is entitled to have that dispute heard at arbitration.[27]

25.4.3 Time limits in arbitration and litigation

The Limitation Act 1980, which requires every legal action to be commenced within a certain period, applies to arbitration as well as to litigation. If a case goes to court, the question is whether the claimant issued a claim form (formerly known as a writ) within the statutory period. If it goes to arbitration, the question is whether notice requiring arbitration was served on the other party within that time. What must be appreciated is that a claimant who satisfies one of these requirements is not automatically regarded as having satisfied the other. To avoid the danger of being out of time for one or other of these procedures, the claimant should both issue a claim form and serve notice of arbitration, while making clear which procedure the claimant really wishes to follow.

Quite apart from the Limitation Act 1980, construction contracts frequently contain their own provisions as to both the earliest and latest time in which disputes can be taken to arbitration. For example, JCT SBC 05 clause 1.10 makes the Final Certificate conclusive evidence of a number of matters in any proceedings commenced more than 28 days after its issue (see Section 18.2.1).

Such contractual time limits will not operate in cases of 'deliberate concealment' of a breach of contract.[28] Furthermore, the courts have jurisdiction under section 12 of the Arbitration Act 1996 to extend time for the commencement of proceedings. However, this can only be done where unforeseen circumstances have arisen and it would be just to extend time, or where the conduct of one party means that it would be unjust to enforce the strict time limit against the other. This is clearly a more restrictive test than the general 'hardship' provision contained in the 1950 Act, and it seems that the courts will not seek to interpret it in a liberal

---

[27] *Halki Shipping Corp v Sopex Oils Ltd* [1997] 3 All ER 833; *Davies Middleton & Davies Ltd v Toyo Engineering Corporation* (1997) 85 BLR 59.
[28] *Crestar Ltd v Carr* (1987) 37 BLR 113.

way. Thus where, due to an administrative oversight, a contractor failed by a mere eight days to commence arbitration within the three-month period required by ICE 6 clause 66, the Court of Appeal held that there was no reason to extend time.[29]

Problems have arisen under the form of wording formerly used in ICE contracts. ICE 6 clause 66(6) provided that, if an engineer's decision on a dispute was to be challenged by arbitration, this must be done within three months. Similarly, if the engineer had failed for one month to give a decision, which had been requested by the contractor (three months if a Certificate of Substantial Completion had been issued), the matter must again be taken to arbitration within three months. If this were not done, the decision (or lack of it) would become unchallengeable.

Since the 2004 amendment to clause 66 of ICE 7, the engineer's decision is no longer a condition precedent to the referral of a dispute to arbitration (see Section 24.5). Before this amendment, however, an unfortunate situation could occur regarding the engineer's decision: while the ICE 7 conditions specifically applied only to those 'disputes' that arose when one of the parties served a 'Notice of Dispute' on the engineer, they did not require such a notice, or indeed an engineer's decision, to be served in any particular form (so long as it is in writing). As a result, it was sometimes difficult to know whether an engineer's response to a contractor's demands constituted a 'decision' or not. Indeed, the wording of the 5th edition was even less clear on this matter.[30]

Regarding the form ICE 7, as amended in 2004, the parties can refer a dispute to arbitration directly, without a decision by the engineer. However, if an adjudicator has decided upon the dispute, the referral to arbitration has to be made within three months of the adjudicator's decision (ICE 7 clause 66B(3)). Without a timely 'notice to refer' (to arbitration), the adjudicator's decision becomes final and binding (to this issue see also Section 25.1.2).

---

[29] *Harbour and General Works Ltd v Environment Agency* [1999] BLR 409.
[30] See, for example, *ECC Quarries Ltd v Merriman Ltd* (1988) 45 BLR 90.

References

Abrahamson, M. (1984) Risk management. *International Construction Law Review*. **1**(3), 241–64.

Allott, T. (Ed.) (1998) *Common arrangement of work sections for building works.* London: Construction Project Information Committee.

Andrews, J. (1983) The age of the client. *Architect's Journal.* 13 July, 32–3.

Arden, M. (1996) *Privity of contract: contracts for the benefit of third parties: item 1 of the sixth programme of law reform: the law of contract.* Law Com; no. 242. London: HMSO.

Banwell, G.H. (1964) *The placing and management of contracts for building and civil engineering works.* London: HMSO.

Bartlett, A., Lloyd, H. and Palmer, N.E. (2002) *Emden's construction law.* London: Lexis Nexis Butterworths.

Bennett, J. and Jayes, S. (1998) *The seven pillars of partnering.* London: Thomas Telford.

Bennett, J. and Peace, S. (2006) *Partnering in the construction industry. A code of practice for strategic collaborative working.* Oxford: Butterworth-Heinemann.

Bowley, M. (1966) *The British construction industry.* Cambridge: Cambridge University Press.

Bresnen, M. (1990) *Organizing construction: project organization and matrix management.* London: Routledge.

Bresnen, M. and Marshall, N. (2000) Building partnerships: case studies of client-contractor collaboration in the UK construction industry. *Construction Management and Economics*, **18**(7), 819–32.

Bridgewater, M. and Hemsley, A. (2006) NEC 3: a change for the better or a missed opportunity? *International Construction Law Review*, **23**(1), 39–58.

Brown, J.C. (1988) Will the JCT's new management contract stand the test of time? (Part 2) *Building Technology & Management.* April/May, 10–11.

Brown, H.J. and Marriott, A.L. (1999) *ADR principles and practice.* 2nd ed. London: Sweet and Maxwell.

Bunni, N.G. (1985) *The spectrum of risks in construction.* Report of the Standing Committee on Professional Liability. Lausanne: Fédération Internationale des Ingénieurs-Conseils.

Cecil, R. (1988) *RIBA indemnity research: risk avoidance.* London: RIBA Publications.

Chappell, D. (2006) *MW05 contract administration guide.* London: RIBA Publishing.

Chapman, R.J. (1998) *An investigation of the risk of changes to key project personnel during the design stage.* Unpublished PhD Thesis, School of Construction Management and Engineering, University of Reading.

Charmer, K. (1990) Definitions of project management. *Bulletin of the Association of Project Managers.* **2**(7), March, 18–19 and **3**(1) June, 13–14.

Cherns, A.B. and Bryant, D.T. (1984) Studying the client's role in construction management. *Construction Management and Economics.* **2**(2), 177–84.

CIRIA (1984) *A client's guide to management contracts in building.* Special Pubn. No. 33. London: Construction Industry Research and Information Association.

Conditions of Contract Standing Joint Committee (2004) Amendment to the ICE Conditions of Contract (July 2004), London: Institution of Civil Engineers.

Construction Industry Board (1997) Code of Practice for the Selection of Main Contractors. London: Thomas Telford.

Draper, M. (2007) Dispute adjudication boards. *Construction Law,* **18**(3), 32.

Egan, J. (1998) *Rethinking construction: the report of the Construction Task Force to the Deputy Prime Minister, John Prescott, on the scope for improving the quality and efficiency of UK construction.* London: Department of the Environment, Transport and the Regions Construction Task Force.

Eggleston, B. (2006) *The NEC 3 engineering and construction contract. A commentary.* 2nd ed. Oxford: Blackwell Science.

Elliott, P. (1972) *The sociology of the professions.* London: Macmillan

FIDIC (2001) *FIDIC contracts guide (Construction, Plant and Design-Build and EPC/Turnkey Contract).* Geneva: Fédération Internationale des Ingénieurs-Conseils.

Gaitskell, R. (2005) Trends in dispute resolution. *Arbitration.* **71**(4), 288–99.

Gray, C. and Hughes, W.P. (2000) *Building design management.* London: Butterworth-Heinemann.

Greenwood, D.J. (1993) *Contractual arrangements and conditions of contract for the engagement of specialist engineering contractors for construction projects.* London: CASEC.

Gruneberg, S.L. and Hughes, W.P. (2005) Understanding construction consortia: theory, practice and opinions. *RICS Research Papers.* **6**(1), 1–55.

Gutman, R. (1988) *Architectural practice: a critical view.* New York: Princeton Architectural Press.

Hackett, M., Robinson, I. and Statham, G. (2007) *The Aqua Group guide to procurement, tendering and contract administration.* London: Blackwells.

Haswell, C.K. and de Silva, D.S. (1989) *Civil engineering contracts: practice & procedure.* 2nd ed. London: Butterworth.

Hawk, D. (1996) Relations between architecture and management. *Journal of Architectural and Planning Research.* **13**(1), 10–33.

Hayes, R. (1986) Who carries the risk? Management contracting – yesterday, today and tomorrow. *Building Technology & Management.* June, 42–5.

HM Treasury (2003) *The green book: appraisal and evaluation in central government.* London: The Stationery Office.

Honeyman, S. (Chairman) (1991) *Construction management forum: report and guidance.* Centre for Strategic Studies in Construction, University of Reading.

Hughes, W.P. (1989) *Organizational analysis of building projects.* Unpublished PhD Thesis, Department of Surveying, Liverpool Polytechnic.

Hughes, W.P. (1997) Construction management contracts: law and practice. *Engineering, Construction and Architectural Management.* **4**(1), 59–79.

Hughes, W.P., Gray, C. and Murdoch, J.R. (1997) *Specialist trade contracting: report.* London: CIRIA Publications.

Hughes, W.P. and Greenwood, D.G. (1996) The standardization of contracts for construction. *International Construction Law Review.* **13**(2), 196–206.

Hughes, W.P. and Hillebrandt, P.M. (2003) Construction industry: historical overview and technological change. *In:* Mokyr, J. (Ed), *The Oxford Encyclopedia of Economic History,* Oxford: Oxford University Press, Vol 1, 504–512.

Hughes, W.P., Hillebrandt, P., Greenwood, D.G. and Kwawu, W.E.K. (2006) *Procurement in the construction industry: the impact and cost of alternative market and supply processes.* London: Taylor and Francis

Hughes, W.P., Hillebrandt, P. and Murdoch, J.R. (1998) *Financial protection in the UK building industry: bonds, retentions and guarantees.* London: Spon.

Hughes, W.P. and Murdoch, J.R. (2001) *Roles in construction projects: analysis and terminology.* Birmingham: Construction Industry Publications.

Huxtable, P.J.C. (1988) *Remedying contractual abuse in the building industry.* Technical Information Service No. 99, Ascot: Chartered Institute of Building.

Institution of Civil Engineers (1991) *CESMM3: civil engineering standard method of measurement.* 3rd ed. London: Thomas Telford.

Institution of Civil Engineers (1993) *New engineering contract.* London: Thomas Telford.

Institution of Civil Engineers (1995) *Engineering and construction contract.* 2nd ed. London: Thomas Telford.

Institution of Civil Engineers (2005a) *ICE dispute resolution board procedure.* London: Thomas Telford.

Institution of Civil Engineers (2005b) *NEC 3 engineering and construction contract.* 3rd ed. London: Thomas Telford.

Institution of Civil Engineers (2005c) *NEC 3 engineering and construction contract guidance notes.* 3rd ed. London: Thomas Telford.

Institution of Civil Engineers (2005d) *NEC 3 professional services contract.* 3rd ed. London: Thomas Telford.

Institution of Civil Engineers (2005e) *NEC 3 professional services contract: guidance notes and flow charts.* 3rd ed. London: Thomas Telford.

Joint Contracts Tribunal (1987a) *Practice note MC/1: management contracts under the JCT documentation.* London: RIBA Publications.

Joint Contracts Tribunal (1990) *The use of standard forms of building contract.* London: RIBA Publications.

Joint Contracts Tribunal (1995) *Mediation on a building contract or sub-contract dispute.* London: RIBA Publications.

Joint Contracts Tribunal (2001) *Practice note 5 (Series 2): deciding on the appropriate JCT form of main contract,* London: RIBA Publications.

Joint Contracts Tribunal (2002a) *Guide to construction management documentation.* London: RIBA Publications.

Joint Contracts Tribunal (2002b) *Practice note 6: main contract tendering.* London: RIBA Enterprises Ltd.

Joint Contracts Tribunal (2005a) *Design and build contract guide (DB/G).* London: Sweet and Maxwell.

Joint Contracts Tribunal (2005b) *Standard building contract guide (SBC/G).* London: Sweet and Maxwell.

Latham, M. (1994) *Constructing the team: final report of the government/industry review of procurement and contractual arrangements in the UK construction industry.* London: HMSO.

Lawrence, P.C. and Lorsch, J.W. (1967) *Organization and environment: managing differentiation and integration.* Massachusetts: Harvard University Press.

Lupton, S (2006) *Guide to SBC 05.* London: RIBA Publishing.

Mosey, D. (2003) *ACA Standard form of contract for project partnering.* PPC2000 amended 2003. Bromley: Association of Consultant Architects.

Murray, M. and Langford, D. (Eds.) (2003) *Construction reports 1944–98.* Oxford: Blackwell.

NJCC (1994a) *Code of procedure for single stage selective tendering.* London: National Joint Consultative Committee for Building.

NJCC (1994b) *Code of procedure for two stage selective tendering.* London: National Joint Consultative Committee for Building.

NJCC (1994c) *Joint venture tendering for contracts in the United Kingdom.* London: National Joint Consultative Committee for Building.

NJCC (1995) *Code of procedure for selective tendering for design & build.* London: National Joint Consultative Committee for Building.

Office of Government Commerce (2007) *Achieving excellence in construction procurement guide.* London: Office of Government Commerce.

Pain, J. and Bennett, J. (1988) JCT with contractor's design form of contract: a study in use. *Construction Management and Economics,* **6**, 307–337.

Parris, J. (1985) *Default by sub-contractors and suppliers.* London: Collins.

Phillips, R. (2004) *Guide to RIBA forms of appointment. 2004 revisions.* 2nd ed. London: RIBA Enterprises.

Property Advisors to the Civil Estate (1998a) *GC/Works/5 General conditions for the appointment of consultants.* London: The Stationery Office

Property Advisors to the Civil Estate (1998b) *GC/Works/5 model forms.* London: The Stationery Office.

Property Advisors to the Civil Estate (1998c) *GC/Works/1 model forms and commentary.* London: The Stationery Office.

Royal Institute of British Architects (2004) *SFA/99: standard form of agreement for the appointment of an architect.* Updated April 2004. London: RIBA Enterprises.

Royal Institute of British Architects (2005a) *Code of professional conduct.* London: RIBA Publications.

Royal Institute of British Architects (2005b) *Code of professional conduct: guidance Note 3.* London: RIBA Publications.

Royal Institution of Chartered Surveyors (1998) *SMM7: standard method of measurement of building works.* 7th ed. London: RICS Books.

Schwarz, C. (Ed.) (1993) *The Chambers dictionary.* Edinburgh: Chambers Harrap Publishers.

Scott, W.R. (1981) *Organizations: rational, natural and open systems.* Englewood Cliffs, NJ: Prentice-Hall.

Seppala, C.R. (1995) The new FIDIC International civil engineering subcontract. *International Construction Law Review.* **12**(1), 5–22.

Shash, A.A. (1993) Factors considered in tendering decisions by top UK contractors. *Construction Management and Economics,* **11**(2), 111–18.

Shearer, H. (2004) *ICE introduces measured changes*, Construction Law, **15**(9), 20–22.

Skinner, D.W.H. (1981) *The contractor's use of bills of quantities.* CIOB Occasional Paper No. 24. Ascot: Chartered Institute of Building.

Smith, N.J., Merna, T. and Jobling, P. (2006) *Managing risk in construction projects.* 2nd ed. London: Blackwell.

Spiers, G.S. (1983) *The standard form of contract in times of change.* Ascot: Chartered Institute of Building.

Thompson, A. (1999) *Architectural design procedures*, 2nd ed. London: Arnold.

Thompson, J.D. (1967) *Organizations in action.* New York: McGraw-Hill.

Tjosvold, D. (1992) *The conflict-positive organization.* Massachusetts: Addison-Wesley.

Uff, J. (Ed.) (1997) *Construction contract reform: a plea for sanity.* London: Construction Law Press.

Uff, J. (2003) *Duties at the legal fringe: ethics in construction law.* Society of Construction Law Paper 115. London: Society of Construction Law.

Uff, J. (2005) *Construction law: law and practice relating to the construction industry.* 9th ed. London: Sweet and Maxwell.

Wakita, O.A. and Linde, R.M. (2002) *Professional practice of architectural working drawings.* 3rd ed. London: Wiley.

Wallace, I.N.D. (1986) *Construction contracts: principles and policies in tort and contract.* London: Sweet and Maxwell.

Wallace, I.N.D. (1995) *Hudson's building and engineering contracts.* 11th ed. London: Sweet and Maxwell.

Woodward, J. (1965) *Industrial organization: theory and practice.* London: OUP.

Author Index

Subject Index

An environmentally friendly book printed and bound in England by www.printondemand-worldwide.com